Advances in Sustainability Science and Technology

The book series aims at bringing together valuable and novel scientific contributions that address the critical issues of renewable energy, sustainable building, sustainable manufacturing, and other sustainability science and technology topics that have an impact in this diverse and fast-changing research community in academia and industry.

The areas to be covered are

- Climate change and mitigation, atmospheric carbon reduction, global warming
- Sustainability science, sustainability technologies
- Sustainable building technologies
- Intelligent buildings
- Sustainable energy generation
- Combined heat and power and district heating systems
- Control and optimization of renewable energy systems
- Smart grids and micro grids, local energy markets
- Smart cities, smart buildings, smart districts, smart countryside
- Energy and environmental assessment in buildings and cities
- Sustainable design, innovation and services
- Sustainable manufacturing processes and technology
- Sustainable manufacturing systems and enterprises
- Decision support for sustainability
- Micro/nanomachining, microelectromechanical machines (MEMS)
- Sustainable transport, smart vehicles and smart roads
- Information technology and artificial intelligence applied to sustainability
- Big data and data analytics applied to sustainability
- Sustainable food production, sustainable horticulture and agriculture
- Sustainability of air, water and other natural resources
- Sustainability policy, shaping the future, the triple bottom line, the circular economy

High quality content is an essential feature for all book proposals accepted for the series. It is expected that editors of all accepted volumes will ensure that contributions are subjected to an appropriate level of reviewing process and adhere to KES quality principles.

The series will include monographs, edited volumes, and selected proceedings.

Nasrin Aghamohammadi · Mat Santamouris
Editors

Urban Overheating: Heat Mitigation and the Impact on Health

Springer

Editors
Nasrin Aghamohammadi
Department of Social and Preventive
Medicine, Faculty of Medicine
University of Malaya
Kuala Lumpur, Malaysia

Mat Santamouris
School of Built Environment
UNSW Sydney
Sydney, NSW, Australia

ISSN 2662-6829 ISSN 2662-6837 (electronic)
Advances in Sustainability Science and Technology
ISBN 978-981-19-4706-3 ISBN 978-981-19-4707-0 (eBook)
https://doi.org/10.1007/978-981-19-4707-0

This Springer imprint is published by the registered company Springer Nature Singapore Pte Ltd.
The registered company address is: 152 Beach Road, #21-01/04 Gateway East, Singapore 189721, Singapore

Preface

Urban Heat Island (UHI), an iconic consequence of urbanization and local climate condition, often affects air pollution, energy use and public health. Hence, better understanding of the temporal dynamics of UHI in a specific location driven by varying urbanization intensities is essential for sustainable urban planning to mitigate urban heating phenomenon under a changing climate. Specifically, UHI and air pollution are the two major problems of the urban environment in Malaysia which are expected to be amplified during severe haze episodes. Overheating with direct solar radiation annually in cities undergoes a rapid urbanization that triggered UHI phenomenon. World experiences severe air pollution episodes every year due to transboundary haze originating from industrial activities, open burning as well as forest fires. There is a need on different studies conducted to explore the synergy between UHI and air pollution and their impact on health. It is also noteworthy that both UHI and heat-trapping air pollutants have raised the consumption of energy in terms of air-conditioners use in the urban environment. In return, high-energy consumption is combusted to heat and adds heat to the cities. Such phenomenon also affects the quality of life of the urban dwellers by reducing outdoor thermal comfort levels and causing heat stress. Therefore, the interaction between UHI and air pollution as well as their impact on air-conditioning and public health during should be studied concurrently to materialize the creation of sustainable and liveable future cities. Meanwhile, this book will propose some urban planning approaches via the establishment of city forests and cooling pavers in the UHI hotspots of the selected study areas through simulation approaches for UHI mitigation.

The present book reviews and reports the recent progress and knowledge on the specific impact of current and projected urban overheating as well as of the urban mitigation technologies on mortality and morbidity and urban vulnerability. It presents recent data and developments on the topic reported by some of the more distinguished researchers in this area. In parallel, it discusses new findings related to the characteristics and the magnitude of urban overheating and reports and analyses the recent knowledge on the synergies between urban heat island and heatwaves. This book helps to advance our understanding on the interaction between UHI and overheating as well as their impact on energy demand and public health globally. Exploring the interaction between UHI and energy consumption requires the understanding on the dynamics of UHI intensity and air pollution index in different land-use and how such interactions may vary in different cities in the World. Moreover, this book will focus on different cities in Australia, Austria, Belgium, Brazil, Canada, Cyprus, Greece, Hong Kong, India, Asia, Spain, UK and USA.

We have the privilege of a number of world-renown experts from various countries contributing and providing support in different chapters of the book. We are absolutely indebted to mentors, colleagues, and particularly researchers who have contributed to this book. In order to realize the goal of a low-carbon city, it is our aim that the findings of this book will be applied by epidemiologists, engineers, scientists, and urban designers for the energy demands and urban planning of future cities.

Kuala Lumpur, Malaysia
Sydney, Australia

Nasrin Aghamohammadi
Mattious Santamouris

Contents

Editors and Contributors

About the Editors

Nasrin Aghamohammadi is Associate Professor in Environmental Health Engineering in faculty of Medicine, University of Malaya. Her core expertise is on urban health, climate change adaptation, Urban Heat Island Phenomenon, and health impact. She is currently leading a grand challenge project in Malaysia on Urban Heat Island and climate change addressing Sustainable Development Goals for future cities. Urban Heat Island Phenomenon In Tropical Urban Area is one of the recent grand challenges led by her funded by University of Malaya which consists of four interdisciplinary wings from build environment, social behavior, computer science, and engineering. She has been Investigator on more than 14 research projects. She has supervised 37 Master and 12 Ph.D. students and has published several books/chapters, and 58 papers on public health, air pollution, environmental pollution, modelling, and review papers in peer-review publications with an H-index of 19. She is Member of Institution of Chemical Engineers (IChemE) in UK, Member of Institution of the Charted Institute of Environmental Health (CIEH) in UK, Member of Environmental Management and Research Association (ENSEARCH) in Malaysia, Member of the International Association of Engineers (IAENG) in Hong Kong, and Member of Centre for Occupational and Environmental Health (COEHUM) in University Malaya and Centre for Epidemiology and Evidence-Based Practice in University of Malaya.

Mattious Santamouris is Scientia, Distinguished, Professor of High Performance Architecture at UNSW; past Professor in the University of Athens, Greece; Visiting Professor: Cyprus Institute, Metropolitan University London, Tokyo Polytechnic University, Bolzano University, Brunnel University, Seoul University, and National University of Singapore; Past President of the National Center of Renewable and Energy Savings of Greece; Editor in Chief of the *Energy and Buildings Journal*; Past Editor in Chief of the *Advances Building Energy Research*; Associate Editor of the *Solar Energy Journal*; Member of the Editorial Board of 14 Journals; Editor of the

Series of Book on Buildings, published by Earthscan Science Publishers; Editor and Author of 17 international books published by Elsevier, Earthscan, Springer, etc.; Author of 348 scientific articles published in journals; Reviewer of research projects in 29 countries including USA, UK, France, Germany, Canada, Sweden, etc.; ranked by the Standford researcher list as No. 1 Scientist on Building and Construction for 2018 and 2019 and No. 679 in the list of all sciences; and highly cited Researcher, 1%, by Clarivate for 3 continuous years.

Contributors

Aghamohammadi Nasrin Department of Social and Preventive Medicine, Faculty of Medicine, Centre for Epidemiology and Evidence-Based Practice, University of Malaya, Kuala Lumpur, Malaysia;
Centre for Energy Sciences, University of Malaya, Kuala Lumpur, Malaysia

Armendáriz-López J. F. Facultad de Ciencias de la Ingeniería y La Tecnologia, Universidad Autonóma de Baja California, Tijuana, México

Ascione Fabrizio Department of Industrial Engineering, DII—Università degli Studi Di Napoli Federico II, Napoli, Italy

Begou Paraskevi Department of Physics, Laboratory of Meteorology and Climatology, University of Ioannina, Ioannina, Greece

Berardi Umberto Department of Architectural Science, Faculty of Engineering and Architectural Science, Ryerson University, Toronto, ON, Canada

Carlos Bartesaghi-Koc School of Architecture and Built Environment, Faculty of Science, Engineering and Technology, The University of Adelaide, Adelaide, Australia

Carlosena Alicia Faculty of Medicine and Health, University of New South Wales (UNSW), Sydney, NSW, Australia

Carlosena Laura Engineering Department, Public University of Navarre (UPNA), Pamplona, Spain

Cureau Roberta Jacoby CIRIAF Interuniversity Research Centre On Pollution and Environment Mauro Felli, University of Perugia, Perugia, Italy

Dardir Mohamed Department of Architectural Science, Faculty of Engineering and Architectural Science, Ryerson University, Toronto, ON, Canada

Davies Mike Institute for Environmental Design and Engineering, University College London, London, UK

De Masi Rosa Francesca Department of Engineering—DING, Università degli Studi del Sannio, Benevento, Italy

Ehsan Sharifi School of Architecture and Built Environment, Faculty of Science, Engineering and Technology, The University of Adelaide, Adelaide, Australia

Farid Nik Daliana Nik Department of Social and Preventive Medicine, Faculty of Medicine, University of Malaya, Kuala Lumpur, Malaysia

Fong Chng Saun Institute for Advanced Studies, University of Malaya, Kuala Lumpur, Malaysia

González-Trevizo M. E. Facultad de Ingeniería, Arquitectura y Diseño, Universidad Autonóma de Baja California, Ensenada, México

Heaviside Clare Institute for Environmental Design and Engineering, University College London, London, UK

Hua Junyi School of International Affairs and Public Administration, Ocean University of China, Qingdao, China

Jamei Elmira Institute for Sustainable Industries and Liveable Cities, Victoria University, Melbourne, VIC, Australia

Kassomenos Pavlos Department of Physics, Laboratory of Meteorology and Climatology, University of Ioannina, Ioannina, Greece

Kumar Prashant Department of Civil and Environmental Engineering, Faculty of Engineering and Physical Sciences, Global Centre for Clean Air Research (GCARE), University of Surrey, Guildford, Surrey, UK

Lau Kevin Ka-Lun Institute of Future Cities, The Chinese University of Hong Kong, Shatin, Hong Kong, China

Luna-León A. Facultad de Arquitectura y Diseño, Universidad Autonóma de Baja California, Mexicali, México

Martínez-Torres K. E. Facultad de Ingeniería, Arquitectura y Diseño, Universidad Autonóma de Baja California, Ensenada, México

Mastellone Margherita Department of Architecture, DIARC, Università degli Studi di Napoli Federico II, Napoli, Italy

Mavrogianni Anna Institute for Environmental Design and Engineering, University College London, London, UK

Milojevic Ai London School of Hygiene and Tropical Medicine, London, UK

Mohammadi Parshad Agha Department of Civil Engineering, Faculty of Engineering, Science and Research Branch of Azad University, Tehran, Iran

Ng Edward Yan Yung School of Architecture, The Chinese University of Hong Kong, Shatin, Hong Kong, China

Oikonomou Eleni Institute for Environmental Design and Engineering, University College London, London, UK

Paravantis John A. University of Piraeus, Piraeus, Greece

Petrou Giorgos Institute for Environmental Design and Engineering, University College London, London, UK

Pigliautile Ilaria CIRIAF Interuniversity Research Centre On Pollution and Environment Mauro Felli, University of Perugia, Perugia, Italy;
Department of Engineering, University of Perugia, Perugia, Italy

Pisello Anna Laura CIRIAF Interuniversity Research Centre On Pollution and Environment Mauro Felli, University of Perugia, Perugia, Italy;
Department of Engineering, University of Perugia, Perugia, Italy

Pyrgou Andri Department of Civil Aviation, Nicosia, Cyprus

Ramakreshnan Logaraj Institute for Advanced Studies, University of Malaya, Kuala Lumpur, Malaysia

Ren Chao Faculty of Architecture, The University of Hong Kong, Pokfulam, Hong Kong, China

Sadeghi Mahsan Centre for Air Pollution, Energy, and Health Research (CAR), NHMRC Centre of Research Excellence (CRE), Sydney, Australia;
School of Architecture, Design & Planning, The University of Sydney, Sydney, Australia

Sandoval-Félix J. Urban Lab: Laboratorio de Urbano y de Políticas Públicas, Ensenada, México

Santamouris Mat School of Built Environment, University of New South Wales, Sydney, Australia;
Anita Lawrence Chair High Performance Architecture, University of New South Wales, Kensington, South Wales, Australia

Santamouris Mattheos Faculty of Built Environment, University of New South Wales, Sydney, NSW, Australia

Scott Hawken School of Architecture and Built Environment, Faculty of Science, Engineering and Technology, The University of Adelaide, Adelaide, Australia

Shi Yuan Institute of Future Cities, The Chinese University of Hong Kong, Shatin, Hong Kong, China

Symonds Phil Institute for Environmental Design and Engineering, University College London, London, UK

Tapper Nigel School of Geography and Environmental Science, Urban Climate Research, Monash University, Melbourne, VIC, Australia

Tariello Francesco Department of Agricultural, Environmental and Food Sciences, Università degli Studi del Molise, Campobasso, Italy

Taylor Jonathon Faculty of Built Environment, Tampere University, Tampere, Finland

Tsoulou Ioanna Institute for Environmental Design and Engineering, University College London, London, UK

Vanoli Giuseppe Peter Department of Medicine and Health Sciences—Vincenzo Tiberio, Università degli Studi del Molise, Campobasso, Italy

Veronica Soebarto School of Architecture and Built Environment, Faculty of Science, Engineering and Technology, The University of Adelaide, Adelaide, Australia

Wilkinson Paul London School of Hygiene and Tropical Medicine, London, UK

Wilson Jeffrey School of Environment, Enterprise, and Development, Faculty of Environment,, University of Waterloo, Waterloo, ON, Canada

Chapter 1
Urban Overheating and Impact on Health: An Introduction

Anna Mavrogianni, Ioanna Tsoulou, Clare Heaviside, Eleni Oikonomou, Giorgos Petrou, Phil Symonds, Mike Davies, Jonathon Taylor, Ai Milojevic, and Paul Wilkinson

Abstract Anthropogenic climate change is projected to increase ambient temperatures, and the frequency, intensity and duration of extreme heat events worldwide. As a result, recent years have seen an increasing policy and research focus on the potential adverse effects of warm and hot weather on human thermal comfort, well-being, morbidity and mortality. Certain segments of the population, including the very young, the old, the chronically ill and the socially isolated, are likely to be disproportionately affected. Climate change induced warming trends, and associated heat risks are expected to be significantly exacerbated in urban environments due to the Urban Heat Island effect, a well-established phenomenon of inadvertent urban climate modification. Building fabric characteristics are also identified as a key modifier of indoor thermal exposures, particularly in climates where the population tends to spend a significant proportion of their time indoors. This chapter aims to provide an overview of urban heat health risk and vulnerability factors for different geographic and climatic contexts and to summarise the impacts of urban overheating on human health and wellbeing, with a focus on drivers of urban heat risk inequalities.

Keywords Urban environment · Cities · Heat · Overheating · Urban temperature · Urban warming · Urban climate · Urban heat Island · Climate change · Thermal comfort · Health · Wellbeing · Heat-related mortality

---

A. Mavrogianni (✉) · I. Tsoulou · C. Heaviside · E. Oikonomou · G. Petrou · P. Symonds · M. Davies
Institute for Environmental Design and Engineering, University College London, London, UK
e-mail: a.mavrogianni@ucl.ac.uk

J. Taylor
Faculty of Built Environment, Tampere University, Tampere, Finland

A. Milojevic · P. Wilkinson
London School of Hygiene and Tropical Medicine, London, UK

N. Aghamohammadi and M. Santamouris (eds.), *Urban Overheating: Heat Mitigation and the Impact on Health*, Advances in Sustainability Science and Technology,
https://doi.org/10.1007/978-981-19-4707-0_1

1 Introduction

Throughout the centuries, through a series of interconnected physiological, behavioural and cultural processes, humans have acclimatised to prevailing local climatic conditions. Exceedance of the amount of heat or cold exposure an individual can tolerate, however, e.g. during a heatwave event, can lead to severe adverse health effects, even death [1–4]. Excess heat is recognised as a major environmental and occupational hazard [5], and its physiological effects on the human body have been widely studied. Approximately one-third of the global population is currently exposed to intense and prolonged heat episodes [6]. Heat events with large-scale health impacts, such as the 2003 heatwave in Europe and the 2021 heatwave in Canada, resulted in economic losses and more than 70,000 and 800 excess deaths, respectively. As a result of anthropogenic climate change, such heat events will have become common in many parts of the world by the middle of the century; extreme heatwaves previously only experienced every 50 years are now projected to occur once per decade [7].

Urban warming is driven by the combined effects of climate change and the Urban Heat Island (UHI) phenomenon and can alter urban thermal conditions at a faster pace than urban residents can adapt to. Urban overheating can lead to a series of interlinked effects on the thermal comfort, associated energy consumption used for space conditioning (both heating and cooling), health and wellbeing of people living in urban environments [8]. This chapter offers an introduction to the key health impacts of urban overheating, the main urban heat risk determinants and the key parameters affecting the distribution of heat risk across and within cities worldwide.

2 Urban Overheating Exposure Trends

According to the Intergovernmental Panel on Climate Change (IPCC)'s Report *Climate Change 2021: The Physical Science Basis*, which was released in August 2021 [7], unless large-scale, rapid climate change mitigation actions take place worldwide, global temperature change close to 1.5 °C or even below 2 °C will not be possible [7, 9]. Even if the 1.5 °C limit target is reached, an increase in the frequency, severity and duration of record-breaking heatwaves is expected, in conjunction with shortening of cold seasons around the world. If anthropogenic global warming exceeds 2 °C, critical thresholds for heat-related health will be exceeded with increasing frequency [7], with the number of people affected by extreme heat stress across the world possibly increasing 15-fold [10].

According to *The Lancet* [11], the two defining questions of this decade will be about:

1. identifying and quantifying the health consequences of a hotter climate and
2. determining how extreme heat is managed at different scales and by different actors.

Climate change induced overheating risk trends are intensified in cities all over the world due to *urban overheating*: the phenomenon of urban areas experiencing higher ambient temperatures than their rural and suburban surroundings, which has been documented in more than 400 major cities globally [12]. The Urban Heat Island (UHI) effect is a well-established phenomenon of inadvertent local climate modification, whereby urban areas are generally, on average, warmer than nearby suburban and rural locations [13]; this temperature differential is commonly referred to as the *UHI intensity*. The UHI is predominantly a nighttime phenomenon. It is the result of modifications of the urban surface energy balance, including the introduction of humanmade surfaces and the reduction of evaporative surfaces; dense and complex urban geometries, which tend to trap, re-reflect and re-emit solar radiation as heat; and the addition of heat generated by anthropogenic activities. UHI intensities are, thus, higher during days and nights with clear skies characterised by low wind velocities due to the lack of heat dispersion. In most cities around the world, the average UHI intensity is typically between 2 and 4 °C [14–17] but can reach up to 10 °C in larger cities under conditions that are favourable to UHI formation [18], which often coincide with heatwave events.

According to the report *The Future We Don't Want* produced by the C40 Cities initiative [6], the number of cities experiencing extreme high temperatures will have almost tripled by the middle of the century and heatwaves are projected to be one of the deadliest climate risks. A recent analysis of global cause-specific temperature-attributable mortality in nine countries[1] in 2019 showed that 1.7 million deaths were linked to extreme heat and cold through associations with 17 causes of death, mainly cardiorespiratory or metabolic disease, but also suicide and other injuries; 356,000 of these deaths were heat-related [19]. A more recent study estimated that 37% of global heat-related deaths are attributable to climate change and that there is evidence of increased mortality on every continent [20].

It is estimated that approximately 200 million people living in over 350 cities currently face hottest 3-month average maximum temperatures reaching at least 35 °C [6], the majority of them located in Asia and the Middle East. This number is expected to rise to 970 cities, with the most intense heat exposure levels observed in some of the poorest parts of the world. By the 2050s, heatwaves will become common and summers with temperatures above 35 °C will be considered typical summers in many cities around the world, with Asia, Africa, North America and parts of South America affected most.

The significant health risks faced by urban residents are of particular importance in terms of global public health due to current global urbanisation trends: in 2007, the number of people living in urban areas exceeded for the first time the number of those living in rural areas [21]. Notably, urban population increases are expected to be higher in areas that will experience dramatic rises in urban overheating with 90% of urbanisation in the next three decades projected to be concentrated in Asia and Africa. Within these three decades, it is estimated that more than 1.6 billion people (more than 40% of the current global urban population) living in close to

[1] Brazil, Chile, China, Colombia, Guatemala, Mexico, New Zealand, South Africa and the USA.

1000 cities will be frequently experiencing extreme heatwaves. As older people are more susceptible to heat, global population ageing trends will also further magnify risks from excess heat exposure [22–24].

Crucially, there is a series of urban health risks intersecting with overheating risk. These range from poor housing and overcrowding, to ambient air, water, soil and noise pollution [24]. The populations most at-risk often coincide with areas of high UHI intensity, as well as high levels of social deprivation or other types of vulnerability, thus contributing to health inequalities, as will be further discussed in Sect. 5.

There are environmental and social aspects that determine the health, comfort, wellbeing and energy implications for urban overheating. The net effect of urban heat broadly depends on context-specific characteristics of an urban setting [25, 26]: Warmer climates experience net negative health and wellbeing outcomes as a result of urban warming with a rise in ambient temperatures resulting in increased outdoor and indoor heat exposure, thermal discomfort and overheating and potential increased risks of heat-related morbidity and mortality. Other negative effects may include potential reductions in work productivity (especially in relation to outdoor manual labour) and disruptions to critical infrastructure. This further leads to potential increases in cooling demand and air conditioning installation, in particular peak electricity loads (where active cooling systems are available). In temperate or cooler climate regions, however, urban overheating may have positive net annual effects, such as protection from cold weather, reduced space heating demand and associated carbon emissions, reduced cold-related illness and deaths, a potential reduction of winter fuel poverty and longer growing seasons (although more frequent droughts and floods may hamper this). A study in the US, for example, found that the UHI increased heat-related mortality by 1.1 deaths per million population, but decreased cold-related mortality by 4.0 deaths per million [27]. In the West Midlands, UK, a highly urbanised area, the UHI offsets around 15% of cold-related mortality in winter [28]. It has, however, been estimated that in many regions such benefits will be outweighed by the increase in heat-related morbidity and mortality [3, 29]. This highlights the importance of adopting a systems thinking approach towards the assessment of urban overheating impacts, to allow the holistic consideration of interconnected factors, their interactions and contributions to health outcomes.

3 Urban Overheating Impacts on Human Health and Wellbeing

3.1 Impacts on Urban Health

The World Health Organisation [30] states that excess heat affects human health *directly* and *indirectly* (Fig. 1).

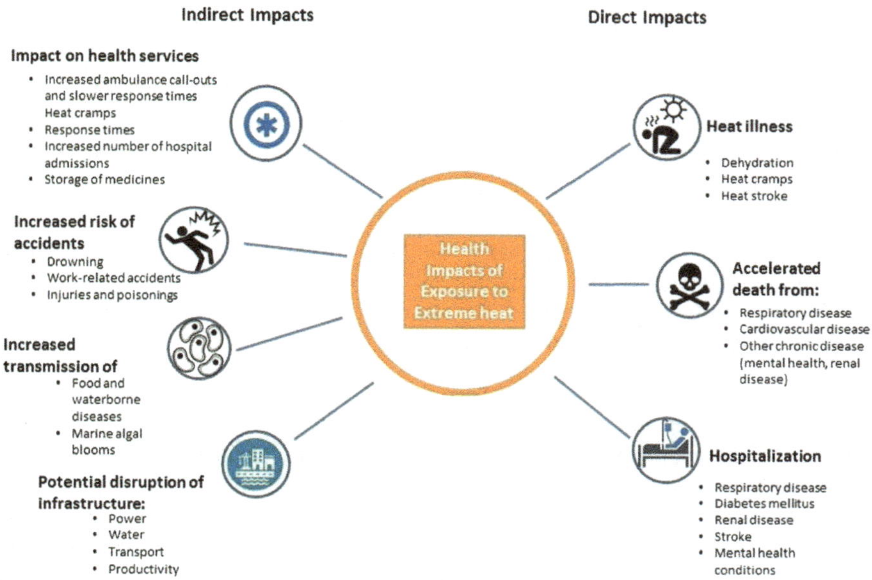

Fig. 1 Direct and indirect heat health impacts. *Source* [30]

Direct heat health impacts are typically the result of heat gain generation in the human body resulting from both ambient heat exposure and internal metabolic heat generation. When an individual's exposure to heat rapidly increases to levels higher than they can tolerate or are accustomed to, the human body's ability to thermoregulate is compromised and the following adverse health effects may occur:

1. Heat illness, including dehydration and electrolyte imbalance, and clinical symptoms of heat stress, such as heat stroke, heat exhaustion, heat syncope, heat cramps and hyperthermia.
2. Accelerated death due to underlying respiratory or cardiovascular disease or other chronic conditions (such as renal disease or mental health conditions). This could begin to be observed at relatively low temperature thresholds and not only as a result of extreme heat exposure.
3. Hospitalisation due to respiratory disease, diabetes, rental disease, stroke or mental health conditions.

Indirect heat health impacts are associated with potential large-scale knock-on effects of heat illness, in particular in relation to the capacity and responsiveness of critical public health and social care infrastructure. Such indirect health effects may include:

1. The potential increased burden on health services and public health infrastructure, for example, due to slower response times or capacity to deal with urgent conditions, as a result of heat-related ambulance call-outs and hospital admissions.

2. Potential disruption to other types of infrastructure (including power, water and transport).
3. Potential adverse effects on medicine low temperature storage capacity.
4. Increased risk of accidents, in particular in work environments. For example, negative impacts of heat on sleep quality or overall cognitive performance may indirectly cause accidents in the workplace.
5. Increased transmission of food and waterborne diseases.
6. Synergies with concurrent risks, for example, exposure to air pollutants that may be at high concentrations during heat events, such as ozone.
7. Conflicts with concurrent risks, for example, increased time spent indoors due to shielding to avoid Covid-19 infection.

With regard to synergistic effects between air pollution and heat, there is a well-documented relationship between elevated concentrations of harmful air pollutants, such as ground level ozone and particulate matter, and increases in urban temperatures. The UHI modifies urban air flows, and coupled with high air pollutant emissions in urban environments, thereby increases concentration of pollutants close to the ground. Tropospheric ozone is also generated by photochemical reactions in ambient air, which are accelerated through chemical interactions with nitrogen oxides and hydrocarbons in the presence of sunlight. Ozone is considered as a modifier in the temperature–mortality relationship, as it is highly toxic with adverse effects on the human cardiovascular and respiratory system [31, 32]. Future ozone concentration increases driven by increased precursor emissions could pose further human health risks [33]. In middle-income countries where power generation is closer to cities, another indirect health effect of urban overheating is the negative health impact resulting from increased exposure to pollutants generated by power plants needed to support a rise in cooling demand due to urban overheating.

The onset of the Covid-19 pandemic in early 2020 highlighted the importance of indoor environmental quality for health. Attention was also given by many public health professionals and researchers on the overlapping risk factors for Covid-19 and heat-related illness [34–36]. It was hypothesised that people who were identified as more vulnerable to Covid infection were more likely to stay indoors due to their vulnerability and/or as a result of self-isolating, thus increasing their exposure to indoor heat. Both Covid and heat can have severe impacts on the lungs, heart and kidneys and result in systemic inflammation. They also share risk determinant factors, such as being older (above 65 years of age), or suffering from certain chronic health conditions, such as respiratory, cardiovascular, cerebrovascular and chronic obstructive pulmonary and kidney disease, or high blood pressure. Future research is required in order to assess the combined effect of Covid-19 and heat exposure and ensure that trying to protect from one, risks are not increased for the other.

3.2 Geographical Variation of Urban Heat Health Risk

The type, magnitude and severity of urban heat-related health impacts largely depend on a series of geographically varying parameters:

- The levels of *acclimatisation* of the local population, *preparedness* and *resilience* of the physical and social infrastructure in each city.
- The *timing* (e.g. whether a heat event occurs early or later in the summer season), *intensity* and *duration* of the heat event itself.
- Other climate parameters, such as wind, humidity and ambient air pollution that have synergistic effects with heat. For example, low wind speeds are common during heatwaves and may make the existing UHI more intense and simultaneously increase air pollutant concentrations; even moderate temperatures could cause thermal discomfort when combined with high humidity, which is particularly important for cities in hot and humid regions.

It is worth noting that these parameters are interconnected; for example, populations tend to be better prepared to cope with warmer weather later in the summer or after several consecutive hot days or periods, both at an individual level (thermal comfort expectations and adaptive capacity) as well as at population and community level. Whilst there is no universal definition of what meteorological conditions constitute a *heatwave* event, the magnitude of a heat event is crucial, with increases in mortality, morbidity and hospital admissions associated with even small deviations from the baseline thermal conditions of a given region and season. In some instances, heat results in forward *mortality displacement* of frail individuals (also known as *early harvesting*), i.e. people of poor health status dying sooner as a result of a hot spell [37, 38]. This phenomenon has been observed in some cities, specifically during the first days of heatwaves [39] but not others [40], another indication of the complex underlying factors driving urban heat-related mortality. Heat deaths are also recorded a few days following a hot spell, often when external temperatures have decreased (*lag time* effects). With regard to the diurnal temporal variation of heat exposure, although peak heatwave temperatures are recorded during the daytime, nighttime temperatures can be detrimental to health if they remain elevated during a heatwave. This is particularly the case if the heatwave persists for a number of days and there is no nighttime respite from the heat; this is of fundamental importance in cities, as the UHI is predominantly a nighttime phenomenon. High heat-related mortality observed during the 2003 heatwave in Paris was partly attributed to high nighttime temperatures [39].

There are complex interrelationships between the confounding factors outlined above and urban heat-related health risk. Key methodological approaches in epidemiological research employed to explore heat-related health risk are as follows [41, 42]:

- The *heat episode analysis* approach, which calculates excess mortality on the basis of historical baseline trends.

- The *regression model* approach, which regresses time-series data to generate the baseline values by controlling for confounding factors such as season.
- Case-crossover studies, which may be an alternative to time regression approaches, in which a case is the day of death of each individual and proximate days (e.g. ±7 days) are controls.
- Case–control studies, which compare the effects on urban and rural populations.
- Case-only approaches, which quantify the modification effect of several risk factors on the heat–mortality relationship.

At the population level, there are well-established epidemiological relationships between excess hot (or cold) temperatures and mortality, developed based on time regression analyses. As mentioned above, ambient temperature is both a determinant of mortality in its own right and a confounder of other determinants, such as air pollution [43]. *Excess mortality* refers to the short-term increase in the number of deaths recorded in a region during a period of time that exceeds the baseline average historic rates typically observed in the area at this time of the year [44]. It is worth noting that excess mortality could be observed for a variety of reasons; this method can be used to offer an approximate estimate of deaths related to heat but cannot directly attribute them. *Heat* (or *cold*)*-related death* is defined as death in which the exposure of an individual to relatively high (or low) temperatures either caused or contributed to death [5]. The precise threshold above which excess heat (or cold)-related mortality is observed varies geographically.

In a recent comprehensive review of the state of the art in urban over-heating research, Santamouris [12] summarised the quantitative relationship between external air temperature increase and heat-related morbidity and mortality across various cities worldwide. It is pointed out that comparisons across epidemiological studies can be challenging due to the inconsistent use of urban overheating and morbidity/mortality indicators, for example, different metrics of ambient temperature and diverse metrics of illness (e.g. calls to health practitioners, hospital admissions, emergency department visits, etc.). Urban areas are affected the most due to the UHI increment of ambient temperature increase; a time-series study analysing death rates across England and Wales between 1993 and 2003 indicated that daily mortality was linked with temperature across all regions but the effect was stronger in London [45]. During the 1995 heatwave, a 9% increase of mortality was observed in London compared to other areas. During the 2003 heatwave, 2139 excess deaths were recorded in England (a 17% increase from the baseline); the corresponding figure in London was 616 (42% increase [46]).

In the majority of cities around the world, a U- or V-shaped curve characterises the relationship between external air temperature and mortality. The graphs in Fig. 2 were generated in the context of a study that quantified the total mortality burden attributable to non-optimum ambient temperature (heat and cold) [2]. Ambient temperature and daily mortality data from thirteen cities in four different continents featuring diverse populations, levels of acclimatisation, urbanisation, building stocks and public health infrastructure were plotted. The cumulative exposure–response curves overlaid on the data of each graph represent the best linear unbiased predictions of the temperature–mortality relationship. Although the specific heat- and cold-related mortality thresholds vary between cities, a common theme emerging is that whilst cold-related mortality risk increases gradually and linearly, the heat mortality–temperature slope is usually steeper with heat-related mortality risk increasing more rapidly and, in some cases, nonlinearly. This corroborates findings in other urban areas [47]. It also highlights the critical risks of urban overheating currently facing most cities worldwide, and the importance of prioritising climate resilient, responsive public health infrastructure.

Whilst there is a lot of intra-climate region variation, there is a correlation between temperature-related thresholds and latitude, and an overall trend emerges with heat-related mortality threshold temperatures being higher in hotter than in colder climates (Fig. 3); this is primarily attributed to higher population acclimatisation to warmer environmental conditions, but potentially also building stocks, public health and infrastructure better adapted to heat [37, 48].

Despite the potential positive effects of urban overheating on winter mortality and morbidity, in particular under climate change, reducing excess heat exposure should still be a public health priority in colder regions for a number of reasons:

- Heat-related health risks in colder climates are often linked to social infrastructure and risk perception and are, thus, largely avoidable.
- Population-level acclimatisation to a warmer climate may shift overall susceptibility to heat, with higher cold and heat-related morbidity and mortality thresholds likely to be observed in the future. This will subsequently reduce the current protective role of urban heat.

It is important to note that excess heat-related mortality is not limited to heat-wave events or 'extreme' heat episodes; heat deaths occur even at moderately high temperatures [2, 29]. As shown in Fig. 4, the majority of deaths in the examined cities are associated with moderate cold. The comparison of heat-related mortality rates due to moderate vs. extreme heat is less clear but it is evident that exposure to moderately warm weather is responsible for a significant number of deaths in most cities studied.

During hot spells, similar trends have been observed between outdoor air temperature and emergency hospital admissions but the relationship is less strong or not even observed, possibly due to: (1) deaths occurring before the individuals are admitted to hospital or before the death is classified as heat-related and (2) renal and respiratory disease being the primary reason for admissions, but not cardiovascular disease [5] which is not as strongly associated with hospitalisation [49].

3.3 Urban Heat Health Risk Determinants

The identification of high heat risk groups is the first key step that will allow the prioritisation of public health interventions for less heat tolerant and vulnerable

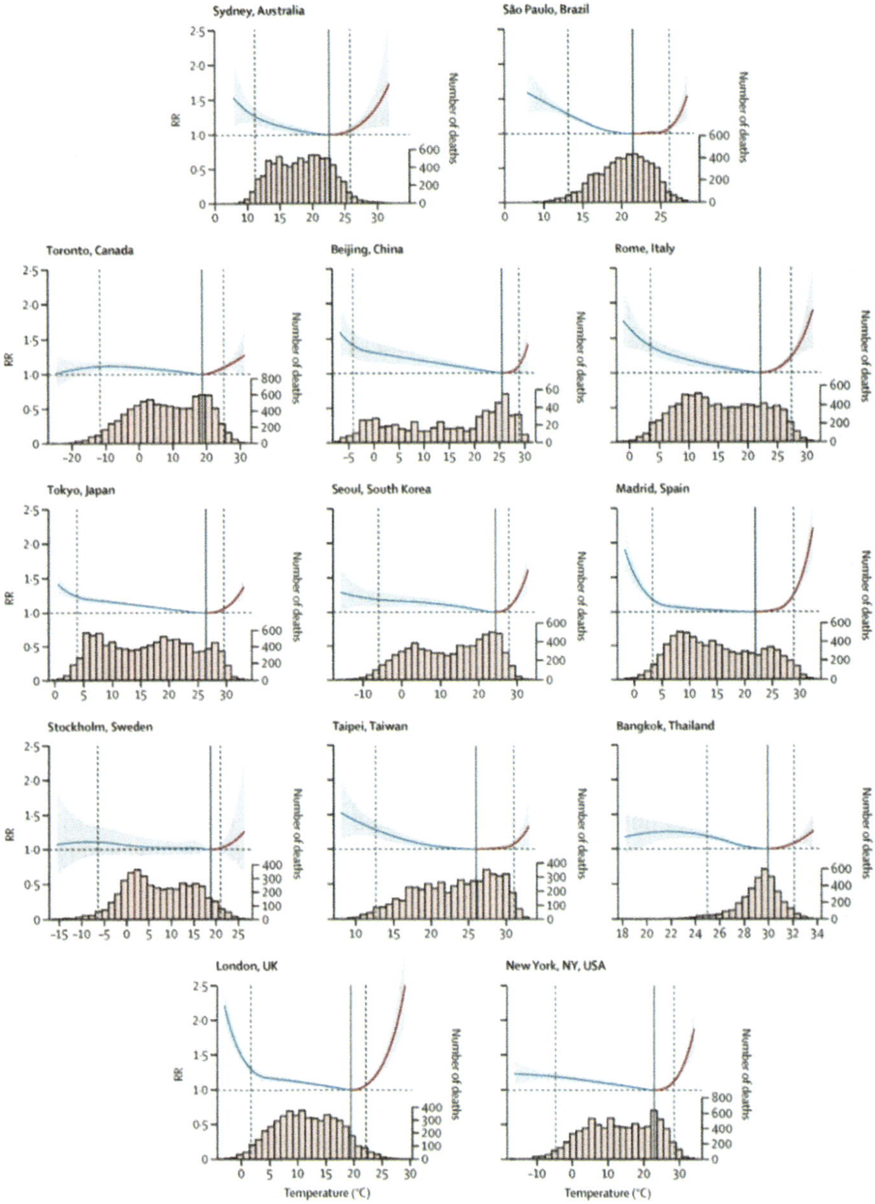

Fig. 2 Overall cumulative exposure–response associations in 13 cities. Exposure–response associations as best linear unbiased prediction (with 95% empirical C.I., shaded grey) in representative cities of the 13 countries, with related temperature distributions. Solid grey lines are minimum mortality temperatures, and dashed grey lines are the 2.5th and 97.5th percentiles. RR = relative risk. *Source* [2]

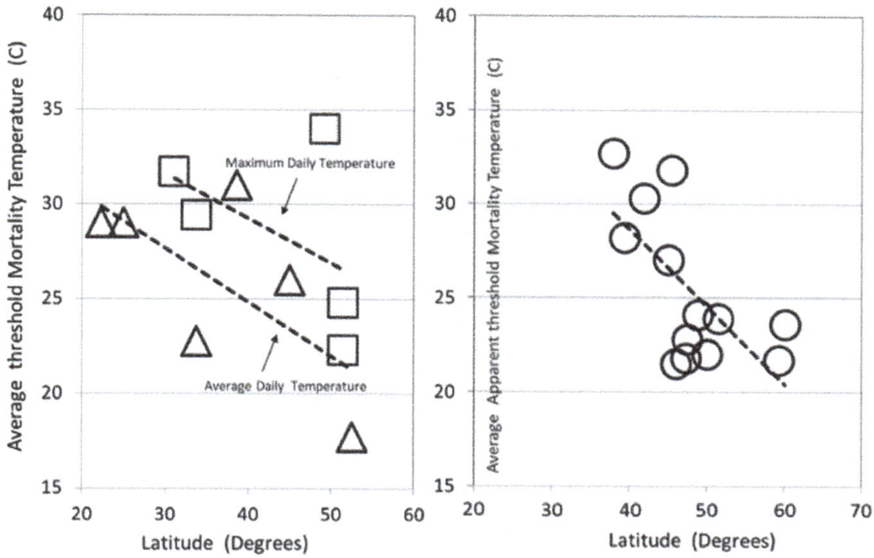

Fig. 3 Correlation between the average threshold mortality temperature and latitude, squares refer to the maximum daily temperature and triangles to the average daily temperature. *Source* [12]

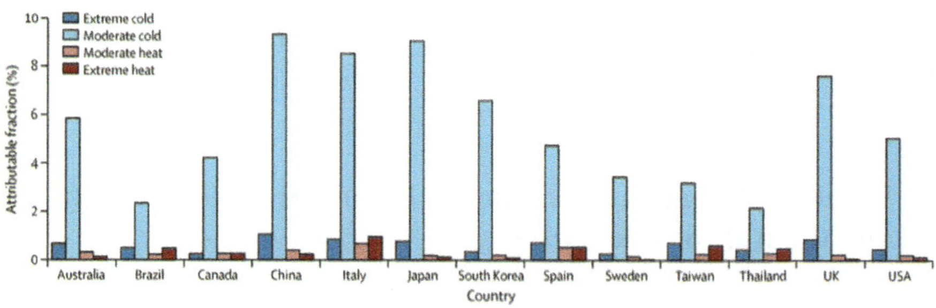

Fig. 4 Fraction of all-cause mortality attributable to moderate and extreme hot and cold temperature by country. Extreme and moderate high and low temperatures were defined with the minimum mortality temperature and the 2.5th and 97.5th percentiles of temperature distribution as cutoffs. *Source* [2]

individuals among a population. WHO (2009) classifies heat determinant factors into three main areas:

- exposure to heat,
- sensitivity and
- ability to adapt.

These factors may be, in turn, determined by environmental, physiological, social and psychological parameters. Another way to classify heat vulnerability risk factors is by dividing them into:

- *intrinsic* (i.e. age, disease, disability) and
- *extrinsic* (i.e. indoor environment, behaviour, lifestyle) [5].

There is overwhelming epidemiological evidence that heat-related vulnerability is correlated with age, with people above 65 years of age, but also the very young, being at higher risk [50]. This is because of changes occurring in the human thermoregulatory system as a result of ageing. Babies and young children also have limited thermoregulatory ability, and they may be at high risk of dehydration during hot weather.

Any underlying health condition that may affect the thermoregulatory system is another major heat risk factor, mainly for people with respiratory, cardiovascular, cerebrovascular, renal conditions, individuals using medication affecting thermoregulation, diabetes fluid/electrolyte and other neurological disorders, obesity and chronic diseases. Heat vulnerability also increases for people living with mental health conditions, such as dementia, depression and Parkinson's disease [5, 50].

Social isolation may potentially contribute to heat-related morbidity and mortality risk [51, 52]. Low-income individuals or people with limited access to public health support systems or community support networks may not have the adaptive capacity and resources necessary to respond to heat–health warnings or afford to adapt their surroundings and protect themselves from excess heat exposure.

Gender has been identified as a potential heat-related morbidity and mortality risk factor, with women usually being more severely affected by heat exposure [44, 50, 53]. However, it is unclear whether this association should be attributed to physiological or sociocultural factors.

4 Building Design as a Modifier of Urban Heat Health Risk and the Role of Energy Use

The design of buildings plays a key role in heat-related health impacts because it has a significant modifying effect on heat exposure. This modifying effect is particularly important for settings where people tend to spend their majority of time indoors, i.e. mid- and high-latitude cities and for some of the more vulnerable population demographics, such as older individuals or people suffering from chronic health conditions. Whilst there has been a plethora of studies quantifying the impact of *outdoor* temperatures on health and associated thresholds, as presented in Sect. 3.2, the role of *indoor* temperatures, and indoor environments more generally, on human health is less well understood.

There is some evidence that links summer heat-related mortality risk with thermally heavyweight or uninsulated structures, top floor flats and dwellings with limited means of ventilation [54]. Such characteristics are potentially more common in urban than rural areas, for example, in high-rise or mid-rise purpose-built blocks of flats, where only single-sided ventilation is possible for many flats. This relationship was particularly explored following analysis of heat-related deaths that occurred during the 2003 and 2006 European heatwaves:

- In Paris and cities around it, an odds ratio of 5:1 between non-insulated and insulated dwellings and an odds ratio of 4:1 for top floor bedrooms were calculated for heat-related mortality risk during the catastrophic 2003 heatwave [55]. Mortality risk also appeared to reduce with higher green coverage around homes.
- In London, an exploration of the interrelationship between housing characteristics, location of a dwelling within the UHI and heat-related mortality during the 2006 heatwave found that the quality of housing contributed more to mortality than the UHI local temperature variation, with 23.5 deaths per million of population attributed to dwelling characteristics as opposed to 6.1 deaths per million of population attributed to the UHI [56].

Summertime overheating and associated mortality may be higher for the older occupants of vulnerable settings such as nursing and care homes. During the 2003 heatwave in London, excess deaths of adults over 75 living in nursing homes were about 10% higher compared to those living at home [45]. Likewise, a study of daily mortality in England and Wales between 1993 and 2003 found that the elderly living in nursing and care homes were most vulnerable to heat mortality [57]. Besides older occupants' susceptibility to heat, building characteristics of these settings may also play a part in this. For instance, care homes have historically prioritised heating instead of cooling systems, aiming to keep residents warm throughout the year, whilst additional considerations, such as safety, have further led to restricting window opening and therefore natural ventilation [58]. Consequently, care home managers and staff may be less prepared to deal with a heatwave.

Housing quality is likely to be correlated with socioeconomic characteristics, which makes disentangling the influence of individual factors on urban heat risk challenging. More affluent population groups are more likely to:

- have access to health advice and support,
- inhabit larger dwellings with access to cooler outdoor spaces or more variation of indoor temperatures (e.g. rooms of different orientations or transitional, shaded, semi-outdoor spaces),
- be in close proximity to cooler, greener areas or green infrastructure more broadly, or be able to afford adjustments of their immediate environment, for example, through housing refurbishment (e.g. installation of shading systems) or the purchase and operation of active cooling systems.

Despite reduction in winter thermal discomfort and heating needs, urban overheating in warmer parts of the world will pose a significant thermal and energy burden on people, buildings and communities. The projected urban heat trends are likely to test the overheating resilience of urban populations in currently cooling-dominated climates: existing passive cooling strategies, such as ventilation and shading, may prove inadequate for the projected unprecedented temperature increases and heatwave events. Whilst it may be easier to incorporate comfort cooling in an already mechanically ventilated building, naturally ventilated buildings may be particularly vulnerable to excess heat as the cooling potential of ventilation may be compromised as external temperatures become increasingly warmer. Unless concerted efforts are made to promote low carbon, low energy cooling solutions, when combined with the aggressive marketing of air conditioning at a large scale, this will lead to overreliance on active cooling and a rise in energy consumption, or adverse health effects for lower income groups that cannot afford it. On the other hand, the existing building stock and design of cities in currently heating-dominated climates are not well suited to extreme heat events as building envelopes are optimised to maximise the penetration of solar heat gains and daylight and minimise heat losses and are, thus, unprepared for warmer weather. This will result in large parts of the building stock performing poorly in the future, unless adapted. In cities located in cooler climates, lessons from warmer countries will need to be embedded in urban- and building-level transformations to increase their climate resilience. In warmer countries, further innovation will be required to alleviate urban overheating risk under extreme heat conditions, taking into account other health confounders and exposures, such as air pollution.

Figure 5 visualises the *cooling energy penalty*, i.e. the mean cooling demand increase caused by urban overheating, typically calculated as the cooling load % difference between an urban and a rural reference location. Meta-analysis of existing studies in thirteen cities estimated that cooling loads in urban areas are approximately 13.1% higher compared to their rural counterparts (Santamouris [12]). As a short-term measure, air conditioning systems can reduce vulnerability to heat. However, they increase energy consumption, fuel costs and potentially summer fuel poverty [59, 60] and, when not powered by renewable energy sources, carbon emissions and peak electricity loads, thus putting a significant strain on electricity grids and requiring the construction of additional power plants. Air conditioning units also dissipate waste heat in urban heat canyons, further increasing the UHI, hence creating a vicious circle of overheating and cooling demand.

More sustainable and equitable urban cooling solutions should, therefore, be deployed across building stocks to mitigate the negative health impacts of urban overheating. The next section explores the determinants of intra-city variation of heat health risks and resulting health inequalities.

5 Urban Overheating and Inequalities

Within a city, the health risk due to overheating varies spatially reflecting potentially large variations in socioeconomic as well as environmental factors. UHIs are highly heterogenous, and local temperature fluctuates spatially across most cities as a function of underlying variations in microclimate characteristics. Because of this and the numerous confounders affecting urban health, mapping intra-city inequalities in urban heat health is complex; however, patterns do emerge in many urban settings that link social deprivation with increased risk of heat exposure, morbidity and mortality [61, 62].

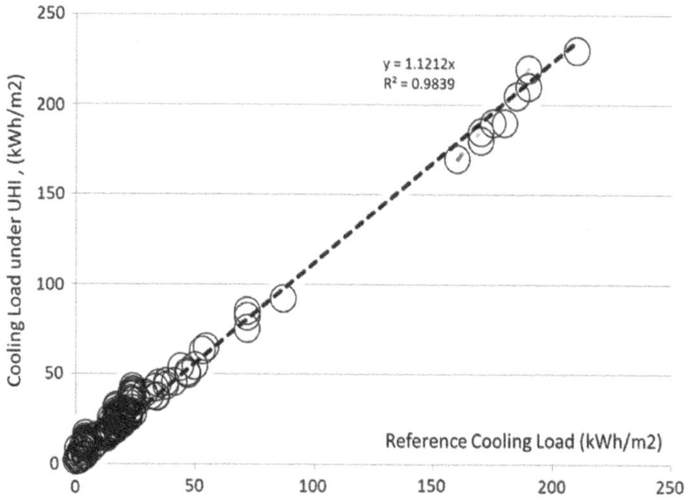

Fig. 5 Correlation between the calculated cooling energy demand under rural (reference) and urban (UHI) conditions. *Source* [12]

In most cities, there is a positive correlation between land surface temperature and the proportion of lower income, ethnic minority, lower education level, older population groups and crime rates [63]. The heat-related mortality risk of people living in parts of the city characterised by higher local temperatures has nearly 6% higher risk of mortality compared to those living in the cooler areas of the same city. The heat-related mortality risk of people living in urban neighbourhoods with lower levels of green infrastructure is 5% higher compared to the residents of greener areas [64]. On average, the thermal centre of most UHIs is located in the centre of the city, where built density is high and humanmade surfaces dominate over green or natural surfaces. In many cases, this coincides with large concentrations of socially disadvantaged populations.

The cooling energy penalty is also higher in these areas, as shown in Fig. 5, and therefore, inner-city populations face increased energy costs compared to the average population, which could contribute to fuel poverty in the summer. Inner-city deprivation has been found to influence heat susceptibility in some settings but not in others, again possibly reflecting interlinked factors, such as penetration of air conditioning or quality of housing [50, 61].

As mentioned in the previous section, belonging to a lower income group or being socially isolated is likely to correlate with other heat risk factors, such as suffering from one of the chronic health conditions mentioned in Sect. 3.1 or co-morbidities. Such individuals may also be characterised by lack of physical and social resources to cope with heat, and limited adaptive thermal capacity, as discussed in Sect. 4. For example, they may not be able to afford adjustments to their immediate environment, such as the purchase of an air conditioning system, or have access to cooler indoor or outdoor spaces. Social isolation is another factor, which often may be high among older demographics.

6 Summary

Urban heat exposure at the regional and city level, as a result of anthropogenic greenhouse gas emissions and the UHI phenomenon, respectively, already poses major challenges to populations worldwide and is projected to have increasing detrimental impacts on human health and wellbeing. This chapter provided an overview of health and wellbeing impacts of urban overheating, key determinant factors and drivers behind geographical variation of intra-city health risk and health inequalities. Older people are among the most severely affected demographic and are particularly at risk from urban heat if already chronically ill, socially deprived or inner-city residents. The distribution of heat vulnerability and health burden across a city is determined by the combination of (1) background climate conditions, (2) underlying socioeconomic structures and (3) the heterogeneity in urban form, building stock characteristics and human activities, which in turn influence the interplay of thermal processes within the urban environment.

Despite the potential benefit of ambient temperature increases for some temperate and colder climatic contexts (such as reductions in space heating demand and fuel poverty during winter), urban overheating is projected to have negative net effects on urban health, and residents may be ill prepared for unprecedented heat events. Systems thinking is required to assess the complex interactions between the human and physical determinants of urban overheating effects, and a transformative, health-focussed, multi-domain, multi-sectoral and multi-actor approach towards the development of climate resilient cities is, therefore, required; this should not only aim to protect human health from the potentially catastrophic impacts of urban overheating, but also promote broader co-benefits for human wellbeing, economies and communities.

References

1. Basu R (2009) High ambient temperature and mortality: a review of epidemiologic studies from 2001 to 2008. Environ Health: Glob Access Sci Source 8(1). https://doi.org/10.1186/1476-069X-8-40
2. Gasparrini A, Guo Y, Hashizume M (2015) Mortality risk attributable to high and low ambient temperature: a multicountry observational study. Lancet 386:369–375. https://doi.org/10.1016/S0140-6736(14)62114-0
3. Vardoulakis S, Dear K, Hajat S, Heaviside C, Eggen B, McMichael AJ (2015) Comparative assessment of the effects of climate change on heat- and cold-related mortality in the United Kingdom and Australia. Environ Health Perspect 122(12):1285–1292. https://doi.org/10.1289/ehp.1307524
4. Baccini M, Biggeri A, Accetta G, Kosatsky T, Analitis A, Anderson HR, Bisanti L, D'Ippoliti D, Danova J (2008) Heat effects on mortality in 15 European cities. Epidemiology 19(5):711–719. https://doi.org/10.1097/EDE.ObO
5. Kovats R, Hajat S (2008) Heat stress and public health: a critical review. Annu Rev Public Health 29(1):41–55. https://doi.org/10.1146/annurev.publhealth.29.020907.090843

6. C40 Cities, & UCCRN (2018) The future we don't want: how climate change could impact the World's greatest cities. Retrieved from https://www.c40.org/what-we-do/scaling-up-climate-action/adaptation-water/the-future-we-dont-want/
7. IPCC (2021) Summary for policymakers. In Masson-Delmotte V, Zhai P, Pirani A, Connors SL, Péan C, Berger S, Caud N, Chen Y, Goldfarb L, Gomis MI, Huang M, Zhou B (eds) Climate Change 2021: the physical science basis. contribution of working group I to the Sixth Assessment Report of the Intergovernmental Panel on Climate Change
8. Oke TR (1987) Boundary layer climates, 2nd edn. Routledge, Abingdon, UK
9. IPCC (2018) Global warming of 1.5 °C, summary for policymakers. Geneva, Switzerland. Retrieved from http://www.ipcc.ch/pdf/special-reports/sr15/sr15_spm_final.pdf
10. MetOffice (2021) One billion face heat-stress risk from 2 °C rise. Retrieved from https://www.metoffice.gov.uk/about-us/press-office/news/weather-and-climate/2021/2c-rise-to-put-one-in-eight-of-global-population-at-heat-stress-risk
11. The Lancet (2021) Health in a world of extreme heat. Lancet 398(641). https://doi.org/10.1016/S0140-6736(21)01860-2
12. Santamouris M (2020) Recent progress on urban overheating and heat island research. Integrated assessment of the energy, environmental, vulnerability and health impact. Synergies with the global climate change. En Build 207. https://doi.org/10.1016/j.enbuild.2019.109482
13. Oke TR (1982) The energetic basis of the urban heat island. Q J Roy Meteorol Soc 108(455):1–24
14. Smith CL, Webb A, Levermore GJ, Lindley SJ, Beswick K (2011) Fine-scale spatial temperature patterns across a UK conurbation. Clim Change 109(3–4):269–286. https://doi.org/10.1007/s10584-011-0021-0
15. Gedzelman SD, Austin S, Cermak R, Stefano N, Partridge S, Quesenberry S, Robinson DA (2003) Mesoscale aspects of the urban heat Island around New York City. Theoret Appl Climatol 75(1–2):29–42. https://doi.org/10.1007/s00704-002-0724-2
16. Oke TR (1973) City size and the urban heat Island. Atmos Environ 7(8):769–779. https://doi.org/10.1016/0004-6981(73)90140-6
17. Bohnenstengel SI, Evans S, Clark PA, Belcher SE (2011) Simulations of the London urban heat island. Q J Roy Meteorol Soc 137(659):1625–1640. https://doi.org/10.1002/qj.855
18. Santamouris M (2015) Analyzing the heat island magnitude and characteristics in one hundred Asian and Australian cities and regions. Sci Total Environ 512–513:582–598. https://doi.org/10.1016/j.scitotenv.2015.01.060
19. Burkart KG, Brauer M, Aravkin AY, Godwin WW, Hay SI, He J, Iannucci VC, Larson SL, Lim SS, Liu J, Murray CJ, Stanaway JD (2021) Estimating the cause-specific relative risks of non-optimal temperature on daily mortality: a two-part modelling approach applied to the Global Burden of Disease Study. Lancet, 398(10301):685–697. https://doi.org/10.1016/S0140-6736(21)01700-1
20. Vicedo-Cabrera AM, Scovronick N, Sera F, Royé D, Schneider R, Tobias A, Astrom C, Guo Y, Honda Y, Hondula DM, Abrutzky R, Gasparrini A (2021) The burden of heat-related mortality attributable to recent human-induced climate change. Nat Clim Change 11(6):492–500. https://doi.org/10.1038/s41558-021-01058-x
21. UN DESA (2018) World urbanization prospects: The 2018 revision, key facts. New York, USA. https://doi.org/10.1017/CBO9781107415324.004
22. Campbell-Lendrum D, Corvalán C (2007) Climate change and developing-country cities: implications for environmental health and equity. J Urban Health 84(Suppl. 1):109–117. https://doi.org/10.1007/s11524-007-9170-x
23. McMichael AJ, Wilcox BA (2009) Climate change, human health, and integrative research: a transformative imperative. EcoHealth 6(2):163–164. https://doi.org/10.1007/s10393-009-0262-9
24. Patz JA, Campbell-Lendrum D, Holloway T, Foley JA (2005) Impact of regional climate change on human health. Nature 438(7066):310–317. https://doi.org/10.1038/nature04188
25. O'Neill MS, Ebi KL (2009) Temperature extremes and health: Impacts of climate variability and change in the United States. J Occup Environ Med 51(1):13–25. https://doi.org/10.1097/JOM.0b013e318173e122

26. Huang C, Barnett AG, Wang X, Vaneckova P, Fitzgerald G, Tong S (2011) Projecting future heat-related mortality under climate change scenarios: a systematic review. Environ Health Perspect 119(12):1681–1690. https://doi.org/10.1289/ehp.1103456
27. Lowe SA (2016) An energy and mortality impact assessment of the urban heat island in the US. Environ Impact Assess Rev 56:139–144. https://doi.org/10.1016/j.eiar.2015.10.004
28. Macintyre HL, Heaviside C, Cai X, Phalkey R (2021) Comparing temperature-related mortality impacts of cool roofs in winter and summer in a highly urbanized European region for present and future climate. Environ Int 154. https://doi.org/10.1016/j.envint.2021.106606
29. Hajat S, Vardoulakis S, Heaviside C, Eggen B (2014) Climate change effects on human health: projections of temperature-related mortality for the UK during the 2020s, 2050s and 2080s. J Epidemiol Community Health 68(7):641–648. https://doi.org/10.1136/jech-2013-202449
30. WHO (2021) Heat and health. World Health Organisation (WHO). Retrieved from https://www.who.int/news-room/fact-sheets/detail/climate-change-heat-and-health
31. Lai LW, Cheng WL (2009) Air quality influenced by urban heat island coupled with synoptic weather patterns. Sci Total Environ 407(8):2724–2733. https://doi.org/10.1016/j.scitotenv.2008.12.002
32. Reid CE, Snowden JM, Kontgis C, Tager IB (2012) The role of ambient ozone in epidemiologic studies of heat-related mortality. Environ Health Perspect 120(12):1627–1630. https://doi.org/10.1289/ehp.1205251
33. Heal MR, Heaviside C, Doherty RM, Vieno M, Stevenson DS, Vardoulakis S (2013) Health burdens of surface ozone in the UK for a rang of future scenarios. Environ Int 61:36–44. https://doi.org/10.1016/j.envint.2013.09.010
34. Martinez GS, Linares C, de'Donato F, Diaz J (2020) Protect the vulnerable from extreme heat during the COVID-19 pandemic. Environ Res 187:13–15. https://doi.org/10.1016/j.envres.2020.109684
35. Wilhelmi OV, Howe PD, Hayden MH, O'Lenick CR (2021) Compounding hazards and intersecting vulnerabilities: experiences and responses to extreme heat during COVID-19. Environ Res Lett 16(8). https://doi.org/10.1088/1748-9326/ac1760
36. Bose-O'Reilly S, Daanen H, Deering K, Gerrett N, Huynen MMTE, Lee J, Karrasch S, Matthies-Wiesler F, Mertes H, Schoierer J, Shumake-Guillemot J, Nowak D (2021) COVID-19 and heat waves: New challenges for healthcare systems. Environ Res 198. https://doi.org/10.1016/j.envres.2021.111153
37. Hajat S, Kosatky T (2010) Heat-related mortality: a review and exploration of heterogeneity. J Epidemiol Community Health 64(9):753–760. https://doi.org/10.1136/jech.2009.087999
38. Toulemon L, Barbieri M (2008) The mortality impact of the August 2003 heat wave in France: investigating the "harvesting" effect and other long-term consequences. Popul Stud 62(1):39–53. https://doi.org/10.1080/00324720701804249
39. Hajat S, Armstrong BG, Gouveia N, Wilkinson P (2005) Mortality displacement of heat-related deaths: a comparison of Delhi, São Paulo, and London. Epidemiology 16(5):613–620. https://doi.org/10.1097/01.ede.0000164559.41092.2a
40. Le Tertre A, Lefranc A, Eilstein D, Declercq C, Medina S, Blanchard M, Chardon B, Fabre P, Filleul L, Jusot JF, Pascal L, Ledrans M (2006) Impact of the 2003 heatwave on all-cause mortality in 9 French cities. Epidemiology 17(1):75–79. https://doi.org/10.1097/01.ede.0000187650.36636.1f
41. Kovats S, Hajat S (2008) Heat stress and public health: a critical review. Annu Rev Public Health 29:41–55. https://doi.org/10.1146/annurev.publhealth.29.020907.090843
42. Heaviside C, Macintyre H, Vardoulakis S (2017) The urban heat Island: implications for health in a changing environment. Curr Environ Health Rep 4(3):296–305. https://doi.org/10.1007/s40572-017-0150-3
43. Armstrong B (2006) Models for the relationship between ambient temperature and daily mortality. Epidemiology 17(6):624–631. https://doi.org/10.1097/01.ede.0000239732.50999.8f
44. Basu R, Samet JM (2002) Relation between elevated ambient temperature and mortality: a review of the epidemiologic evidence. Epidemiol Rev 24(2):190–202. https://doi.org/10.1093/epirev/mxf007

45. Kovats RS, Johnson H, Griffith C (2006) Mortality in southern England during the 2003 heat wave by place of death. Health Stat Q 29:6–8
46. Johnson H, Kovats RS, McGregor GR, Stedman JR, Gibbs M, Walton H, Cook L, Black E (2005) The impact of the 2003 heatwave on mortality and hospital admissions in England. Health Stat Q 25(7):6–11
47. McMichael AJ, Wilkinson P, Kovats RS, Pattenden S, Hajat S, Armstrong B, Vajanapoom N, Niciu EM, Mahomed H, Kingkeow C, Kosnik M, Nikiforov B (2008) International study of temperature, heat and urban mortality: the "ISOTHURM" project. Int J Epidemiol 37(5):1121–31. https://doi.org/10.1093/ije/dyn086
48. Milojevic A, Armstrong BG, Gasparrini A, Bohnenstengel SI, Barratt B, Wilkinson P (2016) Methods to estimate acclimatization to urban heat island effects on heat-and cold-related mortality. Environ Health Perspect 124(7):1016–1022. https://doi.org/10.1289/ehp.1510109
49. Turner LR, Barnett AG, Connell D, Tong S (2012) Ambient temperature and cardiorespiratory morbidity: a systematic review and meta-analysis. Epidemiology 23(4):594–606. https://doi.org/10.1097/EDE.0b013e3182572795
50. Stafoggia M, Forastiere F, Agostini D, Biggeri A, Bisanti L, Cadum E, Caranci N, de'Donato F, De Lisio S, De Maria M, Michelozzi P, Perucci CA (2006) Vulnerability to heat-related mortality: a multicity, population-based, case-crossover analysis. Epidemiology 17(3):315–323. https://doi.org/10.1097/01.ede.0000208477.36665.34
51. Harlan SL, Brazel AJ, Darrel Jenerette G, Jones NS, Larsen L, Prashad L, Stefanov WL (2007) In the shade of affluence: the inequitable distribution of the urban heat island. Res Soc Probl Public Policy 15(07):173–202. https://doi.org/10.1016/S0196-1152(07)15005-5
52. Fouillet A, Rey G, Wagner V, Laaidi K, Empereur-Bissonnet P, Le Tertre A, Frayssinet P, Bessemoulin P, Laurent F, De Crouy-Chanel P, Jougla E, Hémon D (2008) Has the impact of heat waves on mortality changed in France since the European heat wave of summer 2003? A study of the 2006 heat wave. Int J Epidemiol 37(2):309–17. https://doi.org/10.1093/ije/dym253
53. Yu W, Vaneckova P, Mengersen K, Pan X, Tong S (2010) Is the association between temperature and mortality modified by age, gender and socio-economic status? Sci Total Environ 408(17):3513–3518. https://doi.org/10.1016/j.scitotenv.2010.04.058
54. Vandentorren S, Bretin P, Zeghnoun A, Mandereau-Bruno L, Croisier A, Cochet C, Ribéron J, Siberan I, Declercq B, Ledrans, M (2006). August 2003 heat wave in France: risk factors for death of elderly people living at home. Eur J Public Health 16(6):583–591. https://doi.org/10.1093/eurpub/ckl063
55. Salagnac JL (2007) Lessons from the 2003 heat wave: a French perspective. Build Res Inf 35(4):450–457. https://doi.org/10.1080/09613210601056554
56. Taylor J, Wilkinson P, Davies M, Armstrong B, Chalabi Z, Mavrogianni A, Symonds P, Oikonomou E, Bohnenstengel SI (2015) Mapping the effects of Urban Heat Island, housing, and age on excess heat-related mortality in London. Urban Clim 14:517–528
57. Hajat S, Kovats S, Lachowycz K (2007) Heat-related and cold-related deaths in England and Wales: who is at risk? Occup Environ Med 64(2):93–100. https://doi.org/10.1136/oem.2006.029017
58. Gupta R, Howard A, Davies M, Mavrogianni A, Tsoulou I, Oikonomou E, Wilkinson P (2021) Examining the magnitude and perception of summertime overheating in London care homes. Build Serv Eng Res Technol 42(6):653–675. https://doi.org/10.1177/01436244211013645
59. Sanchez-Guevara C, Núñez Peiró M, Taylor J, Mavrogianni A, Neila González J (2019) Assessing population vulnerability towards summer energy poverty: case studies of Madrid and London. Energy Build 190. https://doi.org/10.1016/j.enbuild.2019.02.024
60. Kolokotsa D, Santamouris M (2015) Review of the indoor environmental quality and energy consumption studies for low income households in Europe. Sci Total Environ 536:316–330. https://doi.org/10.1016/j.scitotenv.2015.07.073
61. Ellena M, Breil M, Soriani S (2020) The heat-health nexus in the urban context: a systematic literature review exploring the socio-economic vulnerabilities and built environment characteristics. Urban Clim 34. https://doi.org/10.1016/j.uclim.2020.100676

62. Reid CE, O'Neill MS, Gronlund CJ, Brines SJ, Brown DG, Diez-Roux AV, Schwartz J (2009) Mapping community determinants of heat vulnerability. Environ Health Perspect 117(11):1730–1736. https://doi.org/10.1289/ehp.0900683
63. Huang G, Cadenasso ML (2016) People, landscape, and urban heat island: dynamics among neighborhood social conditions, land cover and surface temperatures. Landscape Ecol 31(10):2507–2515. https://doi.org/10.1007/s10980-016-0437-z
64. Schinasi LH, Benmarhnia T, De Roos AJ (2018) Modification of the association between high ambient temperature and health by urban microclimate indicators: a systematic review and meta-analysis. Environ Res 161:168–180. https://doi.org/10.1016/j.envres.2017.11.004

Chapter 2
A Global Synthesis of Heat-Related Mortality in Overheated Cities

Nasrin Aghamohammadi, Logaraj Ramakreshnan, Chng Saun Fong, and Prashant Kumar

Abstract In the twenty-first century, megacities around the world are vulnerable to climate change, and its effect on urban mortality is exacerbated by extreme heat events (EHE) and urban heat islands (UHI). Since climate projections tend to exclude the EHE and UHI, their impact on urban health and urban mortality could be underestimated. This present study aims to provide a systematic synthesis of the available evidence of the impact of UHI on urban mortality across the globe. A literature survey was performed on research articles published by Web of Science and Google Scholar, and relevant peer-reviewed articles were included to investigate the relationship between all-cause mortality with UHI episodes in the megacities around the world. The evidence pertaining to the all-cause mortality based on field survey, retrospective time series analysis, and models were extracted for expert judgement. The results suggest that the UHI contributed to the total heat-related mortality during the major heatwave episodes in the world. Effects were found to vary with the cause of death, age, gender, geographical settings, and sociodemographic status in the reported studies. Comprehension of the main determinants of heat-related mortality and the projected trend of this association in the rapidly expanding urban regions is prudential to inform preparedness and targeted interventions across the globe.

Keywords Extreme heat event · Heatwave · Urban health · Urban heat island · Urban mortality

N. Aghamohammadi (✉)
Department of Social and Preventive Medicine, Faculty of Medicine, Centre for Epidemiology and Evidence-Based Practice, University of Malaya, 50603 Kuala Lumpur, Malaysia
e-mail: nasrin@ummc.edu.my

Centre for Energy Sciences, University of Malaya, 50603 Kuala Lumpur, Malaysia

L. Ramakreshnan · C. S. Fong
Institute for Advanced Studies, University of Malaya, 50603 Kuala Lumpur, Malaysia

P. Kumar
Department of Civil and Environmental Engineering, Faculty of Engineering and Physical Sciences, Global Centre for Clean Air Research (GCARE), University of Surrey, Guildford, Surrey GU2 7XH, UK

N. Aghamohammadi and M. Santamouris (eds.), *Urban Overheating: Heat Mitigation and the Impact on Health*, Advances in Sustainability Science and Technology,
https://doi.org/10.1007/978-981-19-4707-0_2

1 Introduction

Rising temperatures due to urbanization and population explosion have accelerated the global warming effect that has threatened both human and environmental health globally [1]. People living in urban areas worldwide face a variety of health risks due to various pollutions caused by industries, transport, commercial, and anthropogenic activities [2, 3]. This is of particular concern to the public health community, rapid urbanization has placed about 55% of the global population in urban areas, and this is projected to increase up to 60% by 2030 [4]. If the trend continues, the health risks for urban populations will continue to increase in the future, particularly in the developing world.

Urban areas can experience ambient temperatures appreciably warmer than surrounding rural areas, a phenomenon known as the urban heat island (UHI) effect [5]. This phenomenon is mainly attributed to the built environment landscapes, urban materials such as building facades, roads, and other paved areas that absorb and store more heat than natural landscapes as well as waste heat generated by energy processes in the buildings, transport systems, commercial and industrial operations, and other anthropogenic activities [6]. Human activities in changing natural land covers can lead to changes in thermal capacities, albedo coefficients, heat conductivity, and moisture contents of the surroundings [7, 8]. Human health effects associated with rising temperatures are expected to increase significantly by the mid-to-late century. A large body of work now estimates an increase in mean global temperature from pre-industrial averages of more than 2 °C by the late century under mid-range emissions scenarios [9]. A smaller but growing body of work has sought to estimate the effects of projected warming on heat-related mortality.

By employing health impact functions derived from epidemiological studies of historical warm season mortality rates, recent work projects an increase in annual heat-related mortality of between 3500 and 27,000 deaths in the United States by mid-century [10]. Studies focussed on individual cities estimate an increase in annual heat-related mortality by a factor of 2–7 by the mid-to-late twenty-first century [9, 10]. Medical and health researchers have shown that fatalities associated with high-surrounding temperatures are most commonly due to heat's negative effect on the cardiovascular system that caused respiratory and cardiovascular diseases [11]. In an attempt to control one's internal temperature, the body's natural instinct is to circulate large quantities of blood to the skin. However, performing this protective measure against overheating actually harms the body by inducing extra strain on the heart. This excess strain has the potential to trigger a cardiac event in individuals with chronic health problems, such as older adults [12]. In a study, Frumkin [13] showed that the relationship between mortality and temperature creates a J-shaped function, showing a steeper slope at higher temperatures in the city centres. Similarly, Avashia et al. [14] reported that air temperature and mortality associations for summer's atmospheric temperature (AT) and land surface temperature (LST) indicate a 'J' ('hockey-stick') shaped curve in Ahmedabad, India, as shown in Fig. 1. This suggests that the lowest point on the curve, indicating the lowest heat, represents a low-mortality rate, which

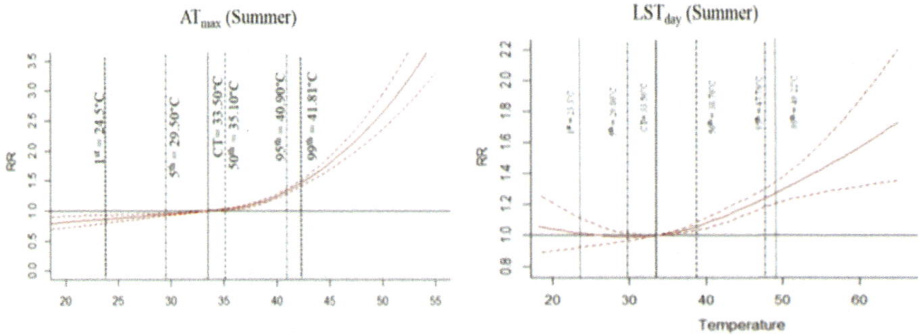

Fig. 1 Temperature and mortality risks in Ahmedabad, illustrating a 'J'/'hockey-stick' curve. This suggests that the lowest point on the curve (lowest heat) represents a low-mortality rate, which increases as the urban heat increases. *Source* [14]

increases as the urban heat increases. Records show that more casualties have resulted from urban heat and heatwaves in the developed countries than hurricanes, floods, and tornadoes together.

1.1 The Mechanism of Heat and Mortality

The core temperature of a human is 37 °C. The human body is acclimatized to this temperature and local climates in terms of its physiological functions. All the humans have absolute limits to the amount of heat exposure that they can tolerate. In fact, healthy adults have heat regulatory mechanisms to cope with temperature increases up to a particular threshold. When the external environment is hot, the body maintains core temperature by losing heat via radiant, convective, and evaporative heat loss by vasodilatation and perspiration [15]. If the surrounding temperature is the same or higher than the body temperature, the effectiveness of this mechanism is remarkably reduced. This eventually led to a clinical condition called heat stress that describes a spectrum of heat-related illness that occur when the body is unable to maintain its core temperature (37 °C) [16]. Heat stress ranges from heat cramps to heat exhaustion. When the core temperature rises above 40.5 °C., it can lead to heatstroke that can lead to multiple organ dysfunction. Heatstroke occurs due to the failure of the hypothalamus, the region of the brain that coordinates our physiological response to excessive heat [17]. At extremely hot environment, irreversible brain damage can occur with neurological signs such as lack of coordination, consciousness, and seizures. Besides, the skin will become hot and dry due to less sweating to release heat. This elevates the heart rate and breathing and lowers the blood pressure. This also damages the body cells as well as impairs heart and kidneys' functions that can lead to multiple organ dysfunction. This eventually accelerates the mortality risk as people with severe heatstroke can die within a few hours or several days from organ failure. The studies have demonstrated that the heat caused mortality risk increases with age, illness, and medical conditions [18]. In terms of the epidemiological studies, the heat-mortality function at a community-level is assessed using routinely collected data to measure short-term (day-to-day) associations between daily mortality counts and ambient temperature levels after controlling for confounding factors (e.g.: health conditions, air pollution, etc.) [19].

1.2 Heat and UHI as a Menace to Urban Health Across the Globe

In general, increasing heat in the surrounding environment is identified as one of the known threats to humankind. For example, during prolonged extreme heat events often reported as heatwaves, in July 2009 in British Columbia and July 2010 in Quebec, public health officials estimated that there were 156 and 280 excess deaths, respectively, from heat-related causes [20]. In Montreal, 66 people died due to a heatwave in July 2018 when temperatures spiked as high as 35.3 °C [21]. Although heat-related deaths and illnesses are largely preventable [22], extreme heat is a leading cause of weather-related morbidity and mortality globally [23]. In the United States, extreme heat is the primary cause of weather-related mortality [22]. In Canada, high temperatures have been found to be a significant health risk, with high temperatures in summer associated with excess mortality in many Canadian cities [24].

The most common illnesses provoked during extreme heat, which often lead to fatalities, are respiratory and cardiovascular diseases due to the extensive strain on the heart from blood circulation to the skin as a cooling function. This excess strain on the heart becomes calamitous for those with chronic heart problems and the elderly when it reaches a critical point that can invoke a cardiac event. Danger also exists for those with congestive heart failure (CHF), which is a circumstance where the heart cannot pump sufficient blood to the rest of the body, leaving vital organs without enough oxygen or essential nutrients. Increased temperatures exacerbate this condition, leaving the body unable to feed vital organs enough blood or regulate body temperature leading to overheating and possibly death [11]. Studies show that the groups most affected by extreme heat are in fact those that are most likely to suffer from the above conditions [25]. These groups include those between the ages of 65–74, people unmarried or living alone, the population previously hospitalized for chronic pulmonary disease, and people suffering from psychiatric disorders. In addition, women and unmarried people that fall in the 75-f age group were also more vulnerable to heat. The lack of self-dependence and weakened body health in these groups are the biggest reasons for their being the most affected by increases in temperature. The population between the ages of 65 and 74 is more likely to have a history of suffering from heart disorders or be in a weakened state from ageing, which causes their susceptibility to fluctuations in heat. Single subjects, as well as those with psychiatric disorders living alone, may not be aware of the adverse effects the excess heat has on their bodies [25].

The UHI effect is another factor known to magnify health impacts during extreme heat events. The UHI can contribute to thermal discomfort and a range of morbidity and mortality outcomes [26]. Increased temperature adds extra pressure to human physiology and makes the body more vulnerable to stress. Residents living in a UHI region or in its vicinity are at increased health risk [27]. Therefore, the UHI is one of the major urban planning problems, which requires monitoring as well dedicated strategies to reduce its impact [28]. As an example, Gosling et al. [29] reported that an increasing urban temperature has a direct impact on the increase in mortality rate. Other studies have shown that citizens in UHI areas suffer from heat-related illnesses such as digestive diseases, nervous system issues, insomnia, depression, and mental illnesses [30]. UHIs have also directly led to a rise in infectious diseases. Along with this, in a systematic review by Thompson et al. [31] increased risks of mental health-related admissions and emergency department visits in higher temperatures areas were found. The findings of Jenerette et al. [32] also demonstrated that the symptoms of heat-related illness were correlated with parcel-scale surface temperature patterns during the daytime in an urban ecosystem. Wong et al. [22] reported that more psychological and social health issues are associated with UHI, such as depression and the restriction of social activities. Further associations between UHI and physical, psychosomatic, and psychological (PPP) health symptoms [33] and somatic symptoms [34] were also presented in the tropical settings. Indeed, the UHI phenomenon compounds the potential effects of global-scale climate change on heat-related mortality amongst urban populations. Time series analyzes of climatic trends in cities find large urbanized regions to be warming at a higher rate than proximate rural areas, with many cities warming at more than twice the mean global rate. The combined effects of UHI formation and the global greenhouse effect are projected to significantly increase the number of extreme heat events in urbanized regions. At present, the extent to which the UHI effect may further increase heat-related mortality is not well established. Therefore, this chapter presents a synthesis of the association between UHI and mortality based on the published articles.

2 Methodology

A literature search was conducted from September to October 2021 using electronic databases such as Web of Science and Google Scholar to obtain research evidence on the UHI and mortality across the countries. A Boolean search using a combination of keywords including 'urban heat', 'heat island', 'urban heat island', 'mortality', and 'death' was conducted without any restrictions on the time to retrieve peer-reviewed original research articles from each of the databases. All types of quantitative studies that empirically assessed the relationship between the UHI phenomenon and mortality were eligible for inclusion in this review. However, the literature search was limited to research articles published in English which are available in open source and also retrievable in full-text PDF format.

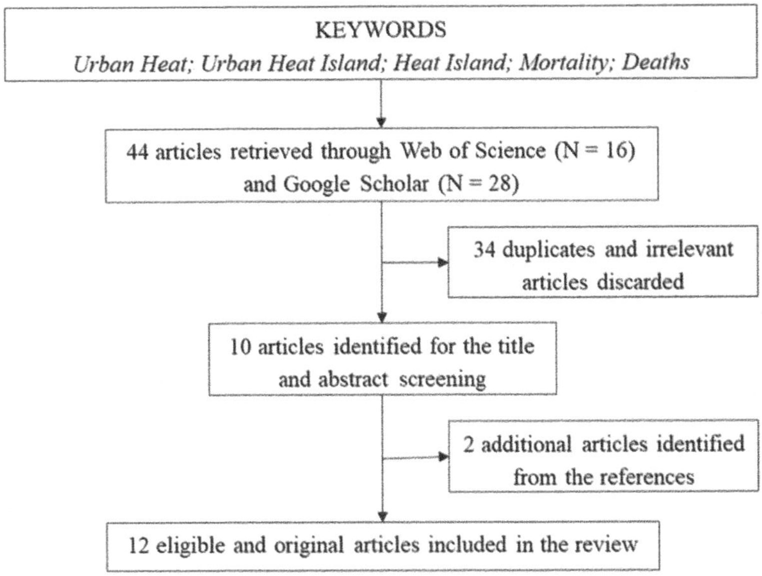

Fig. 2 Flow chart demonstrating the selection process of papers

The search strategy yielded 44 articles from Web of Science ($N = 16$) and Google Scholar ($N = 28$). Finally, about ten original articles were identified for a full-text review after discarding the duplicates and irrelevant articles through title and abstract screening. References and citations of the finalized eligible articles were checked, and two articles were added to ensure the inclusion of all the relevant articles in this review. From the remaining original articles ($N = 12$), the information pertaining to the UHI and mortality across the countries was extracted for expert judgement. Other than these studies, relavant manuscripts are also reviewed to discuss the concept of UHI, heat-related mortality, its mechanism, and risk factors as well as mitigation measures in different parts of this chapter to facilitate the understanding of the readers. All the analysis and findings reviewed in this article are purely based on the author's interpretation of the reported findings in each paper. The flow of paper selection is illustrated in Fig. 2.

3 Association Between UHI and All-Cause Mortality Across the Globe

Relatively, few studies have explored the association between UHI and heat-related mortality. Most of the existing studies concentrated on investigating the relationship between heat waves or extreme heat events (EHE) and mortality, with some studies looking into the variations in the vulnerability of the population from such factors as population age or socioeconomic deprivation [35, 36]. By using mortality case-crossover analyzes from 1993 until 2006, Milojevic et al. [37] examined the degree of acclimatization to heat- and cold-related mortality in relation to UHI anomalies in London, United Kingdom. The findings revealed that there is little difference in the relative risk of mortality on hot versus normal days across UHI decile groups. A 1 °C UHI anomaly multiplied the risk of heat death by 1.004 (95% CI = 0.950, 1.061) compared with the expected value of 1.070 (95% CI = 1.057, 1.082) if there were no acclimatization. The corresponding actual versus expected UHI interaction for the cold was 1.020 (95% CI = 0.979, 1.063) and 1.030 (95% CI = 1.026, 1.034), respectively. Glutting [38] investigated the relationship between micro-UHIs (MUHI) and mortality in the city of Barcelona, Spain, using Landsat 7 thermal infrared imagery overlaid them onto at-home deaths between 2000 and 2003. In the period 2000–2003, at-home deaths in MUHI were associated with 15% greater odds of dying on hot days than at-home deaths outside of MUHI.

In another study conducted in London, the United Kingdom, Taylor et al. [39] examined the spatial distribution of mortality risk from high temperatures based on weather data for 2.65 million dwellings in a building stock database from 26 May to 19 July 2006. They estimated UHI values for individual buildings varied from 20.5 to 23.0 °C (mean of 22.5 °C) during the study period (28.2–31.5 °C, with a mean of 30.2 °C during the hot spell). Houses located in the centre of the UHI could face an additional 1.2 times increase in excess relative risk for an individual over 85 years old, relative to the average London UHI risk under the hot spell conditions. They also reported that the ward-mean maximum daily living room temperatures (MMDTs) when the 24.8 °C outdoor temperature-mortality threshold is exceeded the range from 26.4 to 28.3 °C (mean of 27.1 °C), whilst ward-mean UHI values showed a negative skew, with a range from 20.8 to 22.9 °C (mean of 22.5 °C). The results indicated that the increase in mortality inwards where the average building location UHI and MMDT is greater than the London-wide average, thus presenting positive anomalies and an increase in relative risks. The findings also revealed that the suburbs of London are predicted to have higher all-cause mortality due to the higher proportion of elderly people in outer compared with inner London.

By establishing a city-wide temperature versus daily mortality function on the basis of a case-crossover design for June–August 1990–2003, Smargiassi et al. [40] explored whether people located in micro-UHI are at higher risk of mortality during hot summer days in Montreal, Canada. They found that the association between temperature and mortality is curvilinear, with a linear component close to zero, and then increased gradually until about 20 °C, after which the increase in relative risks was exponential. On days with an average daily ambient temperature of 26 °C compared with days with an average of 20 °C, the odds ratio for heat-induced mortality was 1.16 (lag 0 days, $OR_{26\,°C vs 20\,°C} = 1.04 \times 1.05 \times 1.06$; 95% CI = 1.12, 1.19). ORs of dying at elevated daily ambient temperatures were greater for the 2-day mean (0–1 days) than for lag 0 or lag 1 day alone. On the other hand, Lowe [41] assessed the UHI impact on energy and mortality on an annual basis in New York, United States of America. The mortality impacts are addressed from both the heat and cold perspectives. They identified that the impact of the UHI can be seen as the increase in the death rate of 1.1 people per million. The impact of the UHI also can be seen as a decrease in the cold-related death rate of 4.0 deaths per million.

In an attempt of analyzing the impact of high temperatures, heatwaves, and UHI on the cardiovascular and respiratory mortality of people over 65 years of age for the years 2002–2012 in Athens, Greece, Paravantis et al. [42] identified that the correlation of the daily cardiovascular and respiratory mortality count with various temperature measures confirmed a U-shaped exposure–response curve, with fewer deaths in the range of moderate temperatures. They identified that the fatalities are the lowest around maximum temperature values of 20–30 °C and increase below and above these values. In addition, the highest daily mortality counts are observed between 35 and 40 °C. At high and very high temperatures, the mortality increased by 20–35%, correspondingly with a one-day lag for the maximum temperature effect on mortality. Heaviside et al. [43] quantified the attribution of the UHI to heat-related mortality in the West Midlands during the heatwave of August 2003 by comparing health impacts based on two modelled temperature simulations. The results suggest that the UHI contributed around 50% of the total heat-related mortality during the 2003 heatwave in the West Midlands. In a comparison of the UHI contribution to mortality expected from a range of projected temperatures based on the UKCP09 Climate Projections, a typical heatwave in 2080 could be responsible for an increase in mortality of around 3-times the rate in 2003 (278 vs. 90 deaths) when including changes in population, population weighting, and the UHI effect in the West Midlands.

In an investigation regarding the influence of micro-UHI on daily mortality in Hong Kong from June–September, 2001–2009, Goggins et al. [44] used Poisson generalized additive models (GAMs) to identify that the mean temperatures (lags 0–4) above 29 °C and low-mean wind speeds (lags 0–4) were significantly associated with higher daily mortality, and these associations were stronger in areas with high UHII. A 1 °C rise above 29 °C was associated with a 4.1% (95% CI = 0.7%, 7.6%) increase in natural mortality in areas with high UHII but only a 0.7% (95% CI = −2.4%, 3.9%) increase in low-UHII areas. Lower mean wind speeds (5th percentile vs. 95th percentile) were associated with a 5.7% (95% CI = 2.7, 8.9) mortality increase in high-UHII areas versus a 0.3% (95% CI = −3.2%, 2.6%) change in low-UHII areas.

Besides, Zhu et al. [45] used city-specific exposure–response functions to multiple temperatures and population projections under different climate and urbanization scenarios to assess the non-optimum temperature-related mortality burdens in China from 2000 to 2050. The findings underscored that temperature-related deaths will decrease from 1.19 million in 2010 to 1.08–1.17 million in 2050, except for the most populous scenario. Excess deaths attributable to non-optimal temperatures under representative concentration pathway 8.5 (RCP8.5) were 2.35% greater than those under RCP4.5. Simultaneously, the net effects of UHIs are beneficial in the historical periods, preventing 3493 (95% CI = 22, 6964) deaths in 2000. Nevertheless, UHIs will cause an additional 6951 (95% CI = −17,637, 31,539, SSP4-RCP4.5) to 17,041 (95% CI = -10,516, 44,598, SSP5-RCP8.5) deaths in 2050. Moreover, the findings highlighted that the mortality burden exhibited strong spatial variations, with heavy burdens concentrated in the hotspots, including Beijing-Tianjin Metropolitan Region,

Yangtze River Delta, Chengdu-Chongqing City Group, Guangzhou, Wuhan, Xi'an, Shandong, and Henan. These hotspots should be priority areas for the allocation of the national medical resources to provide effective public health interventions.

Macintyre et al. [46] quantified the UHI intensity in wintertime for a heavily urbanized UK region (West Midlands, including Birmingham) using a regional weather model and used a health impact assessment (HIA) to estimate the associated impact on cold-related mortality. The findings showed that the population-weighted mean winter UHI intensity was +2.3 °C in Birmingham city centre and comparable with that of summer. The results suggest a potential protective effect of the wintertime UHI, equivalent to 266 cold-related deaths avoided (~15% of total cold-related mortality over ~11 weeks). When including the impacts of climate change, the results suggest that the number of heat-related deaths associated with the summer UHI will increase from 96 (in 2006) to 221 in the 2080s, based on the RCP8.5 emissions pathway. The protective effect of the wintertime UHI is projected to increase only slightly from 266 cold-related deaths avoided in 2009 to 280 avoided in the 2080s.

By using an empirical temperature-mortality relationship for London, Milojevic et al. [47] revealed that the overall number of heat-related deaths in London during the period of study was 118.3 (95% CI = 105.8, 130.7). The percentage of heat deaths attributable to the UHI effect, which varies with temperature conditions, was substantial for the non-extreme heat of the study period: 37.7% in 'outer London', 46.6% in 'inner London', and 47.2% in 'central London'. In winter, the UHI acts to reduce cold-related mortality. In a study to understand current temperature-related mortality impacts in Ahmedabad city, India, Avashia et al. [14] used satellite images (MODIS from NASA), temperature data from India Meteorological Department (IMD) and daily all-cause mortality from Ahmedabad Municipal Corporation between 2001 and 2015 to create a distributed lag non-linear model. They underpinned that the association between temperature and mortality persisted across the lag period of 0–5 days. However, the effect size was observed to decrease with increasing lag. For the summer maximum temperature association, as the temperature rises beyond the 95th percentile ATmax, the relative risk (RR) values increase at a higher rate. At the 99th percentile temperature of 44 °C, the RR is 1.64, whilst at 47 °C, it crosses the RR of 2. The study also infers that with an increase in built-up spaces by 1% in the land use mix, the relative risk of heat-related mortality increases by 0.59 points at 40 °C and by 0.78 points at 45 °C.

The summary of the aforementioned studies was provided in Table 1 by highlighting the locations, methods, and main findings.

Table 1 Summary of the reviewed studies on UHI and mortality

Study	Locations	Methods	Findings
Milojevic et al. [37]	London, UK	Mortality case-crossover analyzes from 1993 until 2006	• 1 °C UHI anomaly multiplied the risk of heat death by 1.004 (95% CI = 0.950, 1.061) (interaction rate ratio) compared with the expected value of 1.070 (1.057, 1.082) if there were no acclimatization • UHI interaction for cold was 1.020 (0.979, 1.063) versus 1.030 (1.026, 1.034)
Glutting [38]	Barcelona, Spain	Remote sensing using Landsat 7 thermal infrared imagery overlaid onto at-home deaths between 2000 and 2003	• At-home deaths in MUHI were associated with 15% greater odds of dying on hot days than at-home deaths outside of MUHI
Taylor et al. [39]	London, UK	Statistical analyzes on the relationship between mean maximum temperature and mortality using risk estimate proposed by Armstrong et al. [48] from May 26–July 19, 2006	• UHI values for individual buildings varied from 20.5 °C to 23.0 °C (mean of 22.5 °C) during the study period (28.2–31.5 °C, with a mean of 30.2 °C during the hot spell) • Houses located in the centre of the UHI could face an additional 1.2 times increase in excess relative risk for an individual over 85 years old • Ward-mean maximum daily living room temperatures (MMDTs) when the 24.8 °C outdoor temperature-mortality threshold is exceeded range from 26.4 °C to 28.3 °C (mean of 27.1 °C), whilst ward-mean UHI values showed a negative skew, with a range from 20.8 to 22.9 °C (mean of 22.5 °C)

(continued)

Table 1 (continued)

Study	Locations	Methods	Findings
Smargiassi et al. [40]	Montreal (Canada)	Mortality case-crossover design for June–August 1990–2003	• The association between temperature and mortality is curvilinear, with a linear component close to zero, and then increased gradually until about 20 °C, after which the increase in relative risks was exponential • On days with an average daily ambient temperature of 26 °C compared with days with an average of 20 °C, the odd ratio for mortality (total non-accidental mortality) was 1.16 (lag 0 days, $OR_{26\,°C versus 20\,°C} = 1.04 \times 1.05 \times 1.06$; 95% CI = 1.12–1.19) • ORs of dying at elevated daily ambient temperatures were greater for the 2-day mean (0–1 days) than for lag 0 or lag 1 days alone
Lowe [41]	New York, United States	Correlation between UHI and mortality dataset over 12 months	• The impact of the UHI is an increase in the death rate of 1.1 people per million • The impact of the UHI is a decrease in the cold-related death rate of 4.0 deaths per million
Paravantis et al. [42]	Athens, Greece	Regression time series analysis and the estimation of exposure-response curves between mortality and temperature for the years 2002 to 2012	• Fatalities are lowest around maximum temperature values of 20–30 °C and increase below and above these values • The highest daily mortality counts are observed between 35 and 40 °C • At high and very high temperatures, the mortality increased by 20–35%, correspondingly with a one-day lag for the maximum temperature effect on mortality

(continued)

Table 1 (continued)

Study	Locations	Methods	Findings
Heaviside et al. [43]	West Midlands, UK	Health impact assessment–heat-related mortality during the heatwave	• The UHI contributed around 50% of the total heat-related mortality during the 2003 heatwave in the West Midlands • In a comparison of the UHI contribution to mortality expected from UKCP09 Climate Projections, a typical heatwave in 2080 could be responsible for an increase in mortality of around 3-times the rate in 2003 (278 vs. 90 deaths)
Goggins et al. [44]	Hong Kong	Poisson generalized additive models (GAMs) using meteorological variables and mortality dataset from June–September, 2001–2009	• Mean temperatures (lags 0–4) above 29 °C and low-mean wind speeds (lags 0–4) were significantly associated with higher daily mortality in areas with high UHII • 1 °C rise above 29 °C was associated with a 4.1% (95% CI = 0.7%, 7.6%) increase in natural mortality in areas with high UHII • Lower mean wind speeds (5th percentile vs. 95th percentile) were associated with a 5.7% (95% CI = 2.7, 8.9) mortality increase in high-UHII areas
Zhu et al. [45]	China	City-specific exposure–response functions under different climate and urbanization scenarios from 2000 to 2050	• The temperature-related deaths will decrease from 1.19 million in 2010 to 1.08–1.17 million in 2050 • The UHIs will cause an additional 6951 (95% CI = −17,637, 31,539, SSP4-RCP4.5) to 17,041 (95% CI = −10,516, 44,598, SSP5-RCP8.5) deaths in 2050 • The mortality burden exhibited strong spatial variations, with heavy burdens concentrated in the hotspots, including Beijing-Tianjin Metropolitan Region, Yangtze River Delta, Chengdu-Chongqing City Group, Guangzhou, Wuhan, Xi'an, Shandong, and Henan

(continued)

Table 1 (continued)

Study	Locations	Methods	Findings
Macintyre et al. [46]	UK region (West Midlands and Birmingham)	Health impact assessment to estimate the associated impact of UHI on cold-related mortality	• The population-weighted mean winter UHI intensity was +2.3 °C in Birmingham city centre • Results suggest a potential protective effect of the wintertime UHI, equivalent to 266 cold-related deaths avoided (~15% of total cold-related mortality over ~11 weeks) • When the impacts of climate change are considered, the results suggest that the number of heat-related deaths associated with the summer UHI will increase from 96 (in 2006) to 221 in the 2080s, based on the RCP8.5 emissions pathway
Milojevic et al. [47]	London, UK	An empirical temperature-mortality relationship for London from 26 May until 21 June 2006	• An overall number of heat-related deaths in London was 118.3 (95% CI = 105.8, 130.7) • The percentage of heat deaths attributable to the UHI effect was substantial for the non-extreme heat of the study period: 37.7% in outer London, 46.6% in inner London, and 47.2% in central London
Avashia et al. [14]	Ahmedabad, India	Linear modelling using satellite images, temperature data, and daily all-cause mortality between 2001 and 2015	• The association between temperature and mortality persisted across the lag period of 0–5 days. However, the effect size was observed to decrease with increasing lag • For the summer maximum temperature association, as the temperature rises beyond the 95th percentile ATmax, the relative risk (RR) values increase at a higher rate. At the 99th percentile temperature of 44 °C, the RR is 1.64, whilst at 47 °C, it crosses the RR of 2, respectively

4 Strengthening the Public and Institutional Preparedness and Response to UHI and Heat-Related Events Across the Globe

It is eminent that the health impacts of the UHI encompass a vast range of consequences, especially on human mortality. Although the increase in heat-related deaths due to the UHI has been well documented for several decades, research into this area has been continual and has increased in recent years as heat-related deaths are often used as an impetus to develop strategies to mitigate the UHI [49, 50]. Data on heat-related deaths clearly indicate that most fatalities occur during heat waves [51]. It is also clear that the frequency and longevity of heatwaves are exacerbated by the UHI [52].

Researchers and public health professionals can play an important role in advancing UHI-reduction actions [21]. Whilst the power to regulate land uses and the built environment typically rests with municipal planning departments, public health professionals working in health authorities, provincial governments, universities, and non-profit organizations can be proactive and seek to influence make modifications in the built environment to create liveable and climate-friendly cities. The researchers and public health professionals can support municipalities and other partners in advancing UHI-reduction actions by assisting in building a strong understanding and capacity amongst the urban design professionals, providing input in the municipal planning and development process as well as implementing physical measures to reduce UHIs in the construction of buildings used for delivering health services [53]. On top of that, such actions can also be facilitated by administering grants that provide funds to other sectors to reduce UHIs in their buildings and operations [54].

Public health officials should work towards building a proper understanding and capacity amongst urban design professionals, architects, engineers, municipal staff, and the public about the risks that UHIs pose to human health and the multiple co-benefits of UHI-reduction actions. These can be done in the form of partnerships, advocacy, education, and decision support system by presenting persuasive, evidence-based arguments for how actions may protect public health. Public health authorities can provide input through municipal planning and development processes through public consultation processes to help shape the design of communities, buildings, and infrastructure to reduce adverse health impacts and improve health outcomes. Researchers and public health professionals can undertake Health Impact Assessments (HIA) and subsequently share their outcomes through the formal engagement process to develop municipal plans and policies [43].

Importantly, UHI adaptation efforts and resources should be targeted primarily to those areas in a city or region where the most exposed and most vulnerable populations reside. Besides, community participation and actions are pivotal to reducing UHIs, such as planting trees [55], installing cool roofs [56], and less use of heat-emitting devices such as air-conditioners [57], cool pavements [58], integrating, or installing climate-sensitive applications [59] into the premises. Such measures cannot only lower heat-related illnesses and deaths but also help facilitate active living and reduce greenhouse gas emissions. At the municipal level, the trigger for actions that help reduce urban temperatures is frequently an issue other than UHIs.

5 Conclusion

There is a wealth of existing research on the health impacts of the UHI phenomenon. However, the impact of UHI on mortality still needs more investigation, especially in tropical regions. This review has only focussed on recent literature, which aims to quantitatively assess the UHI impacts on mortality, and how these impacts may be avoided through various mitigation techniques such as natural ventilation [60], green infrastructure [8], reforestation [61], cooling urban materials [62], and nature-based solutions [63]. Even though there is a limited study on this research matter, the results suggest that UHI exacerbates the mortality rate amongst the vulnerable population, especially amongst older adults and those with existing comorbidities. Climate-friendly urban planning measures designed to mitigate UHI effects may lessen mortality, especially when coupled with unfavourable summertime meteorological conditions.

References

1. Ma X, Zhang L, Guo M, Zhao J (2021) The effect of various urban design parameter in alleviating urban heat island and improving thermal health—A case study in a built pedestrianized block of China. EnvironSci Pollut Res 1–20
2. Hu L, Luo D, Wang L, Yu M, Zhao S, Wang Y, Mei S, Zhang G (2021) Levels and profiles of persistent organic pollutants in breast milk in China and their potential health risks to breastfed infants: a review. Sci Total Environ 753:142028
3. Sahani J, Kumar P, Debele S, Emmanuel R (2022) Heat risk of mortality in two different regions of the United Kingdom. Sustain Cities Soc 80:103758
4. United Nations, Department of Economic and Social Affairs, Population Division (2018) Accessed at https://population.un.org/wup/ on 27 Oct 2021
5. Ulpiani G (2021) On the linkage between urban heat island and urban pollution island: three-decade literature review towards a conceptual framework. Sci Total Environ 751:141727
6. Degirmenci K, Desouza KC, Fieuw W, Watson RT, Yigitcanlar T (2021) Understanding policy and technology responses in mitigating urban heat islands: a literature review and directions for future research. SustainCities Soc 102873
7. Khan MS, Ullah S, Chen L (2021) Comparison on land-use/land-cover indices in explaining land surface temperature variations in the city of Beijing, China. Land 10(10):1018
8. Tiwari A, Kumar P, Kalaiarasan G, Ottosen TB (2021) The impacts of existing and hypothetical green infrastructure scenarios on urban heat island formation. Environ Pollut 274:115898
9. Stone B Jr, Vargo J, Liu P, Habeeb D, DeLucia A, Trail M, Hu Y, Russell A (2014) Avoided heat-related mortality through climate adaptation strategies in three US cities. PLoS ONE 9(6):e100852

10. Voorhees AS, Fann N, Fulcher C, Dolwick P, Hubbell B, Bierwagen B, Morefield P (2011) Climate change-related temperature impacts on warm season heat mortality: a proof-of-concept methodology using BenMAP. Environ Sci Technol 45(4):1450–1457
11. Wong KV, Paddon A, Jimenez A (2013) Review of world urban heat islands: many linked to increased mortality. J Energy Res Technol 135(2):022101
12. Wong LP, Alias H, Aghamohammadi N, Aghazadeh S, Sulaiman NMN (2017) Urban heat island experience, control measures and health impact: a survey among working community in the city of Kuala Lumpur. Sustain Cities Soc 35:660–668
13. Frumkin H (ed) (2016) Environmental health: from global to local. Wiley
14. Avashia V, Garg A, Dholakia H (2021) Understanding temperature related health risk in context of urban land use changes. Landsc Urban Plan 212:104107
15. Bouchama A, Abuyassin B, Lehe C, Laitano O, Jay O, O'Connor FG, Leon LR (2022) Classic and exertional heatstroke. Nat Rev Dis Primers 8(1):1–23
16. Laitano O, Leon LR, Roberts WO, Sawka MN (2019) Controversies in exertional heat stroke diagnosis, prevention, and treatment. J Appl Physiol 127(5):1338–1348
17. Chauhan NR, Kapoor M, Singh LP, Gupta RK, Meena RC, Tulsawani R, Nanda S, Singh SB (2017) Heat stress-induced neuroinflammation and aberration in monoamine levels in hypothalamus are associated with temperature dysregulation. Neuroscience 358:79–92
18. Åström DO, Bertil F, Joacim R (2011) Heat wave impact on morbidity and mortality in the elderly population: a review of recent studies. Maturitas 69(2):99–105
19. Hajat S, Kosatky T (2010) Heat-related mortality: a review and exploration of heterogeneity. J Epidemiol Community Health 64(9):753–760
20. Bustinza R, Lebel G, Gosselin P, Bélanger D, Chebana F (2013) Health impacts of the July 2010 heat wave in Quebec, Canada. BMC Public Health 13(1):1–7
21. Beaudoin M, Gosselin P (2016) An effective public health program to reduce urban heat islands in Québec, Canada. Rev Panam Salud Publica 40:160–166
22. Luber G, McGeehin M (2008) Climate change and extreme heat events. Am J Prev Med 35(5):429–435
23. Hondula DM, Barnett AG (2014) Heat-related morbidity in Brisbane, Australia: spatial variation and area-level predictors. Environ Health Perspect 122(8):831–836
24. Casati B, Yagouti A, Chaumont D (2013) Regional climate projections of extreme heat events in nine pilot Canadian communities for public health planning. J Appl Meteorol Climatol 52(12):2669–2698
25. Schifano P, Cappai G, De Sario M, Michelozzi P, Marino C, Bargagli AM, Perucci CA (2009) Susceptibility to heat wave-related mortality: a follow-up study of a cohort of elderly in Rome. Environ Health 8(1):1–14
26. Heaviside C, Macintyre H, Vardoulakis S (2017) The urban heat island: implications for health in a changing environment. Curr Environ Health Rep 4(3):296–305
27. Tomlinson CJ, Chapman L, Thornes JE, Baker CJ (2011) Including the urban heat island in spatial heat health risk assessment strategies: a case study for Birmingham, UK. Int J Health Geogr 10(1):1–14
28. Peng J, Xie P, Liu Y, Ma J (2016) Urban thermal environment dynamics and associated landscape pattern factors: a case study in the Beijing metropolitan region. Remote Sens Environ 173:145–155
29. Gosling SN, Lowe JA, McGregor GR, Pelling M, Malamud BD (2009) Associations between elevated atmospheric temperature and human mortality: a critical review of the literature. Clim Change 92(3):299–341
30. Tan J, Zheng Y, Tang X, Guo C, Li L, Song G, Zhen X, Yuan D, Kalkstein AJ, Li F, Chen H (2010) The urban heat island and its impact on heat waves and human health in Shanghai. Int J Biometeorol 54(1):75–84
31. Thompson R, Hornigold R, Page L, Waite T (2018) Associations between high ambient temperatures and heat waves with mental health outcomes: a systematic review. Public Health 161:171–191

32. Jenerette GD, Harlan SL, Buyantuev A, Stefanov WL, Declet-Barreto J, Ruddell BL, Myint SW, Kaplan S, Li X (2016) Micro-scale urban surface temperatures are related to land-cover features and residential heat related health impacts in Phoenix, AZ, USA. Landsc Ecol 31(4):745–760
33. Aghamohammadi N, Fong CS, Idrus MHM, Ramakreshnan L, Sulaiman NM (2021) Environmental heat-related health symptoms among community in a tropical city. Sci Total Environ 782:146611
34. Aghamohammadi N, Fong CS, Idrus MHM, Ramakreshnan L, Haque U (2021) Outdoor thermal comfort and somatic symptoms among students in a tropical city. Sustain Cities Soc 72:103015
35. Reid CE, O'neill MS, Gronlund CJ, Brines SJ, Brown DG, Diez-Roux AV, Schwartz J (2009) Mapping community determinants of heat vulnerability. Environ Health Perspect 117(11):1730–1736
36. Xu Y, Dadvand P, Barrera-Gómez J, Sartini C, Marí-Dell'Olmo M, Borrell C, Medina-Ramón M, Sunyer J, Basagaña X (2013) Differences on the effect of heat waves on mortality by sociodemographic and urban landscape characteristics. J Epidemiol Community Health 67(6):519–525
37. Milojevic A, Armstrong BG, Gasparrini A, Bohnenstengel SI, Barratt B, Wilkinson P (2016) Methods to estimate acclimatization to urban heat island effects on heat-and cold-related mortality. Environ Health Perspect 124(7):1016–1022
38. Glutting JP (2013) Excess heat-related mortality in micro-urban heat islands: a case-only study in barcelona
39. Taylor J, Wilkinson P, Davies M, Armstrong B, Chalabi Z, Mavrogianni A, Symonds P, Oikonomou E, Bohnenstengel SI (2015) Mapping the effects of urban heat island, housing, and age on excess heat-related mortality in London. Urban Climate 14:517–528
40. Smargiassi A, Goldberg MS, Plante C, Fournier M, Baudouin Y, Kosatsky T (2009) Variation of daily warm season mortality as a function of micro-urban heat islands. J Epidemiol Community Health 63(8):659–664
41. Lowe SA (2016) An energy and mortality impact assessment of the urban heat island in the US. Environ Impact Assess Rev 56:139–144
42. Paravantis J, Santamouris M, Cartalis C, Efthymiou C, Kontoulis N (2017) Mortality associated with high ambient temperatures, heatwaves, and the urban heat island in Athens, Greece. Sustainability 9(4):606
43. Heaviside C, Vardoulakis S, Cai XM (2016) Attribution of mortality to the urban heat island during heatwaves in the West Midlands, UK. Environ Health 15(1):49–59
44. Goggins WB, Chan EY, Ng E, Ren C, Chen L (2012) Effect modification of the association between short-term meteorological factors and mortality by urban heat islands in Hong Kong. PloS One 7(6):e38551
45. Zhu D, Zhou Q, Liu M, Bi J (2021) Non-optimum temperature-related mortality burden in China: addressing the dual influences of climate change and urban heat islands. Sci Total Environ 782:146760
46. Macintyre HL, Heaviside C, Cai X, Phalkey R (2021) The winter urban heat island: Impacts on cold-related mortality in a highly urbanized European region for present and future climate. Environ Int 154:106530
47. Milojevic A, Wilkinson P, Armstrong B, Davis M, Mavrogianni A, Bohnenstengel S, Belcher S (2011) Impact of London's urban heat island on heat-related mortality. Epidemiology 22(1):S182–S183
48. Armstrong BG, Chalabi Z, Fenn B, Hajat S, Kovats S, Milojevic A, Wilkinson P (2011) Association of mortality with high temperatures in a temperate climate: England and Wales. J Epidemiol Community Health 65(4):340–345
49. Dong W, Liu Z, Zhang L, Tang Q, Liao H, Li XE (2014) Assessing heat health risk for sustainability in Beijing's urban heat island. Sustainability 6(10):7334–7357
50. Van Der Hoeven F, Wandl A (2015) Amsterwarm: mapping the landuse, health and energy-efficiency implications of the Amsterdam urban heat island. Build Serv Eng Res Technol 36(1):67–88

51. Centers for Disease Control and Prevention (2013) Heat illness and deaths—New York City, 2000–2011. Morb Mortal Wkly Rep 62(31):617–621
52. Laaidi K, Zeghnoun A, Dousset B, Bretin P, Vandentorren S, Giraudet E, Beaudeau P (2012) The impact of heat islands on mortality in Paris during the August 2003 heat wave. Environ Health Perspect 120(2):254–259
53. Hidalgo J, Lemonsu A, Masson V (2019) Between progress and obstacles in urban climate interdisciplinary studies and knowledge transfer to society. Ann NY Acad Sci 1436(1):5–18
54. Evola G, Gagliano A, Fichera A, Marletta L, Martinico F, Nocera F, Pagano A (2017) UHI effects and strategies to improve outdoor thermal comfort in dense and old neighbourhoods. Energy Procedia 134:692–701
55. Wang Y, Akbari H (2016) The effects of street tree planting on urban heat island mitigation in Montreal. Sustain Cities Soc 27:122–128
56. Yang J, Pyrgou A, Chong A, Santamouris M, Kolokotsa D, Lee SE (2018) Green and cool roofs' urban heat island mitigation potential in tropical climate. Sol Energy 173:597–609
57. Kondo K, Mabon L, Bi Y, Chen Y, Hayabuchi Y (2021) Balancing conflicting mitigation and adaptation behaviours of urban residents under climate change and the urban heat island effect. Sustain Cities Soc 65:102585
58. Wang C, Wang ZH, Kaloush KE, Shacat J (2021) Cool pavements for urban heat island mitigation: a synthetic review. Renew Sustain Energy Rev 146:111171
59. Yang J, Wang ZH, Kaloush KE (2015) Environmental impacts of reflective materials: is high albedo a 'silver bullet' for mitigating urban heat island? Renew Sustain Energy Rev 47:830–843
60. Ahmed T, Kumar P, Mottet L (2021) Natural ventilation in warm climates: the challenges of thermal comfort, heatwave resilience and indoor air quality. Renew Sustain Energy Rev 138:110669
61. Valmassoi A, Gharbia S, Stibe S, Di Sabatino S, Pilla F (2017) Future impacts of the reforestation policy on the amospheric parameters: a sensitivity study over Ireland. Procedia Comput Sci 109:367–375
62. Kakoniti A, Georgiou G, Marakkos K, Kumar P, Neophytou MKA (2016) The role of materials selection in the urban heat island effect in dry mid-latitude climates. Environ Fluid Mech 16(2):347–371
63. Kumar P, Debele SE, Sahani J, Aragão L, Barisani F, Basu B, Bucchignani E, Charizopoulos N, Di Sabatino S, Domeneghetti A, Edo AS, Zieher T (2020) Towards an operationalisation of nature-based solutions for natural hazards. Sci Total Environ 731:138855
64. Peng RD, Bobb JF, Tebaldi C, McDaniel L, Bell ML, Dominici F (2011) Toward a quantitative estimate of future heat wave mortality under global climate change. Environ Health Perspect 119(5):701–706
65. Sheridan SC, Lee CC, Allen MJ, Kalkstein LS (2012) Future heat vulnerability in California, Part I: projecting future weather types and heat events. Clim Change 115(2):291–309

Chapter 3
The Impact of Urban Overheating on Heat-Related Morbidity

Pavlos Kassomenos and Paraskevi Begou

Abstract The global urban population has grown rapidly during the last decades. Currently, more than 4 billion people around the world live in urban areas. Consequently, the natural environments are destroyed due to the escalating demands for urban infrastructure. One of the most calamitous effects of urbanization is the complex and multilayered local phenomenon of urban heat island (UHI) which is a consequence of the prevailing weather conditions and the characteristics of urban cities. Therefore, the urban overheating has become an important issue of public health for both developing and developed countries. Indeed, urban residents suffer from the elevated ambient temperature during day and night. As well as they suffer from unseasonal hot weather conditions and meteorological hazards such as extreme weather events and heat waves. In general, higher outdoor and indoor air temperatures have direct impact on human's health and well-being. Specifically, they can cause heat stress, sleep disorders and cardiovascular and cardiorespiratory diseases. The purpose of the present chapter is to present a comprehensive review about the existing literature with reference to the impact of urban overheating on heat-related morbidity. Moreover, this chapter aims to examine the indirect effects of high ambient temperature on urban systems to evaluate their impacts on the overall environmental quality of cities and public health.

Keywords Public health · Indoor temperature · Outdoor temperature · UHI · Well-being · Mitigation approach

P. Kassomenos (✉) · P. Begou
Department of Physics, Laboratory of Meteorology and Climatology, University of Ioannina, Ioannina, Greece
e-mail: pkassom@uoi.gr

P. Begou
e-mail: p.begou@uoi.gr; paraskevib@cc.uoi.gr

N. Aghamohammadi and M. Santamouris (eds.), *Urban Overheating: Heat Mitigation and the Impact on Health*, Advances in Sustainability Science and Technology,
https://doi.org/10.1007/978-981-19-4707-0_3

1 Introduction

In accordance with the World Meteorological Organization (WMO) Statement on the State of the Global Climate in 2015–2019, the years 2015 to 2019 were the five warmest years on record. Specifically, the average global temperature for 2015–2019 was 1.1 ± 0.1 °C warmer than pre-industrial period (1850–1900) and is the warmest of any equivalent period on record [1]. Also, the 5-year period 2015–2019 is 0.21 ± 0.08 °C warmer than the average for the 5-year period 2011–2015. Moreover, the global mean land-surface air temperature for 2015–2019 was approximately 1.7 °C above the pre-industrial levels and 0.3 °C warmer than the average for the 5-year period 2011–2015 and the global mean sea-surface temperature for the 5-year period 2015–2019 was approximately 0.83 °C above the pre-industrial levels and 0.13 °C warmer than the average for the 5-year period 2011–2015 [1]. During the period 2015–2019, the heat waves were the deadliest meteorological hazard. Many regions around the world have experienced heat waves with unprecedented frequency and intensity, while the climate models project that the weather extremes and heat waves will break the previous records during the period 2051–2080 [2].

In this context, urban areas face a particular challenge. Cities concentrate on several risks health hazards for residents and new health challenges emerge [3]. The high levels of air and noise pollution which adversely affect human health and cause a substantial burden of morbidity outcomes/diseases and premature deaths are common urban health challenges.

The urban climate is defined as a local climate that is modified by interactions between the built-up area (including waste heat and the emission of air pollutants) and regional climate [4] can alter the meteorological variables (air temperature, humidity, wind speed and direction, precipitation rate) in urban areas compared to rural areas [5]. The UHI is the most common phenomenon of the urban climate which mainly results in temperature difference between urban centre and rural areas. Especially, during the night-time, urban and suburban areas experience elevated temperatures compared to the rural surroundings. Highly developed urban areas, which are mainly characterized by impervious surfaces and lack of vegetation, have reduced shading by vegetation canopy and evaporative cooling by evapotranspiration, which in turn, intensify the warming of urban air. These characteristics contribute to higher surface and air temperatures in urban areas. Therefore, the urban form and geometry, the properties of urban materials, the reduced vegetation and anthropogenic heat emissions all contribute to the UHI effect [6]. The urbanization can strengthen the UHI effect by altering small-scale meteorological processes (e.g., land and sea breezes, katabatic winds) and modifying synoptic scale meteorology while the regional-scale weather patterns can also influence the development of UHI [7].

In addition, the impacts of extreme heat and heat waves exacerbate the UHI phenomenon within the urban areas, causing heat stress and heat-related health problems in urban residents [8]. In fact, a 'vicious circle' begins given that the hot weather aggravates the UHI phenomenon in terms of magnitude, intensity and diurnal variation while concurrently the impacts of hot weather are exacerbated under the influence of UHI.

Many studies investigate the synergistic interactions of UHI and heat waves or the effect of heat waves on the intensification of UHI [9–11]. Apparently, the impacts of the global warming combined with the effects of UHI adversely affect health and well-being of urban residents [12]. The evidence from the epidemiological studies and statistical analyses which investigate the relationship between ambient temperature and human health is overwhelming. Both low and high temperatures contribute to the burden of morbidity and mortality. Many hospital admissions, emergency department visits and ambulance dispatches are associated with the exposure to extreme ambient temperatures, especially during the periods of heat waves and cold spells. In general, urban residents are exposed to higher temperatures than residents of surrounding rural areas because of the UHI effect and the urban overheating. Extreme heat events and heat waves have severe impacts on human health. However, the severity of the health outcomes related to heat depends on the timing, intensity and duration of the heat event, personal risk factors, the degree of acclimatization and the adaptability of the local population as well as the institutional and infrastructure systems of a given country (https://www.who.int/news-room/fact-sheets/detail/climate-change-heat-and-health).

Obviously, there is a great burden of heat-related illnesses including heat exhaustion and heatstroke during the extreme heat events. Moreover, several studies found associations between the extreme ambient temperature and the morbidity burden from circulatory, cerebrovascular, cardiovascular and respiratory diseases. Although there are contrasting patterns of hospital admissions and deaths from these diseases during the extreme heat events, many studies based on hospital records found that hot weather results in a rise in hospital admissions for coronary heart disease, acute myocardial infarction, stroke or ischemic stroke, cardiac dysrhythmia and hypotension [13–15]. Also, heat-related respiratory morbidity risk was observed during the extreme heat events [16, 17]. In addition, the exposure to elevated temperature affects mental health and emotional well-being. The sleep time and sleep quality were greatly affected by the elevated temperatures, and as a result, sleep disturbances were found to be associated with high night-time temperatures. Furthermore, studies suggest that heat stress-associated nephropathy may represent an epidemic due to global warming and the frequent and intense heat waves [18]. Also, a recent hypothesis postulates that the increasing prevalence of kidney stones could be attributed to exposure of the urban population to the higher temperatures due to UHI [19]. Indeed, several studies found that during heat waves there was an increase in hospital admission for renal diseases [20]. Moreover, pregnant women and the developing foetus are among the most vulnerable to heat exposure. The exposure to heat jeopardizes the maternal and foetal health and may provoke pregnancy complications or even affect the health of the newborns [21, 22]. Apart from pregnant women, other population groups that are vulnerable to heat exposure are the elderly, infants, and children, people with chronic health problems or disabilities, socially isolated people, low-income people and people who live in poverty, outdoor and manual workers, migrants and refugees, the homeless and people living in slum communities and informal settlements. Also, people taking prescription medications could have higher risk for heat-related illness and certain medications can conduce to more deleterious effects of extreme heat on patients [23].

Also, the UHI phenomenon affects the built environment, the critical infrastructure sectors, the health and social care sectors. The extreme heat events put immense pressure on health care systems and services. Urban areas where most hospitals and healthcare institutions are located experience these events more intensely. Hence, it is of pivotal importance the preparedness of health care and social services coordinated with the Meteorological Services and the Weather Forecasting.

In this chapter, we analyse the recent/state-of-the-art epidemiological evidence on the heat-related morbidity due to extreme heat events and heat waves within the urban areas. Also, we focus on the vulnerable population groups which encounter barriers to reach adequate protection under the menace of urban overheating and require preventive actions against the heat exposure. Finally, we bring up the great challenge posed in the healthcare sector by the extreme heat events.

2 Background Knowledge

2.1 Thermoregulation

Humans are homeotherms and therefore can maintain their core body temperature within narrow limits despite the fluctuations in the ambient temperature. The body can be divided into two regions, namely the 'core' and the 'shell'. The 'core' is made up of the contents of the skull, the thorax, and the abdomen. The 'shell' includes the skin, the subcutaneous tissues, and the limbs [24]. The core body temperature is between 36.1 and 37.8 °C which fluctuates throughout the 24-h with a well-defined circadian rhythm and is maintained within this narrow range to facilitate the optimal functioning of physiological processes. There are also normal fluctuations that occur throughout a lifetime (ageing).

The core body temperature is normally in a state of dynamic equilibrium as a result of a balance between heat loss and heat gain. The heat exchange with the environment occurs via the physical processes of radiation, conduction, convection, and evaporation, while the heat production occurs through the mechanism of muscle contraction, diet-induced thermogenesis and non-shivering thermogenesis as well as hormones are used to increase heat production.

The regulation of body temperature is accomplished through two parallel processes: behavioural and physiological temperature regulation [25]. Behavioural temperature regulation operates largely through conscious behavioural adjustments and may employ any means available, including standing in the shade and wearing light-coloured clothing. Alterations in work rate during self-paced exercise in the heat also constitute behavioural adjustments that contribute to regulate body temperature [25].

Abnormal core temperature deviations will challenge the body's thermoregulatory mechanisms and fluctuations in temperature outside the normal range can prove fatal [26]. The body temperature is controlled by a feedback loop, and the 'set point' is defined by the hypothalamus.

The hypothalamus is responsible for thermoregulation by activating receptors in the skin and visceral organs to facilitate heat loss and maintain the core temperature within the normal values. The core body temperature could deviate outside the normal range of 36.1–37.8 °C during prolonged exercise, fever due to illness and exposure to extreme hot and cold environmental conditions. Prolonged exposure to excessive heat compromises the body's ability to regulate the core temperature and can result in a wide range of physiological impacts for humans, amplifying existing medical conditions and provoking a cascade of heat-related illnesses.

Heat-related illness is a spectrum of clinical syndromes ranging from heat oedema, heat rash, heat syncope and heat cramps to heat exhaustion and heat stroke [27]. Heat stroke is characterized by a core body temperature above 40 °C and is a life-threatening illness with high mortality rates.

3 Studies on the Association Between Heat Exposure and Heat-Related Morbidity

The heat-related morbidity associated with the urban overheating is a consequence of the urban heat island (UHI) effect along with the effects of extreme heat events and heat waves which synergically interact and affect human health and well-being. The UHI itself has a great impact on public health in urban areas. But, obviously, the synergetic effect between heat events and UHI lead to the exacerbation of the UHI magnitude and intensity. Bigger cities seem to experience an amplification of UHI during the periods of extreme heat [28]. Many authors highlight that these synergetic effects increase the heat-related health risk associated with weather conditions and the vulnerability of urban population [9]. For instance, during the severe heat wave in southern Pakistan in June 2015, the temperature ranged from 49 °C in Larkana and Sibi to 45 °C. In Karachi, the maximum temperature of 44.8 °C was recorded on 20th June while the Heat Index rose up to 66.1 °C and 58.3 °C on the 20th of June and the 22nd of June, respectively. According to the technical report on the heat wave conducted by the Ministry of Climate Change, Government of Pakistan [29] the UHI exacerbate the impact of heat wave and was responsible for the large number of casualties in Karachi [29].

In this context, urban residents face certain risk factors associated with elevated ambient temperatures. In particular, these conditions compromise the health status of the vulnerable populations because of their underlying medical conditions. There is a growing body of literature that focuses on the burden of morbidity during the extreme heat events, heat waves and periods of elevated ambient temperature and/or under the influence of UHI effect.

3.1 Vulnerable Population Groups and Heat Exposure

There is an overwhelming base of evidence suggesting that specific groups of people are more vulnerable to extreme heat. Among these groups cited are older adults (aged 65+), infants, and children, people with chronic health problems or disabilities, socially isolated people, low-income people and people who live in poverty and outdoor workers (Table 1). Also, pregnant women are at higher risk of experience complications from heat exposure (pleases see Sect. 3.6). Moreover, people who use medications are more vulnerable to the effects of extreme heat exposure. Heat can contribute to or exacerbate underlying illness as well as be the primary cause of illness or death. Also, we should place emphasis on the fact that urban residents are more vulnerable groups of people compared to the rural residents due to the urban overheating and the effects of UHI. Also, the living standards, lifestyles, and social behaviour of citizens in urban areas adversely affect their health and well-being. As a matter of fact, they may have unhealthier lifestyle behaviours such as smoking and unhealthy or unbalanced diet, a sedentary lifestyle with lack of exercise and physical activity, which lead to a greater risk of obesity [30]. Obese or overweight individuals are more susceptible to heat-related illness and are among the high-risk groups during extreme heat events (Table 1).

3.1.1 Elderly

It is well-documented that people aged 65 years or older are more prone to heat-related health problem. The age-related physiological changes in sweating, skin blood flow and thermoregulatory system as well as others factors such as the presence of acute and chronic diseases place the elderly at higher risk for heat-related illness [31, 32]. There is a combination of concurrent factors associated with the ageing process and the aggregation of comorbidities in the elderly which pose them to high-risk populations for heat-related morbidity and mortality. Many scientific studies have been reported to increase hospital admissions and emergency department visits during heat waves among older adults.

In summer 2006, a severe heat wave affected North America. In California, both the intensity and duration of the heat wave were exceptional, where during the period from July 16 through July 26 the elevated overnight minimum temperatures surpassed previous daily, or even all-time, high minimum temperatures at several reporting stations. Due to the UHI phenomenon, the abnormally high overnight temperatures were exacerbated in developed areas such as the urban centres of California that have warmed faster than the rest of the state [33, 34].

With regard to this heat wave, Knowlton et al. (2009) found that 16,166 excess ED visits and 1182 excess hospitalizations occurred statewide in California from 15 July to 1 August 2006 heat wave period compared with the non-heat wave reference period (8–14 July and 12–22 August 2006) [35]. They observed significantly elevated RRs in both ED visits and hospitalizations for heat-related illness, electrolyte imbalance,

Table 1 Vulnerable population groups to heat exposure

Vulnerable population group	Risk factor
Elderly	They are less aware to the effects of heat exposure and less adaptable to extreme heat They may suffer from chronic illness or take medications Social isolation Living in nursing homes or they are bedridden
Children	They are sensitive to the effects of extreme heat Their safety and protection depend on adults
Pregnant women	A pregnant woman's core body temperature is often elevated
Overweight and obese individuals	They may be more sensitive to extreme heat and have difficulty in thermoregulating
Individuals with chronic medical conditions and disabilities	Heat exposure may exacerbate their condition They may take medications
People with neurological and psychiatric diseases	They are less aware to the effects of heat exposure They may adapt slowly to environmental changes
People take certain prescription medications	Certain medications may increase the risk for heat-related illness or can conduce to more deleterious effects of extreme heat on patients
People living alone	Social isolation
People living in care facilities	Their safety and protection depend on others
Outdoor workers and people in certain occupations	Prolonged exposure to heat under strenuous labour
Individuals of low socioeconomic status Homeless people Migrants and refugees	They have economic constraints or lack of air-conditioners and cooling devices They may not have access to information about heat waves and cooling centres
Tourists	They lack of heat acclimatization and adaptation because they may be from cooler climates

Source Kansas Extreme Weather Workgroup [47]; National Public Health Organization of Greece [48]; Singh et al. [49]

acute renal failure, and nephritis and nephrotic syndrome among each of the three age groups (0–4, 5–64, ≥65 years of age), whereas the elderly (≥65 years of age) were at greatest risk for heat-related illness. Particularly, the RR (95% CI) for ED visits and hospital admissions for heat-related illnesses, in the ≥65 years age group statewide, were RR = 10.87 (8.39–14.31) and RR = 14.23 (9.56–22.08), respectively. In contrast, in the 5–64 age group statewide, the RR (95% CI) for ED visits and hospital admissions for heat-related illnesses were RR = 5.43 (4.83–6.13) and RR = 7.00 (4.90–10.28), respectively. Also, the authors found elevated rates of ED visits in the ≥65 years age group statewide for diabetes (RR = 1.04; 95% CI, 1.02–1.06) and respiratory illnesses (RR = 1.04; 95% CI, 1.02–1.06).

During the 2009 south-eastern Australia heat wave in Victoria, the maximum temperatures were 12–15 °C above normal for much of the state. In Melbourne, the maximum daytime temperature was above 43 °C for three consecutive days from 28–30 January reaching the maximum daytime temperature of 45.1 °C on 30 January 2009. A report based on separate sources in order to assess the health impact of heat wave compared the data of Ambulance Victoria (AV) for the week of the heat wave (26 January to 1 February 2009) with the same calendar dates in 2008. The analysis revealed that the ambulance attendances for heat-related conditions increased during this week to a total of 514 cases, representing over a 34-fold increase in 2009 compared with 2008 (15 cases), while 313 (61%) of the total heat-related cases were in those 75 years or older. Also, the analysis of the Total Emergency Department (ED) Presentations for the week of the heat wave in 2009 compared with mean of 2004–2008 found that there was a 37% increase in those 75 years and older [36]. The findings of the report suggest that the January 2009 heat wave has had a substantial impact on the health of the residents of Victoria, particularly the elderly.

3.1.2 Children

Children are vulnerable to natural disasters and extreme weather events. Given that children's care, safety and protection normally dependent on adults, extreme heat events reveal their vulnerability and pose them to higher health risk [37]. Also, children adjust more slowly than healthy adults to changes in environmental conditions and they are less able to control their surrounding environment. According to many scientific studies, extreme heat events induce a great risk on children's health [38]. They are physiologically less able to regulate their body temperature, their surface area to body weight ratio is higher than in adults and their usual breathing and heart rates are faster than healthy adults. A review on the impact of heat waves on children's health suggests that the key paediatric diseases or conditions significantly affected by heat waves include renal disease, respiratory disease, electrolyte imbalance and fever [39]. Also, the burden of disease from extreme heat in children includes missed school days and impaired cognitive performance [39].

The study conducted by Knowlton et al. [35] on the impact of 2006 California heat wave on the ED visits and hospital admissions have shown that children (0–4 years) were at the greatest risk given that the ED visits in this age group had elevated RRs for heat-related illnesses ED visits on a statewide level (RR = 6.17; 95% CI, 2.58–17.88) [35].

Studies conducted in Brisbane, Australia [40] and Southwestern Ontario, Canada [41] have shown that the extreme temperatures and extreme heat events are associated with an increased incidence of emergency department visits in children.

Xu et al. [40] collected emergency department admission (EDA) data for the main paediatric diseases in Brisbane, Australia for the period of 1st January 2003 to 31st December 2009 in order to estimate the temperature effects on paediatric EDA [40]. Their analysis revealed that children aged 0–4 years were more vulnerable to heat effects and children aged 10–14 years were more sensitive to both hot and cold effects, while male children may be more vulnerable to extreme temperatures than female children. Given that children aged 10–14 may play outdoors more often than younger children, they exposed directly to extreme temperatures. In general, high temperatures (26.5 °C) were statistically significantly associated with many paediatric diseases including intestinal diseases, respiratory diseases, endocrine, nutritional and metabolic diseases, nervous system diseases and chronic lower respiratory diseases.

Wilk et al. [41] reviewed the paediatric emergency department (ED) visits among children (0–17 years) from two academic hospitals in Southwestern Ontario, Canada, from June through August of 2002–2019 [41]. For the purpose of their study, the extreme heat was defined as the 99th percentile of the maximum temperature distribution (33.1 °C) and found to be associated with an overall 22% increase (RR: 1.22; 95% CI: 1.12–1.32) in ED visits due to all causes excluding injuries compared to the reference temperature of 21 °C. Children between the ages of 1 and 12 accounted for 63.5% of all ED visits and the results of the study indicated that for all ED visits larger association estimates were found in this age group (RR: 1.33; 95% CI: 1.20–1.47) than in other age groups.

3.1.3 People with Chronic Health Problems or Disabilities

Several authors suggest that people with existing medical conditions and chronic illness are particularly vulnerable to extreme weather conditions and heat stress [42]. People with hypertension, diseases of the respiratory system or cardiovascular diseases, kidney diseases or diabetes mellitus are at increased risk for heat-related morbidity and mortality.

A recent review conducted by Javorac et al. [43] on the meteorological factors causing the exacerbations of chronic obstructive pulmonary disease (COPD) highlights that not only low temperatures but also high outdoor temperatures can lead to increased hospitalizations and mortality due to worsening of COPD [43]. Also, the review indicates the adverse impact of diurnal temperature variations on chronic respiratory diseases given that studies have found that the number of hospitalizations due to AECOPD increased with increased diurnal temperature range values.

During the heat wave in Portugal in July 2006 which characterized by high temperatures (from 29 to 37 °C) and moderate relative humidity (42% to 71%), Monteiro et al. [44] have shown that from 11 to 18 July 2006 there was an increase in hospital admissions for COPD (12 cases) of 100% relative to the expected number of admissions during the reference period (6 cases) [44].

A study conducted by Hoffmann et al. [45] analysed the medical data and the characteristics of 990 patients, who were hospitalized for AECOPD in Berlin, Germany, in order to explore the relationship between climate and morbidity of COPD patients in an urban environment [45]. Notably, their analysis revealed that patients with

COPD exacerbations during 'hot summer' periods more often had a prior myocardial infarction and the 72% of the patients were active smokers. According to the authors the 'hot summer' defined as the summer periods (June 1st to August 31st) of 2006 and 2010 with average temperatures (20.3 °C) and the hospital admissions per day due to AECOPD increased if the daily minimum outdoor temperature surpassed 18.3 °C.

Xu et al. [46] analysed data on hospitalizations for diabetes mellitus and meteorological variables from 1814 Brazilian cities during the warm seasons from 2000 to 2015 [46]. They found that every 5 °C increase in daily mean temperature was associated with 6% (OR = 1.06; 95% CI: 1.04, 1.07) increase in hospitalization due to diabetes mellitus with lag 0–3 days at a national level. Specifically, the 7.3% (95% CI: 3.5, 10.9) of all hospitalizations due to diabetes mellitus during the warm season could be attributable to heat exposure by assuming a cause–effect relationship, while this fraction was especially in high the age group above 80 years (19.2%, 95% CI:6.5, 29.5).

3.1.4 Other Populations Whose Socioeconomic Status May Make Them More Vulnerable

Climate change and extreme weather events exacerbate social and economic inequalities and affect global economy, people's financial situation, mobility, social relations and access to basic services [50]. On a global scale, people living in low- and middle-income countries are more affected by the impacts of climate change and global warming than the residents in high-income countries. Also, many authors identified the low socioeconomic status as an important factor in enhancing people's vulnerability to climate variability and change. Chakraborty et al. [51] found that in most (72%) cases within a sample of 25 cities in different climate zones in both developed and developing countries, the UHI disproportionately affects residents of lower socioeconomic status [51].

Gronlund [52] reviewed the racial, ethnic, socioeconomic (i.e. education, income and occupation) and sociodemographic characteristics in order to assess the vulnerability in heat-related morbidity and mortality. They indicated that various factors contribute to people's susceptibility such as their physical health, their inadequate access to healthcare facilities and services, their occupations, their living conditions and the quality of their households as well as the access to air condition [52].

Sánchez et al. [53] analysed the location of the vulnerable population living in the city of Madrid, Spain, based on socioeconomic indicators and demographic characteristic and the intensity of the UHI under the influence of a heat wave event on 15th of July 2015 [53]. Their analysis indicated that certain vulnerable neighbourhoods are located in areas with higher UHI intensity, and as a result, people living in these neighbourhoods are less likely to overcome the problems associated with extreme temperatures. According to Jacobs et al. [54], the maximum daytime values of the thermal indices (i.e. HI, WBGT and UTCI) in Delhi (India), Dhaka (Bangladesh) and Faisalabad (Pakistan) indicated dangerously hot conditions [54]. These areas are defined as climate change 'hot spots areas' and inhabitants in these cities disproportionately exposed to climate risks and have higher levels of vulnerability [55].

3.1.5 People Living in Slum Communities and Informal Settlements

A major issue in the context of heat-related vulnerability is the spatial structure in developing cities because it may pose particularly high health risks in the residents. Given that high proportions of low-income residents living in degraded environments and informal settlement communities or slum communities, they might be particularly vulnerable to heat as well as the natural and weather hazards in general. At the same time, the population living in these areas are threatened by several other socioeconomic problems such as informal economy, violence or even malnutrition and are afflicted by infectious and non-communicable diseases. In fact, in developing countries the vulnerability and risks to extreme temperature conditions may be overlooked and heat is not recognized as a significant health hazard in these countries. For instance, studies focus on heat-related morbidity and mortality in Sub-Saharan Africa are generally limited even though it is estimated to bear the highest burden of climate change in terms of DALYs. Amegah et al. [56] review the studies conducted in sub-Saharan Africa and investigate the association between temperature variability and morbidity and mortality [56]. They concluded that even there are limited meteorological stations, poor health surveillance systems and records keeping across sub-Saharan Africa which are obstacles to climate-health research in the region, the ambient temperature was found to be associated with the incidence of several diseases and increased risks of mortality. For example, elevated ambient temperature resulted in increase in cholera incidence, diarrhoea occurrence and meningitis incidence. Furthermore, the UHI phenomenon is documented in the communities of the developing world where the urban sprawl and the informal and unplanned settlements characterized by high-density residential development and high population density, poor building materials, limited household ventilation and lack of access to cooling devices. Also, the population residing in these settlements are vulnerable to heat exposure due to lack of information on heat wave occurrence and the relevant heat-related health risks. These factors contribute to make the population living in these communities highly vulnerable to heat exposure. Pasquini et al. [57] analysed the heat-health vulnerability of informal settlement residents in Dar es Salaam, Tanzania, which has a tropical climate and the warmest months are January and February [57]. For the purpose of their analysis, they conducted a mixed-method exploratory study including interviews with the informal settlement residents of the Vingunguti ward and climate analysis. The high temperatures are common in Dar es Salaam, and there is the perception that the residents have acclimatized in these environmental conditions. However, the majority of informal settlement respondents mentioned that their discomfort from heat is worse at night-time than during the daytime and the heat exposure is high within their homes given that they live in poor-quality housing and have limited ability to afford air-conditioning or even refrigeration. They also lack of amenities and have difficulties to access to healthcare facilities. One of the most important findings is that Vingunguti residents' health, well-being, work and daily lives are affected by the heat exposure as reported by the interviews but that residents lacked knowledge of all the risks of heat on their health. In this context, people living in slum communities may present similar characteristics to those in Dar es Salaam that increase their vulnerability to heat and lead to heat-related morbidity or even death.

3.1.6 People in Certain Occupations

In addition, vulnerable group of people can be considered outdoor workers such as construction workers, street sweepers, traffic police, fire-fighters, security guards, mail and parcel delivery drivers. Therefore, for certain occupations is required to working under extreme environmental conditions or during extreme weather events, while workers are under huge physical strain. A recent literature review from Moda et al. [58] discussed the health impacts of UHI on outdoor workers and their safety and indicated that the protective clothing used by workers is an additional risk factor for developing heat-related illness [58]. They also highlighted that workers have greater chance of having fatigue and physical exhaustion wherever the ambient temperature exceeds 35 °C [58]. Also, the heat stress can lead to reduced work performance and capacity or even increased risk of errors during the working hours and the operation of machinery [59]. In addition, indoor workers are also at risk of heat-related illness if lacking efficient ventilation systems and cooling or air-conditioning equipment.

Harduar Morano et al. [60] analysed a total of 8,315 emergency department (ED) visits and 1051 inpatient hospitalization (IH) data for occupational heat-related illness over a 5-year period (2007–2011) from nine southeast states in the USA [60]. The majority of occupational ED visits and IH for heat-related illness occurred during the hottest months (June–August). But there are also cases during the months of May and September which occurred mainly among the out-of-state workers indicating their insufficient acclimatization to the heat and humid environment in the south-east region. The acclimatization is another important risk factor for heat-related illness, especially for workers. Full heat acclimatization normally requires 15 days or longer depending on individual factors. Therefore, the Occupational Safety and Health Administration (OSHA) suggests gradual acclimatization for new workers and those returning from a prolonged absence: 20% of the workload on the first day, increasing incrementally by no more than 20% each subsequent day.

Moreover, the international standard (ISO 7243:2017) has been applied in order to evaluate the heat stress to which a person is exposed over a working day and protect them assessing of heat stress using the wet-bulb globe temperature (WBGT). There are several 'heat stress indices' and protective guidelines. The WBGT is one of the most widely used in occupational health, though other alternatives indices have been proposed such as the required sweat rate (SWreq) index, the predicted heat strain (PHS) index and the thermal work limit (TWL) [59].

3.1.7 People Taking Prescription Medications

In the context of people's vulnerability to heat, it is important to mention that certain prescribed or instructed medications may increase the risk for heat-related illness or can conduce to more deleterious effects of extreme heat on patients. According to Westaway et al. [61], medicines that may significantly increase the risk of dehydration and heat-related illness during hot weather include diuretics, especially when combined with an angiotensin-converting enzyme (ACE) inhibitor or an angiotensin II receptor blocker (ARB), anticholinergics and psychotropics. Also, in the presence of comorbidities and use of multiple medicines, the risk of dehydration and heat-related illness is increased significantly, especially during hot weather [61].

Sommet et al. [62] studied the adverse drug reactions, in patients older than 70 years, during the exceptional European heat waves of 2003 and 2006 (1 July to 31 August) in France in relation to two reference years without heat waves (2004 and 2005) [62]. Although the authors note that the total number of adverse drug reactions in heat wave years and the reference years were not significantly different, most frequent adverse drug reactions during the years with heat waves were from cardiovascular medications such as diuretics, angiotensin-converting enzyme inhibitors (ACEIs) and ACE receptors blockers (ARBs).

Hausfater et al. [63] performed a multi-centre observational cohort study of febrile patients (with core temperature $\geq$38.5 °C) and mean age of 79 $\pm$ 19 years admitted to 16 emergency departments belonging to the teaching hospital network of the Paris area during the heat wave of 2003 (between 5 and 14 August) [63]. Their analysis demonstrated that the chronic medication with diuretics was one of the prognostic factors for the survival rates of patients with OR = 1.26 (CI:1.04–1.54) [63].

The mechanisms of the medicines which may increase the risk of heat-related illness include reduced vasodilation, decreased sweating, increased heat production, decreased thirst, dehydration and aggravation of heat illness by worsening hypotension in vulnerable patients [64].

3.2 Studies on Heat-Related Illness

Although heat-related illnesses are preventable, the burden of heat-related morbidity and mortality is substantial. Globally, the incidence of heat-related illness in the general population has been increasing to alarming levels during the heat waves and the periods of extreme heat. Many scientific studies report the association between the ambient temperature or biometeorological indices and heat-related illnesses based on hospital records and morbidity statistics. Given that there is a broader identification in the fields of meteorology and environmental health that urban areas have higher risk for heat-related morbidity, the main focus of research in many studies is the population living in urban areas. In many regions around the world, the effect of UHI as a potential risk factor is documented.

In Japan, the remarkable temperature increase due to UHI has become a great threat to public health. In accordance with Japan Meteorological Agency, the annual mean air temperatures in downtown Tokyo, Japan, have already increased about 3 °C in the past 100 years [65]. In Fukuoka City, which is the fastest-growing urban centre in Japan, a study by Toosty et al. [66] determined the potential risk factors for heat stroke (heat stroke emergency transporters to the hospitals) based on the daily temperature and the wet-bulb globe temperature (WBGT) [66]. Their analysis revealed an increased risk of heatstroke associated with the increased daily temperature and WBGT, but also they found that the elderly is a group of people with the highest susceptibility to heat stroke.

Also, numerous Chinese cities are currently suffering from the UHI along with the high levels of air pollution. Bai et al. [67] investigated the associations between extreme heat and daily heat-related illnesses during the summers of 2011–2013 in Ningbo, China, which is an important port city in the country [67]. Their analysis is based on the data on patients diagnosed with heat-related symptoms due to the exposure to extreme high temperatures in summers provided by the national Heat-related Illness Surveillance System of the Chinese Center for Diseases Control and Prevention (China CDC). In the summer of 2013, the city experienced unequalled heat waves with maximum temperatures exceeding 40 °C occurred on 11 days. As for Ningbo, the authors found that from the overall total number of 3862 heat-related illnesses included during June–August over the period 2011–2013, 1260 cases occurred during the heat waves in July and August 2013. For the whole period from 2011 to 2013, more cases (69.6%) occurred on days with maximum temperature higher than 35 °C. Notably, they highlighted that people of all ages are at risks of heat-related illnesses during heat waves because young adults may be exposed more frequently outdoors to heat and affected by chronic physical and mental health conditions. In addition, Gu et al. [68] using the data from the Heat-related Illness Surveillance System in China for the summer of 2013, indicated a total of 5758 heat-related illness cases in 14 provinces across the whole China, while the reported cases were concentrated mainly in urban areas in the middle and lower reaches of Yangtze River, such as Chongqing, Wuhan, Shanghai and Ningbo [68]. With reference to the 2013 summer heat wave in the Eastern China, Wang et al. [69] point out that the difference between urban and non-urban/rural stations indicates that urban warming reached 1.22 °C based on the observational assessment while the WRF-simulated urban warming (1.46 °C) was much more intense [69]. Their findings confirm that the changes caused by urbanization modify heat wave patterns over the cities and the heat wave aggravates urban-related heat stress in the urban areas. Generally, the urbanization impacts over the metropolitan areas increase of the temperatures during the heat waves.

In the USA, where about the 85% of the population lives in metropolitan areas, many parts of the Northern and Western USA and Alaska have experienced warming trends in the annual average air temperatures since the early twentieth century (https://www.epa.gov/climate-indicators/climate-change-indicators-us-and-global-temperature#ref3). In accordance with the Climate Central Report [70], the cities with the five most intense urban heat islands are New Orleans, Newark, New

York City, Houston, and San Francisco and as a result the extreme urban heat directly impacts human health, including heat-related illnesses and deaths [70]. According to the Morbidity and Mortality Weekly Report (MMWR) of the USA published by the *Centers for Disease Control and Prevention* (CDC), during 2001–2010, approximately 28,000 hospitalizations for heat stress illness occurred during the summer months (May–September) in 20 states [71]. A recent MMWR reported that during June 25–30 in 2021, the states of Oregon and Washington, especially the Portland metropolitan area, recorded temperatures reaching 116°F (46.7 °C) [72]. This report indicated that the number of emergency department visits for heat-related illness skyrocketed on June 28 and during June 25–30 in 2021 was 69 times higher than that during the same days in 2019.

Also, Canada experiences temperature pattern changes with the average (mean) annual temperature to be increased by 1.7 °C from 1948 to 2016 (https://www.canada.ca/en/environment-climate-change/services/climate-change/canadian-centre-climate-services/basics/trends-projections/changes-temperature.html). Bassil et al. [73] examined the association between daily temperature and ambulance response calls for heat-related illness in Toronto, Ontario, Canada during the summer (from 1 June to 31 August) of 2005 and found that for every one degree increase in maximum temperature (°C) there was a 29% increase in ambulance response calls while for every one degree increase in mean temperature (°C) there was a 32% increase in ambulance response calls for heat-related illness [73].

In Australia, the heat waves have increased in duration, frequency, and intensity in many parts of the country and the number of extreme heat records has outnumbered extreme cool records for both daytime maximum temperatures and night-time minimum temperatures since 2001 (https://www.climatechangeinaustralia.gov.au/en/changing-climate/climate-extremes/). Therefore, the health impacts of heat waves are a growing public health problem in the country, especially under the unprecedented levels of urban overheating. Toloo et al. [74] found that even Brisbane, Australia, has a subtropical climate during the heat waves (including single days with increased ambient temperature) of the study period (December to February 2000 to 2012) were increased the emergency departments visits for heat-related illnesses and workload [74]. Specifically, the relative risks of heat-related illness visits on heat wave days increased 5.0 times (95% CI: 3.9, 6.4) and 18.5 times (95% CI: 12.0, 28.5) based on HWD1 (two or more successive days with daily maximum temperature $\geq$34 °C) and HWD2 (two or more successive days with daily maximum temperature $\geq$37 °C) compared to non-heat wave days, respectively.

During the fatal heat wave of August 2003 in Europe, there were almost 15,000 heat-related deaths in France and more than 70,000 additional deaths were recorded in twelve European countries with nearly 45,000 additional deaths in August [75]. Most of these additional deaths occurred in urban areas, where registered record-breaking temperatures. A study conducted by Argaud et al. [76], based on the emergency admissions of the department of Edouard Herriot Hospital (a 1100-bed, inner-city university hospital) in Lyon France, found that 83 patients presented with heatstroke among the 1827 emergency admissions (5%) [76]. Notably, most of the admissions

(73 patients [88%]) occurred from 3 to 13 August 2003, when the maximum daytime and minimum night-time temperatures were above 35 °C and 20 °C, respectively.

Smith et al. [77] investigated the impact of a moderate heat wave from 3 to 23 July 2013 on primary care and emergency department visits in England [77]. They used syndromic surveillance data and estimated that the number of general practitioner consultations for heat illness (heat/sunstroke) during the summer (May to September) of 2013 was 1166 (95% CI 1064 to 1268) which was double the rate observed during non-heat wave years.

Hartz et al. [78] analysed the heat-related dispatches (HRD, emergency 911 calls) those calls that emergency responders had coded as 'heat' for the event classification in Chicago, Illinois, between 2003 and 2006 within the time periods of 1 May through 30 September [78]. They investigated the seasonal, monthly and diurnal trends of the emergency calls and the daily patterns in relation to the maximum air temperature and maximum heat index (HI). Also, the spatial analysis of the emergency calls shows that generally the largest concentration of calls occurring in and nearby to the central business district because of the area's high daytime population density and the recreational activities.

In addition, we should mention that Chicago had experienced a fatal heat wave during mid-July 1995 that caused more than 600 deaths and 3300 excess emergency department visits from 12 to 20 July 1995 with the daily temperatures ranged from 33.9 to 40.0 °C, and on 13 July, the heat index peaked at 48.3 °C [79]. Semenza et al. [80] studied the nonfatal health effects of heat wave for the population served by Cook County hospitals (which includes the city of Chicago) [80]. They found that in July 1995, there were 42,304 people admitted to 47 non-VA (Veterans Administration [VA]) hospitals in Cook County compared with 40,910 people admitted in July 1994. Particularly, during the heat wave study period (13–19 July 1995), the number of hospitalizations exceeded the average for non-heat wave weeks by 1,072 (11%).

With reference to this fatal heat wave over the central USA during mid-July 1995, studies have investigated the critical weather factors which aggravate the impact of this heat event. Kunkel et al. [81] indicated the contribution of UHI in Chicago during this heat wave where the maximum temperatures were 1.6 °C higher in the city centre than in neighbouring areas and at night the city centre was 2.0–2.5 °C warmer [81]. Moreover, they found moisture differences between urban and rural areas as reflected by the differences in T_d which partly lead to the unusually high apparent temperatures (daily average values at 38 °C) during July 1995.

3.3 Extreme Heat and Cardiovascular and Respiratory Morbidity

The health effects of extreme heat on circulatory, cerebrovascular and respiratory diseases have been investigated from many scientific studies across several geographical locations. There is evidence that the exposure to extreme heat increases the morbidity in terms of hospitalizations, emergency department visits due to cardiovascular, cerebrovascular and respiratory diseases or even leads to greater risk for mortality attributable to the same causes.

Han et al. [15] conducted a time-series analysis to evaluate the association between ambient temperature and hypotension events based on daily hospital visits using the National Health Insurance Database (NHID) in Korea [15]. The participants of the study were the residents in seven metropolitan cities (Seoul, Busan, Incheon, Daegu, Daejeon, Gwangju and Ulsan) of Korea from 2011 to 2015. During the summer season (from June to August), the highest mean temperatures were observed in Daegu (25.8 °C) and Gwangju (25.3 °C). Their analysis indicated that a 1 °C increase in the mean ambient temperature was associated with 1.1% increase in hypotension hospital visits in lag day 0, whereas a 1 °C increase in mean ambient temperature was associated with 1.9% increase in hypotension hospital visits in cumulative lag 0–7 days. Interestingly, another finding of the study is that the effect of increasing ambient temperature on hypotension hospital visits was higher in younger age groups compared to elderly group (aged 65 or older) with a 1 °C increase in ambient temperature was associated with 2.0% increase in hypotension hospital visits in lag day 0 age group under 25 years.

Michelozzi et al. [82] evaluate the impact of maximum apparent temperature (T_{appmax}) on daily hospital admissions for cardiovascular, cerebrovascular and respiratory causes by age in twelve European cities (Barcelona, Budapest, Dublin, Ljubljana, London, Milan, Paris, Rome, Stockholm, Turin, Valencia and Zurich) during warm period of the year (April–September) for at least a 3-year time period. Obviously, there was a considerable variability in mean T_{appmax} values, ranging from 14.78 °C in Dublin to 29.58 °C in Valencia [82]. The pooled estimates of the effect of T_{appmax} on daily hospital admissions for respiratory causes indicate that for 1 °C increase over the 90th percentile of T_{appmax} with lag 0–3 days, the admissions increased by 14.5% (95% CI, 1.9–7.3) and 13.1% (95% CI, 0.8–5.5) in the all ages in Mediterranean and North-Continental cities, respectively.

Alessandrini et al. [83] collected daily data of emergency ambulance dispatches for people aged 35 or older during the summer periods (May to September) from 2002 to 2006, for the major towns in the Emilia-Romagna region in Italy (Piacenza, Parma, Reggio, Modena, Bologna, Ferrara, Ravenna, Forlì and Cesena) [83]. The meta-analytic estimates of per cent change/increase (%) in ambulance dispatches associated with 1 °C increase in mean apparent temperature between 25 and 30 °C indicated that the largest effects were observed for respiratory diseases in the 35–64 age group (4.56%). For mean apparent temperature above 30 °C, the per cent change in risk was greater with almost 8 and 5% increase in ambulance dispatches for cardiovascular and respiratory diseases, respectively, while the main effect was observed in the 75+ age group.

Findings from several studies found positive associations with the burden of morbidity from respiratory diseases and the ambient temperature. It is well known that during the winter months, especially in the higher latitudes, low temperatures are associated with increased occurrence of respiratory tract infections or the aggravation of symptoms of patients with respiratory diseases. Recent studies have found that extreme heat events and heat waves contribute to the burden of heat-related respiratory diseases. Moreover, other studies that explore the influencing mechanism of UHI intensity and urban landscape pattern on respiratory diseases suggest that urban residents' health is influenced by the increased ambient temperatures and especially the symptoms of patients with COPD or asthma are exacerbated (Huang et al. [84]).

Song et al. [16] found that the association between ambient temperature and respiratory emergency department visits exhibits a U-shaped relationship with the minimum (optimum) temperature at 21.5 °C [16]. To find the association between extreme temperatures and respiratory emergency department visits ED in Beijing, they collected the daily emergency room visits for respiratory diseases of three comprehensive hospitals in the Haidian District of Beijing from 1 January 2009 to 31 November 2012. They found that the greatest relative risk in daily respiratory morbidity that observed in the study indicated during the heat wave with mean temperature exceeding the 99th percentile. Another remarkable finding is that increased relative risks of upper respiratory diseases observed during the heat waves.

Zhao et al. [17] assessed the burden of morbidity from respiratory diseases that are attributable to ambient temperature in a city in China with subtropical climate [17]. They used daily data on hospital outpatient visits for respiratory diseases between January 2013 and August 2017, from one of the premier hospitals in Dongguan. The associations between temperature and morbidity for all respiratory diseases and asthma exhibited inverse J-shaped, but for COPD and bronchiectasis, the curves were U-shaped. In general, compared with the optimal temperature at a lag of 0–7 days, the relative risks of respiratory morbidity were higher for extreme heat (T = 30.3 °C) than the relative risks for extreme cold (T = 12.6 °C). Also, the heat-related respiratory morbidity risk was higher than cold-related risk for the population aged 0–64 years old, whereas the elderly (≥65 years) was more susceptible to low temperatures. Hence, the attributable fraction of hospital outpatient visits for all respiratory diseases attributable to heat is very high for the younger population.

Unexpectedly, most of the studies did not observe a strong association between heat waves or extreme heat events and risk of cardiovascular emergency hospitalization. For instance, Ponjoan et al. [85] conducted a self-controlled case series methodology (SCCS) in Catalonia, Spain, in order to compare hospitalization rates for cardiovascular diseases (including coronary heart disease, stroke or heart failure) during periods of exposure to extreme temperatures with rates during non-exposure periods within the same individuals [85]. They found that the incident rate ratios (IRR) for cardiovascular emergency hospitalization during heat waves were almost equal to 1. However, they found that the incidence of emergency hospitalizations due to cardiovascular disease increased significantly during cold spells compared with the unexposed periods. Ponjoan et al. [85] provide some explanations for the lack of association between cardiovascular morbidity and heat waves [85]. Firstly, the cardiovascular risk factors tend to be lower in the summer. Indeed, Marti-Soler et al. [86] assessed the seasonality of cardiovascular risk factors (CVRF) in a large set of 24 population-based studies including over 230,000 participants in 15 countries (Australia, Belgium, Czech Republic, Denmark, France, Italy, Lithuania, New Zealand, Northern Ireland, Norway, Portugal, Principality of Liechtenstein, Russia,

Spain and Switzerland) and found that CVRFs show a seasonal pattern characterized by higher levels in winter, and lower levels in summer [86]. Secondly, the heat waves have an immediate impact on mortality due to the fact that vulnerable people might die before being admitted into the hospital, resulting in a decrease of cardiovascular hospitalizations. Thirdly, during extreme heat events general public and especially vulnerable population groups may follow the recommendations of the local health authorities and avoid outdoor exposure.

With reference to the impact of heat waves on mortality, many authors indicate contrasting patterns of mortality and morbidity (e.g., hospital admissions) during the hot weather [87, 88]. A recent review on the epidemiological studies that investigate the heat wave-related disease burden of cardiovascular and respiratory diseases propose that heat waves have a 'main effect' due to the independent effects of daily high temperatures and an 'added' effect due to sustained duration of high temperatures which may play a key role in [89].

This is an important issue given that the evidence suggests that there is an excess of circulatory disease attributed to the exposure to high temperatures reported by the studies investigate the relationships between heat exposure and mortality, but not by studies focus on the relationship with morbidity. Kovats et al. [87] suggest that many deaths occur rapidly or in isolated people before they come to anyone's attention or reach to the hospital [87]. As a matter of fact, Bai et al. [67] highlight that heat stroke due to the exposure to extreme weather conditions is an immediate medical emergency and the emergency treatment of heat-related illnesses could minimize the degree of cell damage that portends progression to organ failure [67]. This is an explanation for the evidence that shows excess mortality associated with circulatory diseases during the extreme heat events.

In contrast, Halahavri et al. [90] review the literature on the negative influence of extreme heat on cardiovascular morbidity and report on several studies that provide evidence on the association between high ambient temperature and hospital admissions and emergency department visits for cardiovascular diseases in the USA, Canada, Australia, China and Korea [90].

For instance, in Ontario, Canada, an analysis of all those who were hospitalized for coronary heart disease between 1 January 1996 and 31 December 2013 found a 6% (95% CI: 1–11%) increase in related admissions on days with high temperatures (daily mean temperature >99th percentile) relative to the optimal temperature (noted in the study as daily mean temperature at the 79th percentile). Overall, the 1.20% (almost 16,000) of the 1.4 million coronary heart disease admissions included in the study were attributable to heat (Bai et al. [14]).

In addition, Basu et al. [13] have found that the increased temperatures have same-day effects on emergency room (ER) visits for several diseases outcomes, including cardiovascular and respiratory diseases, in California [13]. They examined the association between mean daily apparent temperature and emergency room visits during the warm seasons (1 May to 30 September) of 2005–2008 in 16 same climate zones of California. The meta-analysis results indicated excess risk (%) for emergency room visits per 5.6 °C (10 °F) increase in apparent temperature for ischemic heart disease (1.7% [95% CI: 0.2–3.3]), acute myocardial infarction (1.7% [95% CI: −0.5

to 4.0]) ischemic stroke (2.8% [95% CI: 0.9–4.7]), cardiac dysrhythmia (2.8% [95% CI: 0.9–4.9]) and hypotension (12.7% [95%: 8.3–17.4]).

Lu et al. [91] using postcode-level hospitalization data in Queensland, Australia, between 1 January 1995 and 31 December 2016 examined the temporal variation of associations between ambient temperature and hospitalizations for cardiovascular diseases and their trend over time [91]. One of the major findings of their study is that the associations between temperature and hospitalizations for cardiovascular diseases showed different patterns in 1995 and 2016. There was an increasing trend in the risks of hospitalization for cardiovascular diseases associated with higher temperatures in 2016, whereas the effect of lower temperatures on the hospitalizations has decreased. In accordance with the authors, the residents in Queensland have not adapted to the impacts of high temperatures.

A recent study by Bao et al. [92] investigated the modification effect of intra-urban landscape characteristics on the association between heat exposure and stroke morbidity based on hospitalization data in Shenzhen, China [92]. Notably, their analysis reveals that high values of night-time land-surface temperature and the proportion of impervious surface might aggravate the harmful effects of heat on stroke morbidity while high values of the normalized difference vegetation index (NDVI) and the proportion of water bodies may alleviate its effect.

In Northern Europe, the studies on the association between extreme heat events and hospitalizations or emergency department visits are generally scarce and there is a lack of knowledge concerning the effects of heat exposure on morbidity. Although North European countries have cool summer temperatures, a recent study conducted in the Helsinki metropolitan area, Finland suggests that heat waves are public health concern, as indicated by the hospital admissions, even in a Northern climate [93]. In particular, the analysis was used different definition of heat waves based on the 90th and 95th percentile cut-off points for mean daily temperature during the summer months (May–August) from 2001 to 2017. According to these definitions, the average temperature during the heat waves varied between 22.6 and 23.9 °C. Interestingly, associations were mainly indicated between heat waves and hospital admissions, especially for respiratory diseases and to a lesser extend for cardiovascular and cerebrovascular diseases in the 18–64 age group.

In addition, heat waves have been associated with cardiac arrest. In particular, out-of-hospital cardiac arrest (OHCA) is a medical emergency because of its unpredictable and time-sensitive nature. Various studies indicate that heat waves and extreme temperatures significantly raise the risk of OHCA.

Kang et al. [94] have investigated the occurrence of OHCA based on emergency medical services records during the periods of heat waves in seven major cities in Korea with more than 1 million residents, between January 2006 and December 2013 [94]. They found that the relationship between maximal ambient temperature and OHCA was J-shaped, especially in Seoul, Busan and Gwangju. Based on the pooled effects of heat, a 1 °C increase in maximal temperature over 28 °C was associated with a 1.3% excess risk of OHCA while using the absolute temperature criteria for the heat waves according to the Korean Meteorological Administration, the heat wave with an absolute daily maximal temperature at least 35 °C ('heat wave alert') was associated with a 16% increased risk of OHCA. Another important finding of Kang et al. [94] was the distribution of OHCA by hour of the day [94]. The peak hours for OHCA during the non-heat wave days were clearly from 7 to 10 AM and were with a secondary peak from 5 to 8 PM, while during the heat wave days the peak hours for OHCA were during the late afternoon from 3 to 5 PM when the temperature is highest.

In addition, Kranc et al. [95] analysed the OHCA events treated by the Magen David Adom (MDA) in Israel during 2016–2017 in relation to the meteorological conditions indicating that the exposure to extreme high temperature is among the risk factors of OCHA events [95]. Interestingly, they indicated that the 61.6% of OHCA calls were made during the daylight hours (7 AM–7 PM) and the 29.1% of OCHA events were occurred in summer (31 May–22 September).

Doan et al. [96] looked at the data of all OHCA patients who are attended by the Queensland Ambulance Service (QAS) paramedics in Brisbane, Australia, between 1 January 2007 and 31 December 2019 [96]. The 'main effect' of heat waves due to the independent effect of daily high temperatures increased the risk of OHCA in the four different definitions of heat waves which are defined as the daily average temperatures at or above a heat threshold (95th, 98th and 99th percentile of the yearly temperature distribution across the study period) for at least two consecutive days. The risk of OHCA increased as the heat thresholds increased and when the heat threshold of 99th percentile was used, the RR increased to 1.48 (95% CI: 1.11–1.96). In fact, the main effect of heat waves which was the intensity of heat wave days was stronger than the added effect which was the sustained duration of heat wave days.

In accordance with Kang et al. [94], the results of the studies which investigate the relationship between OCHA events associated with the exposure to extreme heat could explain the paradoxical findings of studies showing that hospital admissions for cardiovascular causes, in contrast to cardiovascular mortality, are unaffected by heat exposure [94]. That's to say, sudden and unexpected cardiovascular events occur before the patient reach to a hospital. In the case of cardiac arrest, the survival rate is low such as other acute cardiovascular diseases which contribute to the global leading causes of death.

3.4 *Effects of Heat on Mental Health, Well-Being and Cognitive Performance*

Health is not merely the absence of disease or infirmity. Good health is a state of complete physical, social and mental well-being. People's mental health is jeopardizing from the impacts of climate change [97]. According to Wong et al. [98], the increasing ambient temperature severely impacts people's mental health because elevated temperature is associated with mood disorders, anxiety and depression, emotional disturbance and aggressive motivation and behaviour [98]. As a matter of fact, being exposed to extreme heat can lead to physical and psychological fatigue

[99] and the thermal discomfort leads to hostile feelings, aggressive thoughts and/or violent actions [97]. Anderson [100] suggests that there is a linear relationship between ambient temperature and murders and assault rates in USA [100]. Their hypothesis is supported by many authors who investigate the relationship between meteorological variables and aggressive crime. Also, findings highlight that there is an association between climatic conditions and suicide rates. Burke et al. [101] by using comprehensive data from multiple decades for both the USA and Mexico found that suicide rates rise 0.7% in US counties and 2.1% in Mexican municipalities for a 1 °C increase in monthly average temperature [101].

The impact of UHI and urban environment on mental health and well-being of urban residents has been investigated from many studies. Wong et al. [98] assessed the health impact of UHI in Greater Kuala Lumpur, Malaysia, based on interviews conducted between October 2016 and May 2017 among adult urban residents [98]. Apart from respiratory problems and heat-related illness, a high proportion of respondents experienced psychological impacts such as anxiety, depression and aggressive behaviour as well as social impacts like reduced outdoor activities and skip work or school. Moreover, the effects of UHI on physical and mental health of the residents of Tianjin in four periods from 1992 to 2018 were evaluated by Huang et al. [102]. Their results show that the distribution of the effects of UHI on emotional health increased level from 1992 to 2018, while the affected areas gradually expanded from the city centres to the periphery.

In order to examine the impact of ambient temperature on the emotional well-being in the adult U.S. population, Noelke et al. [103] use data from telephone surveys conducted on 350 days per year during the years 2008–2013 in combination with air temperature data on interview dates [103]. The results of the study indicate that increasing temperatures significantly reduce well-being, especially above 32 °C. Particularly, compared to average daily temperatures in the range of 10–16 °C, temperatures above 21 °C reduce positive emotions and increase negative emotions and fatigue. Also, these effects are especially strong among individuals aged 46+ and less educated.

In addition, many authors suggest that several mental illnesses may be sensitive to exposure to high ambient temperatures. Hence, people with mental illness are identified as a heat-susceptible group of the population. In general, people with mental health problems often experience poorer health with higher rates of morbidity and mortality compared with the general population and the way that people experience their mental illnesses can increase their susceptibility of developing poor physical health. Therefore, they are especially vulnerable to extreme heat or heat waves because of their behavioural issues and medications that interfere with physiological homeostasis [104].

We should place emphasis on the fact that two issues concern the effects of heat on psychiatric illness. Firstly, the nature of the psychiatric condition can be a risk factor for heat-related morbidity and mortality. Secondly, heat can be a risk factor for the exacerbation of the condition [104].

The impact of extreme heat on mental health has been extensively investigated from many scientific studies. A review conducted by Thompson et al. [105] on the associations between heat and mental health concluded that among the serious mental health outcomes included in studies are bipolar disorders and exacerbations of schizophrenic symptoms [105]. A recent study by Almendra et al. [106] assessed the short-term impacts of daily mean temperature on hospital admissions with primary diagnosis of mental disorders, suicide and self-inflicted injury between 2008 and 2014 in Lisbon, Portugal. Their statistical analysis also added potential confounders (relative humidity (RH) and air pollutants (PM_{10} and O_3) concentrations) of the relationship between the mental disorders and the ambient temperature [106]. The results showed a significant increase in hospital admissions for the previously mentioned causes above 30 °C on the day of exposure and 27 °C at lag 0–1 and lag 0–2.

Hansen et al. [104] used health outcome data from Adelaide, South Australia, between 1993 and 2006 in order to estimate the effect of heat waves on hospital admissions and mortalities attributed to mental, behavioural, and cognitive disorders [104]. They show that hospital admissions for mental and behavioural disorders increased at temperatures above 26.7 °C, and they also increased by 7.3% during heat waves compared with control periods (non-heat wave periods).

A study conducted by Vida et al. [107] examined the relationship between temperature, humidity and emergency department visits for mental and psychosocial problems in three geographic areas of Québec, Canada, from 1995 to 2007 from May to September [107]. The incidence rate ratio (IRR) for the number of visits at an emergency department at 20 °C (68 °F), 22.5 °C (72.5 °F) and 25 °C (77.0 °F) increased with increasing temperature which means that for a given geographical area and age group, the incidence rate of visits at that temperature was significantly higher than the overall warm season mean.

In order to assess the effects of extreme ambient temperature on hospital emergency room visits for specific mental and behavioural diseases, Wang et al. [108] conducted a time-series study by using morbidity, meteorological and air pollution data from 2002 to 2010 in Toronto, Canada [108]. They found that exposure to high ambient mean temperature is significantly associated with increases in hospital emergency room visits for individuals with mental and behaviour disorders.

Basu et al. [109] postulate that the biological mechanisms for temperature-related increases in mental health-related outcomes involve several factors. In fact, stressful situations from everyday life might be aggravated when temperatures are elevated and particularly for persons with mental health problems is an another risk factor [109]. Lõhmus [110] review the literature on the biological mechanisms which link mental health with heat and identified that there are several factors contribute to the vulnerability of patients with mental disorders during heat waves and extreme heat events [110]. Among these factors are cited their medical treatment with certain psychotropic medications (please see Sect. 3.1) and their impaired thermoregulation and their ability to adapt during periods of extreme heat.

Another important aspect of people's mental health is their cognitive function and performance. The prolonged exposure to high ambient temperatures negatively impact people's cognitive function. A recent study focuses on the effects of outdoor heat stress on cognition performance by examining heat stress incidence effects on the results of verbal and math tests administered within a national survey in China for people above the age of 40 years [111]. They used the cumulative heat stress degree days (HSDD) as a measure of heat stress and they found that heat stress decreases performance on verbal and math test scores up to 3 days.

3.5 Extreme Heat Events and Sleep

In accordance with a systematic literature review on the relationship between climate change and human sleep, the higher ambient temperatures have negative effects of on sleep time and sleep quality [112]. It is well-documented that the thermal environment is an important determinant of sleep quality. In accordance with Okamoto-Mizuno and Mizuno [113], a key determinant of sleep is the thermal environment because even healthy individuals without insomnia are affected by the excessively high or low ambient temperature [113]. As a matter of fact, they indicated that heat exposure affects the sleep stages and pointed out that humidity is one of the most important factors that increase heat stress during sleep. The ambient temperature affects the body's thermoregulatory system, and thermoregulation is intimately connected with the waking–sleeping cycle [114]. Arifwidodo et al. [115] investigated the effects of UHI on urban residents in Bangkok, Thailand, and Bandung, Indonesia, based on a survey questionnaire. They indicated that UHI causes disruption of their daily activities such as working, sleeping and increases the energy consumption for cooling especially during the night to help them sleep. An experiment with ten participants conducted in a non-air-conditioned house in Baoding assessed the effects of the daily maximum hourly outdoor temperature on sleep quality based on a questionnaire [116]. At 36 and 38 °C, the sleep quality was negatively affected because high temperatures led to a decrease in body temperature and an increase of the average heart rate and respiratory rate [116].

Obradovich et al. [117] analysed data from 765,000 US survey respondents from 2002 to 2011, coupled with night-time temperature data in order to investigate the relationship between climatic anomalies, reports of insufficient sleep, and projected climate change [117]. They concluded that as mean monthly night-time temperature anomalies become more positive, nights with insufficient sleep become more frequent. Also, they found that the effect of night-time temperature anomalies is most acute during the summer and among lower-income respondents and the elderly.

Weinreich et al. [118] investigated the association of short-term exposure to PM_{10}, ozone and temperature with sleep-disordered breathing (SDB) in participants (aged 50–80 years) of the Heinz Nixdorf Recall lived in Essen, Bochum and Mülheim (Germany) [118]. They observed that associations for temperature were stronger in summer yielding a 32.4% (95% CI: 0.0–75.3%) increase in apnoea–hypopnoea index (AHI) per 8.6 °C.

Nastos and Matzarakis [119] evaluated the effects of thermal indices (predicted mean vote (PMV), physiologically equivalent temperature (PET) and standard effective temperature (SET*)), the mean radiant temperature (T_{mrt}) and the minimum air temperature (T_{min}) on sleep disturbances in Athens, Greece [119]. They obtained the daily records of hospital admissions for sleep disturbances of the psychiatric emergency unit of Eginition Hospital of Athens for the years 1989 and 1994 and found that the thermal indices and the T_{min} statistically associated with sleep disturbances.

Min et al. [114] investigated the association between ambient night-time temperature and sleep problems assessed by the prescription of hypnotic drugs in the adults who were 20 years old or older in South Korea [114]. For the purpose of this study, they collected the highest temperature data at night (23:00–07:00) for every month from January to December between 2002 and 2015. We found that a nonlinear, U-shaped relationship between night-time temperature and hypnotic medication prescription. With an increase per 1 °C temperature or an increase in a square per 1 °C, the prescription dose of sleeping pills was significantly increased (both $p < 0.05$). At each 5 °C night-time temperature, subjects belonging to low ($\leq$0 °C and 0–5 °C) or high (20–25 °C and $\geq$25 °C) temperature categories had significantly higher doses of sleeping pills than those at the reference temperature (10–15 °C).

3.6 Maternal Health and the Foetal Health Effects Associated with Heat Exposure

A new area of research focuses on the maternal health and the foetal health effects associated with the exposure to environmental hazards and extreme weather events such as heat waves. Kloog [120] highlights that in the context of the global warming and the rapidly urbanization, the urban environments represent an increasing risk for pregnant women [120]. The urban built environments to which women are exposed during pregnancy influence their health and can pose challenges for newborns' health due to the elevated concentrations of air pollutants, the environmental and meteorological conditions of urban centres. Kuehn and McCormick [21] conducted a literature review on studies that investigate the exposure to extreme heat and heat waves and affect maternal, foetal and neonatal health [21]. They indicated that the urban populations are of major concern because of the UHI phenomenon which adversely affects pregnant women.

Heat exposure is among the risk factors that contribute to complications during pregnancy and is associated with adverse pregnancy outcomes. Several epidemiological studies over the past decade have reported associations between high temperatures or heat waves and adverse pregnancy outcomes, including preterm birth (births that occur before 37 gestational weeks), stillbirth and low birth weight [21, 22]. According to the review conducted by Chersich et al. [22], pregnant women might have increased risk of cardiovascular events during heat waves and experience maternal stress and sleep disturbance when they are exposed to high temperatures [22]. Lin et al. [121] observed a U-shaped exposure–response relationship between daily average temperatures and maternal global severity index (GSI) scores at lag 0–1 days, with higher GSI indicating a higher level of emotional stress [121]. The U-shaped curve reached the lowest level in the optimal temperature range between 20 and 25 °C.

Kim et al. [122] studied the effect of exposure to extreme temperature during pregnancy on maternal and infant hospitalizations, using the universe of administrative inpatient discharge records from three US states (Arizona, New York, Washington) [122]. Given that there is a substantial variation in average temperatures across these geographic regions that could generate differences in adaptation responses, the authors model exposure to extreme heat in terms of standard deviations relative to each county's monthly temperature mean. Generally, their analysis shows that extreme heat exposure during second and third trimesters raises the likelihood of hospitalization during pregnancy. An additional day during the second and trimester with average temperature at least three standard deviations above the county's monthly mean raises the likelihood of hospitalization during pregnancy. Also, an additional day with three standard deviations above the county's monthly mean increases the probability of having any complication related to pregnancy at childbirth. Interestingly, they suggested that extreme heat has both immediate and persistent impacts on maternal pregnancy complications given that the heat exposure in the second trimester is associated with hospitalizations in both second and third trimesters. Moreover, they found that the foetal exposure to extreme temperatures influences infants' health and the effect is driven by exposure to extreme heat during the second trimester. In fact, there is increased the risk of a newborn being diagnosed with dehydration and an infant being re-hospitalized to due prenatal jaundice, prenatal haematological disorders, and respiratory conditions including bronchitis, influenza and pneumonia, which are causes mainly linked to dehydration.

In order to investigate the acute extreme heat exposure towards the end of gestation in relation to preterm birth, Ilango et al. [123] studied the cohort of mothers who gave birth to a live infant during the warm season (May through September) in California, 2005–2013 [123]. With the aim of capturing a broad set of periods with extreme heat of varying intensity and duration, they created 12 different definitions of heat waves using varied extremes (75th, 90th, 95th, 98th percentiles) and durations (at least 2, 3 or 4 consecutive days) from estimated maximum temperature data. Their analysis revealed that the mothers who experienced heat waves during the last week of gestation had an increased risk of preterm birth, with increasing associations given increasing severity of extreme heat episode. Specifically, using the definition of a heat wave with the daily average maximum temperature above the 98th percentile and duration of 4 or more days, the risk of preterm birth was 13% (HR = 1.13; 95% CI 1.05, 1.21) higher among mothers who experienced a heat wave during the last week of gestation compared to mothers who did not.

Similarly, a study conducted in Southern District of Israel, which is an area characterized by a semiarid climate zone with long and hot summers, Spotler et al. [124] identified that short-term exposure to high ambient temperatures is associated with an increased risk of late preterm and early term births [124].

Furthermore, Qu et al. [125] conducted a crossover study in order to investigate the association between the extreme heat exposure and the pregnancy complications during the summer months (June–August) and transitional months (May and September) from 2005 to 2013 in New York State [125]. They obtained data from a database covering over 95% of hospital records in NYS for hospital admissions and

ED visits reporting with a principal diagnosis of pregnancy complication. They used 90th percentile as the threshold to define extreme heat and concluded that during the summer months the extreme heat events was significantly associated with 1–2% excess risk of emergency departments visits due to the pregnancy complications with the effects lasted for 5 days and the strongest cumulative effect occurred on 0–3 days after the extreme heat event with 4% excess risk of emergency department visits. Also, the extreme heat events in transitional months were also associated with increased ED visits and hospital admissions for pregnancy complications and the effect mainly occurred on the same day of exposure. Among their findings on the associations between extreme heat events and subtypes of pregnancy complications, they observed an increased risk in threatened or spontaneous abortion in both summer and transitional months.

In addition, Dastoorpoor et al. [126] investigated the relation between the physiological equivalent temperature (PET) index and the adverse pregnancy outcomes (stillbirth, low birth weight, preterm labour, spontaneous abortion, preeclampsia and gestational hypertension) based on data from two big referral hospitals in Ahvaz, Iran, from April 2008 to March 2018 [126]. The PET index varied between 0.8 and 47.3 °C with a mean value of 27.6 ± 11.6 °C during the study period. It should mention that Ahvaz has a desert climate with hot long summers and short mild winters and often experience high temperatures between 45 and 50 °C during the summer. For instance, Ahvaz recorded the temperature of 54 °C on 29 June 2017. Also, in the metropolitan cities of Iran the phenomenon of the UHI is extremely intense due to the population growth, the urban sprawl and the frequent heat waves through the year [11]. Dastoorpoor et al. [126] concluded that high values of PET (45.4 °C) increased the risk of stillbirth with the effect observed until 0–21 days, while high and low PET values were associated with increased risk for low birth weight [126]. Moreover, the cumulative effects of hot thermal stress significantly increased the risk of stillbirth in lags 0 and 0–13 and LBW in lag 0–13.

Overall, the association of heat exposure with maternal and newborn health is important in order to quantify the burden of heat-related morbidity in the context of urban overheating.

3.7 *High Temperatures and Kidney Diseases*

Several studies have investigated the relationship of high temperature with the risk of hospitalization for renal diseases. Glaser et al. [18] suggest that heat stress-associated nephropathy may represent an epidemic due to global warming, the frequent and intense heat waves [18]. In fact, epidemics of chronic kidney disease (CKD) of unknown aetiology emerge in some regions of the world which are not associated with traditional risk factors for CKD and have been linked with recurrent dehydration and heat stress or even strenuous manual labour under very hot conditions [18, 127].

Also, the relationship between high ambient temperature and kidney disorders or renal failure, acute kidney injury, urolithiasis, nephritis, and nephrotic syndrome have been assessed by De Lorenzo and Liaño [128]. A recent study suggests that patients who suffer from kidney failure have increased risk of hospitalization and mortality during the days with extreme heat events based on data from patients with end-stage renal disease undergoing hemodialysis treatment in Boston/Massachusetts, New York City/New York and Philadelphia/Pennsylvania [129]. They underlined the vulnerability of people living within the urban centres because of their exposure to high environmental temperatures due to UHI effect and in the context of climate change.

Hansen et al. [20] analysed the hospital admissions data for principal discharge diagnoses of renal disease (ICD-10 code: N00–N39) during the heat wave periods compared with non-heat wave periods in the Adelaide metropolitan area for the period from 1 January 1995 to 31 December 2006 [20]. Their analysis revealed that during heat waves there was an increase in hospital admission for renal diseases with an incidence rate ratio of 1.100 (95% CI:1.003–1.206) compared with non-heat wave periods. Specifically, the overall incidence rate ratio was 1.255 (95% CI: 1.037–1.519) for acute renal failure during heat wave periods compared with the non-heat wave periods and the admissions peaking during an extreme heat event when the maximum temperature reached 44.3 °C on 14 February 2004 during an 8-day heat wave.

Kim et al. [130] examined the association between ambient temperature and acute kidney injury using data on emergency department visit from 1 January 2010 to 31 December 2014 in Seoul, South Korea [130]. Using a case-crossover study design each patient serves as his/her own control, they found that the risk of acute kidney injury increases as temperature increases in both warm (April–September) and cool (October–March) seasons. The threshold temperature was found to be at 22.3 °C during the warm season with the odds ratio (OR) increasing above this threshold which indicates a severe risk for acute kidney injury associated with ambient temperatures above 22.3 °C.

Goldfarb and Hirsch [19] hypothesized that the increasing prevalence of kidney stones could be attributed to the urbanization process and the exposure of the urban population to the higher temperatures due to UHI [19]. The human migration from rural areas to warmer urban areas, which began in the last century, along with impact of global warming on the observed increasing prevalence rate of nephrolithiasis globally. Although several other factors and aetiologies explain the prevalence of kidney stones in the population, it is worth mentioned that increased ambient temperature is a risk factor for nephrolithiasis due to the stimulation of heat-induced perspiration (Goldfarb and Hirsch [19]).

Abreu Júnior and Ferreira Filho [131] looked at the data on hospitalizations for nephrolithiasis over the period from 1 January 2010 to 31 December 2015 in Brazilian cities with a population of more than 300,000 inhabitants located in tropical and subtropical climate regions [131]. They correlated the hospitalization data with meteorological variables and found positive association between the number of hospitalizations for nephrolithiasis and air temperature. Their analysis revealed that there will be approximately 592 new hospitalizations per month (or 7104 new hospitalizations per year) for nephrolithiasis for an increase of 1°C in the monthly mean temperature in the total population of Brazil.

4 Urban Overheating: Challenges for the Healthcare System

It is an undeniable fact that heat waves and extreme heat events can pose extreme challenges on the healthcare sector. They can cause acute increased demand for health and social care services. Especially, the urban areas where most hospitals and healthcare institutions are located experience these events more intensely. The extreme heat due to weather events increase the pressure on healthcare system because increase the burden on the health systems including the emergency health service, ambulance dispatches, emergency department visits and hospital admissions. Generally, the preparation for heat waves in the healthcare system is much weaker than winter preparation.

During the 2003 heat wave in France, the Assistance Publique-Hôpitaux de Paris, which is the university hospital operating in Paris and its surroundings, reported more than 2600 excess emergency department visits, most of them classified as heat-related, and 1900 excess hospital admissions from 1 to 14 August 2003 [132]. In the Paris area, the maximal temperatures measured above 35 °C during 14 consecutive days from 1 to 14 August 2003, including eight days with temperatures above 40 °C and nine days with minimal temperatures above 35 °C [133]. During this period, the heat-related illnesses were more frequent in elderly people, especially those who lived in urban areas. Also, the number of deaths at home and in nursing homes was doubled [132]. Claessens et al. [133] with reference to the experience of the fatal 2003 heat wave in Paris suggest that the 'heat island' effect should be considered a major risk for patients in developing heat-related conditions [133].

A study of the 2003 European heat wave in the UK in early August indicated that London experienced night-time temperatures of 26–27 °C during the heat wave, and a maximum of 37.9 °C was recorded in London on 10 August. Although all regions in the UK had high proportion of excess deaths, especially for people over the age of 75 years, from 4 to 13 August 2003, the London region had the greatest excess in the over 75 age group with a 59% (CI: 51–67%) increase [134]. This partly indicates the sensitivity of urban areas such as London during the extreme heat events due to UHI phenomenon which can cause London to be up to 10 °C warmer than neighbouring rural areas.

Therefore, it is crucial to take measures to prevent the adverse health effects of heat on the exposed population and particularly for the high-risk group of population. The adverse health outcomes related to heat depending on the duration, the frequency and the intensity of the extreme heat events as well as the personal heat exposure. The magnitude and the severity of these health risks depend on the capability of Heat-Health Warning Systems and services and public health system to address, prepare for and manage these hazardous events. The Heat-Health Warning Systems address the health risks posed by the extreme heat events and help healthcare professionals cope with the periods of extreme temperatures (https://public.wmo.int/en/media/news/wmo-who-issue-guidance-heat-health-warning-systems). However, the health risks of the extreme heat events depend on several other factors which are discussed in this chapter and based on individual's vulnerability and susceptibility to heat exposure.

The heat-related illnesses and the adverse health effects of heat exposure are preventable and treatable. The prevention requires the development of a strategy for disaster risk management and effective early warning systems in order to respond appropriately to a threat and reduce the risk of morbidity, mortality and property loss and damage. According to Matthies et al. [135], the prevention consists of actions at different levels [135]. Among these, actions is the preparedness of health care and social services coordinated with the National Meteorological Services and Heat-Health Warnings Systems. These actions can be integrated in a defined Heat-Health Action Plan (HHAP).

In order to ensure the provision of health care during the extreme heat events, the health personnel in specific departments such as the Cardiology, Respiratory and Pathology Departments should be trained. As well as the emergency departments of hospitals could be prepared and alerted for the upcoming heat events [136]. Hospitals also need to prepare for a large sudden influx of patients during extreme heat events, which exceed typical daily admission rates. A lack of preparedness could lead to inadequate medical care for the patients or even the collapse of the healthcare system. As well as, treatment protocols and guidelines for the prevention and management of heat-related illnesses should be available.

The medical education and training should be modified in the context of climate change in order to adapt to the upcoming extreme weather events and the environmental crisis [137, 138]. Xie et al. [139] point out that primary care providers may adjust therapies to adapt to the health effects result from the climate change such as the heat-related illness [139]. They mentioned that anticipatory guidance and modification of medications are available to patients during the heat waves. For instance, Westaway et al. [23] provide recommendations for the healthcare professionals to reduce the risk of dehydration and heat-related illness and review the patient's medicines and reduce the dose of certain medications in order to eliminate the health risks during the heat waves [23].

According to Kovats and Osborn [140], heat waves affect the functionality of hospitals and care homes. Health services will be vulnerable to an increase in the frequency and intensity of heat waves [140]. Care homes are at risk of overheating, and care management practices may not sufficiently address heat risks.

In addition, there is a great concern that hospitals, clinics, care and nursing homes are at risk of overheating during the extreme heat events. The healthcare buildings and health facilities need special attention to increase their resilience during the extreme heat events and heat waves. The hospital wards encounter high indoor temperatures during the periods of extreme temperatures due to the lack of adequate cooling devices and air-conditioning. A retrospective study of all the patients who died during the 2003 heat wave at the Nantes University Hospital (Centre hospitalier universitaire (CHU) de Nantes) between 6 and 15 August 2003 showed seventeen patients died from hospital-acquired heat stroke (19% of all hospital deaths) [141]. In accordance with the authors, this condition accounted for a 25% increase in hospital mortality over the same period during 2002.

Therefore, the hospital adaptation to extreme weather conditions should ensure a high quality of medical care and thermal comfort for patients. Therefore, the protection of vulnerable patients from extreme climatic conditions is an issue of great importance for the healthcare sector [142]. Given that inpatients are especially susceptible to heat due to their infirmity, the adverse environmental factors may aggravate their health condition. Generally, the scientific studies focus on the impact of overheating on the inpatient's thermal comfort are scarce, though it is crucial to evaluate the adverse effects of hot weather on their health and well-being. However, studies have focused their attention on thermal comfort and heat vulnerability among care home residents given that they are particularly vulnerable to extreme heat events. Gupta et al. [143] examined the magnitude and perception of summertime overheating in two London-based care homes, occupying modern and older buildings [143]. Although the indoor temperatures were found to be high (>30 °C) with bedroom temperatures often higher at night than daytime across both care settings, the interviews and surveys conducted with the residents in both care homes and revealed that the majority of them perceived themselves to be comfortable in the conditions and did not recognize their susceptibility to heat stress. The results of this study raise many concerns about the overall health and well-being of the elderly, inpatients and people with disabled health and infirmity. The healthcare system requires radical change to protect the high-risk groups of population under the menace of heat waves and extreme weather events which become more frequent and intense in the near future.

Apart from the inpatients in the hospitals, the healthcare staff and people employed in hospitals are also at risk of developing heat-related health problems during the shift work and long work hours in the periods of excess heat. In fact, heat consists of an occupational hazard and people who are physically active in high temperatures are at increased risk of heat injury and experience reduced cognitive ability [144]. The hospital design and construction influence thermal comfort and ventilation during heat waves, which in turn, can cause equipment failure, such as failure of essential refrigeration systems, disruption of IT and laboratory services or even the degradation or loss of medicines [140]. For instance, Carmichael et al. [144] reported that during the 2003 heat wave in the UK the operations in the hospitals were cancelled due to IT equipment failure, laboratory work halted because machines failed in the heat and the failure of a freezer destroyed human specimen samples [144].

In addition, we should mention the heat-related costs for the healthcare services result from the extreme heat exposure, which is an additional pressure on the healthcare system and a substantial economic burden. The economic consequences of heat-related health effects are the incurred medical expenses and healthcare costs, and loss of work productivity. Wondmagegn et al. [145] classified the healthcare cost associated with the extreme heat exposure as direct (health interventions) and indirect (productivity) costs [145]. Particularly, the direct healthcare costs consist of two categories, the cost to the healthcare provider and the costs to patients and their family. The healthcare cost associated with staff, training, medical devices, pharmaceuticals, diagnostic procedures outpatient and inpatient services is substantial and constitutes a challenge for the healthcare system [145]

For example, based on the 2005 Nationwide Inpatient Sample (NIS) data there were estimated 6200 hospital stays (hospitalizations) in the USA that were caused by excessive heat exposure due to weather conditions. The average cost of stays related to heat exposure was $6200 per heat-related stay with average length of heat-related stays at 3.2 days [146]. The aggregate costs of hospital stays resulting from exposure to excessive heat were $38.7 million, whereas the aggregate costs of hospital stay that not related to the exposure to excessive heat and cold due to weather conditions were $315.8 billion. Schmeltz et al. [147] explored the healthcare costs associated with hospitalizations for heat-related illness in the USA from 2001 to 2010 based on the NIS data [147]. They found 73,180 hospitalizations for heat-related illnesses while the total number of hospitalizations was 181,094,795 from May to September during the study period. Notably, for the 75.8% of the heat-related hospitalizations the hospital location was urban, which indicates the vulnerability of urban population to extreme heat.

Moreover, Knowlton et al. [148] calculated the health cost of the 2006 California heat wave [148]. The total health costs associated with the heat wave were approximately $5.4 billion based on the premature deaths, hospitalizations, emergency department visits, and outpatient visits [148].

A recent study conducted by Adélaïde et al. [149] performed an assessment of the health-related economic burden of heat waves from 2015 to 2019 in France based on morbidity (emergency department (ED) visits and outpatient clinic visits for selected heat-related adverse effects) and mortality (all-cause excess mortality) data as well as the loss of well-being (Minor Restricted Activity Day (MRAD)) during the heat waves of the period under study [149]. Between 2015 and 2019, their analysis revealed that the economic impact was €25.5 billion, mainly in mortality (€23.2 billion), minor restricted activity days (€2.3 billion), and morbidity (€0.031 billion).

The above-mentioned studies highlight the great challenge that faces the healthcare system in terms of economic burden during the extreme heat events and heat waves. The financial burdens of individuals hospitalized for heat-related illnesses are substantial and constitute an obstacle for the effectiveness and efficiency of the healthcare sector. Also, hospitals should be prepared for a large sudden influx of patients during these extreme weather events. At the same time, the hospitals have a range of unique heat-health challenges due to vulnerable inpatients. Especially, the urban areas aggregate the major healthcare services and facilities within the countries and face a particular challenge because they are severely afflicted by the extreme heat events due to the UHI.

5 Conclusions

The UHI describes the phenomenon where differences in heat fluxes between urban and rural areas modify the urban climate. To the role of anthropogenic activities in the formation of urban overheating is substantial, while the climate change contributes to the exacerbation of the extreme heat events in the cities. The elevated ambient temperatures are associated with certain risk factors, compromising human health and well-being. In this chapter, we explored all the aspects of heat-related morbidity under the influence of UHI during the periods of extreme heat and the literature review showed that the burden of morbidity due to the urban overheating is considerable. Although the heat-related illness is largely preventable, the heat-related morbidity is unavoidable due to the climate change and the projected extreme heat events. The urban areas are more vulnerable under the influence of UHI and people living in highly urbanized cities are at greater risk. The implementation of urban overheating mitigations strategies is crucial to reducing the thermal stress in cities and minimizing the heat-related health risk. Also, Heat-Health Warnings Systems and Heat-Health Action Plans could prevent the weather-related fatalities caused by the extreme high temperatures.

References

1. WMO (2019) The global climate in 2015–2019. World Meteorological Organization
2. Fischer EM, Sippel S, Knutti R (2021) (2021) Increasing probability of record-shattering climate extremes. Nat Clim Change 11:689–695. https://doi.org/10.1038/s41558-021-01092-9
3. World Health Organization (2010) why urban health matters. World Health Organization. https://apps.who.int/iris/handle/10665/70230
4. World Meteorological Organization (1983) Abridged final report, 8th session. Commission for Climatology and Applications of Meteorology, World Meteorological Organization (WMO No. 600), Geneva
5. Christina K, Sari K, Gerd J, Bettina M (2004) Heat-waves: risks and responses. World Health Organization. Regional Office for Europe. https://apps.who.int/iris/handle/10665/107552
6. U.S. Environmental Protection Agency (2008) Reducing urban heat islands: Compendium of strategies. Draft. https://www.epa.gov/heat-islands/heat-island-compendium
7. Revi A, Satterthwaite DE, Aragón-Durand F, Corfee-Morlot J, Kiunsi RBR, Pelling M, Roberts DC, Solecki W (2014) Urban areas. In: Field CB, Barros VR, Dokken DJ, Mach KJ, Mastrandrea MD, Bilir TE, Chatterjee M, Ebi KL, Estrada YO, Genova RC, Girma B, Kissel ES, Levy AN, MacCracken S, Mastrandrea PR, White LL (eds) Climate Change 2014: impacts, adaptation, and vulnerability. Part A: global and sectoral aspects. Contribution of working group II to the fifth assessment report of the intergovernmental panel on climate change. Cambridge University Press, Cambridge, UK and New York, NY, USA, pp 535–612
8. Prashad L (2014) Urban heat Island. In: Njoku EG (ed) Encyclopedia of remote sensing. Encyclopedia of earth sciences series. Springer, New York, NY. https://doi.org/10.1007/978-0-387-36699-9_81
9. Founda D, Santamouris M (2017) 2017) Synergies between urban heat Island and heat waves in Athens (Greece), during an extremely hot summer (2012). Sci Rep 7:10973. https://doi.org/10.1038/s41598-017-11407-6
10. Zhao L, Oppenheimer M, Zhu Q, Baldwin J, Ebi K, Bou-Zeid E, Guan K, Liu X (2017) Interactions between urban heat islands and heat waves. Environ Res Lett. https://doi.org/10.1088/1748-9326/aa9f73
11. Keikhosravi G (2019) The effect of heat waves on the intensification of the heat island of Iran's metropolises (Tehran, Mashhad, Tabriz, Ahvaz). Urban Clim 28:100453. https://doi.org/10.1016/j.uclim.2019.100453

12. Georgiadis T (2017) Urban climate and risks. In: Oxford handbook online. Available online https://www.oxfordhandbooks.com/view/, https://doi.org/10.1093/oxfordhb/9780190699420.001.0001/oxfordhb-9780190699420-e-11?print=pdf
13. Basu R, Pearson D, Malig B, Broadwin R, Green R (2012) The effect of high ambient temperature on emergency room visits. Epidemiology (Cambridge, Mass) 23(6):813–820. https://doi.org/10.1097/EDE.0b013e31826b7f97
14. Bai L, Li Q, Wang J, Lavigne E, Gasparrini A, Copes R, Yagouti A, Burnett RT, Goldberg MS, Cakmak S, Chen H (2018) Increased coronary heart disease and stroke hospitalisations from ambient temperatures in Ontario. Heart (Br Cardiac Soc) 104(8):673–679. https://doi.org/10.1136/heartjnl-2017-311821
15. Han C, Lim YH, Lee KS, Hong YC (2019) Acute effects of ambient temperature on hypotension hospital visits: a time-series analysis in seven metropolitan cities of Korea from 2011 to 2015. Environ Int 131:104941. https://doi.org/10.1016/j.envint.2019.104941
16. Song X, Wang S, Li T, Tian J, Ding G, Wang J, Wang J, Shang K (2018) The impact of heat waves and cold spells on respiratory emergency department visits in Beijing, China. Sci Total Environ 615:1499–1505. https://doi.org/10.1016/j.scitotenv.2017.09.108
17. Zhao Y, Huang Z, Wang S, Hu J, Xiao J, Li X, Liu T, Zeng W, Guo L, Du Q, Ma W (2019) Morbidity burden of respiratory diseases attributable to ambient temperature: a case study in a subtropical city in China. Environ Health: Glob Access Sci Source 18(1):89. https://doi.org/10.1186/s12940-019-0529-8
18. Glaser J, Lemery J, Rajagopalan B, Diaz HF, García-Trabanino R, Taduri G, Madero M, Amarasinghe M, Abraham G, Anutrakulchai S, Jha V, Stenvinkel P, Roncal-Jimenez C, Lanaspa MA, Correa-Rotter R, Sheikh-Hamad D, Burdmann EA, Andres-Hernando A, Milagres T, Weiss I, Johnson RJ (2016) Climate change and the emergent epidemic of CKD from heat stress in rural communities: the case for heat stress nephropathy. Clin J Am Soc Nephrol: CJASN 11(8):1472–1483. https://doi.org/10.2215/CJN.13841215
19. Goldfarb DS, Hirsch J (2015) Hypothesis: Urbanization and exposure to urban heat islands contribute to increasing prevalence of kidney stones. Med Hypotheses 85(6):953–957. https://doi.org/10.1016/j.mehy.2015.09.003
20. Hansen AL, Bi P, Ryan P, Nitschke M, Pisaniello D, Tucker G (2008) The effect of heat waves on hospital admissions for renal disease in a temperate city of Australia. Int J Epidemiol 37(6):1359–1365. https://doi.org/10.1093/ije/dyn165
21. Kuehn L, McCormick S (2017) Heat exposure and maternal health in the face of climate change. Int J Environ Res Public Health 14(8):853. https://doi.org/10.3390/ijerph14080853
22. Chersich MF, Pham MD, Areal A, Haghighi MM, Manyuchi A, Swift CP et al (2020) Associations between high temperatures in pregnancy and risk of preterm birth, low birth weight, and stillbirths: systematic review and meta-analysis. BMJ 2020:371. https://doi.org/10.1136/bmj.m3811
23. Martin-Latry K, Goumy M, Latry P, Gabinski C, Bégaud B, Faure I, Verdoux H (2007) Psychotropic drugs use and risk of heat-related hospitalisation. Eur Psychiatry 22(6):335–338. https://doi.org/10.1016/j.eurpsy.2007.03.007
24. Cooper KE (1996) Basic thermoregulation. In: Greger R, Windhorst U (eds) Comprehensive human physiology. Springer, Berlin, Heidelberg. https://doi.org/10.1007/978-3-642-60946-6_111
25. Périard JD, Racinais S, Sawka MN (2015) Adaptations and mechanisms of human heat acclimation: applications for competitive athletes and sports. Scand J Med Sci Sports 25(Suppl 1):20–38. https://doi.org/10.1111/sms.12408
26. Tansey EA, Johnson CD (2015) Recent advances in thermoregulation. Adv Physiol Educ 39(3):139–148. https://doi.org/10.1152/advan.00126.2014
27. Gauer R, Meyers BK (2019) Heat-related illnesses. Am Fam Phys 99(8):482–489
28. Katavoutas G, Founda D (2019) Response of Urban heat stress to heat waves in Athens (1960–2017. Atmosphere 10(9):483. https://doi.org/10.3390/atmos10090483
29. Ministry of Climate Change, Government of Pakistan (2015) Technical report on Karachi heat wave June 2015, Prepared by: Dr. Qamar uz Zaman Chaudhry, Dr. Ghulam Rasul, Ahmad Kamal, Munir Ahmad Mangrio and Shahbaz Mahmood

30. Chaolin GU (2020). Urbanization. Kobayashi A (ed) International Encyclopedia of human geography, 2nd edn. Elsevier, pp 141–153. ISBN 9780081022962. https://doi.org/10.1016/B978-0-08-102295-5.10355-5
31. Worfolk JB (2000) Heat waves: their impact on the health of elders, geriatric nursing, vol 21, Issue 2, pp 70–77. ISSN 0197-4572. https://doi.org/10.1067/mgn.2000.107131
32. Balmain BN, Sabapathy S, Louis M, Morris NR (2018) Aging and thermoregulatory control: the clinical implications of exercising under heat stress in older individuals. Biomed Res Int 2018:8306154. https://doi.org/10.1155/2018/8306154
33. Edwards LM, Kozlowski D, Bair A, Juskie J, Blier W, O'Hara B (2006) A review of the July 2006 heat wave in California. American Geophysical Union, fall meeting 2006. Abstract id. A13D-0971 https://ui.adsabs.harvard.edu/abs/2006AGUFM.A13D0971E
34. Tamrazian A, Ladochy S, Willis J, Patzert W (2008) Heat waves in southern California: are they becoming more frequent and longer lasting? Yearbook Assoc Pac Coast Geogr 70:59–69. https://doi.org/10.1353/pcg.0.0001
35. Knowlton K, Rotkin-Ellman M, King G, Margolis HG, Smith D, Solomon G, Trent R, English P (2009) The 2006 California heat wave: impacts on hospitalizations and emergency department visits. Environ Health Perspect 117(1):61–67. https://doi.org/10.1289/ehp.11594
36. Department of Health & Human Services (2012) January 2009 heatwave in Victoria: an assessment of health impacts. https://www2.health.vic.gov.au/about/publications/researchandreports/January-2009-Heatwave-in-Victoria-an-Assessment-of-Health-Impacts
37. Stanberry LR, Thomson MC, James W (2018) Prioritizing the needs of children in a changing climate. PLoS Med 15(7):e1002627. https://doi.org/10.1371/journal.pmed.1002627
38. O'Sullivan K, Chisholm E (2020) Baby it's hot outside: balancing health risks and energy efficiency when parenting during extreme heat events. Energy Res Soc Sci 66:101480. https://doi.org/10.1016/j.erss.2020.101480
39. Xu Z, Sheffield PE, Su H, Wang X, Bi Y, Tong S (2014) The impact of heat waves on children's health: a systematic review. Int J Biometeorol 58(2):239–247. https://doi.org/10.1007/s00484-013-0655-x
40. Xu Z, Hu W, Su H, Turner LR, Ye X, Wang J, Tong S (2014) Extreme temperatures and paediatric emergency department admissions. J Epidemiol Community Health 68(4):304–311. https://doi.org/10.1136/jech-2013-202725
41. Wilk P, Gunz A, Maltby A, Ravichakaravarthy T, Clemens KK, Lavigne É, Lim R, Vicedo-Cabrera AM (2021) (2021) Extreme heat and paediatric emergency department visits in Southwestern Ontario. Paediatr Child Health 26(5):305–309. https://doi.org/10.1093/pch/pxaa096
42. Kenny GP, Yardley J, Brown C, Sigal RJ, Jay O (2010) Heat stress in older individuals and patients with common chronic diseases. CMAJ: Can Med Assoc J = journal de l'Association medicale canadienne 182(10):1053–1060. https://doi.org/10.1503/cmaj.081050
43. Javorac J, Jevtić M, Živanović D, Ilić M, Bijelović S, Dragić N (2021) What are the effects of meteorological factors on exacerbations of chronic obstructive pulmonary disease? Atmosphere 12:442. https://doi.org/10.3390/atmos12040442
44. Monteiro A, Carvalho V, Oliveira T, Sousa C (2013) Excess mortality and morbidity during the July 2006 heat wave in Porto, Portugal. Int J Biometeorol 57(1):155–167. https://doi.org/10.1007/s00484-012-0543-9
45. Hoffmann C, Hanisch M, Heinsohn JB, Dostal V, Jehn M, Liebers U, Pankow W, Donaldson GC, Witt C (2018) Increased vulnerability of COPD patient groups to urban climate in view of global warming. Int J Chron Obstruct Pulmon Dis 13:3493–3501. https://doi.org/10.2147/COPD.S174148
46. Xu R, Zhao Q, Coelho M, Saldiva P, Zoungas S, Huxley RR, Abramson MJ, Guo Y, Li S (2019) Association between heat exposure and hospitalization for diabetes in Brazil during 2000–2015: a nationwide case-crossover study. Environ Health Perspect 127(11):117005. https://doi.org/10.1289/EHP5688
47. Kansas Extreme Weather Workgroup (2014) Kansas extreme heat toolkit Topeka, Kansas. Retrieved from http://bit.ly/2ZHE6oy

48. National Public Health Organization of Greece (2019) Guidelines for protection against heat waves
49. Singh R, Arrighi J, Jjemba E, Strachan K, Spires M, Kadihasanoglu A (2019) Heatwave guide for cities. Red Cross Red Crescent Climate Centre
50. Fiona M (2014) Too hot to handle: assessing the social impacts of extreme heat. Paper for turning up the heat: a symposium for SIA practitioners and researchers, 21 Feb 2014
51. Chakraborty T, Hsu A, Manya D, Sheriff G (2019) Disproportionately higher exposure to urban heat in lower-income neighborhoods: a multi-city perspective. Environ Res Lett 14(10):105003. https://doi.org/10.1088/1748-9326/ab3b99
52. Gronlund CJ (2014) Racial and socioeconomic disparities in heat-related health effects and their mechanisms: a review. Curr Epidemiol Rep 1(3):165–173. https://doi.org/10.1007/s40471-014-0014-4
53. Sánchez-Guevara Sánchez C, Núñez Peiró M, Neila González FJ (2017) Urban heat island and vulnerable population. The case of Madrid. In: Mercader-Moyano P (ed) Sustainable development and renovation in architecture, urbanism and engineering. Springer, Cham. https://doi.org/10.1007/978-3-319-51442-0_1
54. Jacobs C, Singh T, Gorti G, Iftikhar U, Saeed S, Syed A, Abbas F, Ahmad B, Bhadwal S, Siderius C (2019) Patterns of outdoor exposure to heat in three South Asian cities. Sci Total Environ 674:264–278. https://doi.org/10.1016/j.scitotenv.2019.04.087
55. De Souza K, Kituyi E, Harvey B, Leone M, Kallur M, Ford J (2015) Vulnerability to climate change in three hot spots in Africa and Asia: key issues for policy-relevant adaptation and resilience-building research. Region Environ Change 15:747–753. https://doi.org/10.1007/s10113-015-0755-8
56. Amegah AK, Rezza G, Jaakkola JJ (2016) Temperature-related morbidity and mortality in Sub-Saharan Africa: a systematic review of the empirical evidence. Environ Int 91:133–149. https://doi.org/10.1016/j.envint.2016.02.027
57. Pasquini L, van Aardenne L, Godsmark CN, Lee J, Jack C (2020) Emerging climate change-related public health challenges in Africa: a case study of the heat-health vulnerability of informal settlement residents in Dar es Salaam, Tanzania. Sci Total Environ 747:141355. https://doi.org/10.1016/j.scitotenv.2020.141355
58. Moda HM, Filho WL, Minhas A (2019) Impacts of Climate change on outdoor workers and their safety: some research priorities. Int J Environ Res Public Health 16(18):3458. https://doi.org/10.3390/ijerph16183458
59. Kjellstrom T, Holmer I, Lemke B (2009) Workplace heat stress, health and productivity—An increasing challenge for low and middle-income countries during climate change. Glob Health Action 2:2047. https://doi.org/10.3402/gha.v2i0.2047
60. Harduar Morano L, Bunn TL, Lackovic M, Lavender A, Dang GT, Chalmers JJ, Li Y, Zhang L, Flammia DD (2015) Occupational heat-related illness emergency department visits and inpatient hospitalizations in the southeast region, 2007–2011. Am J Ind Med 58(10):1114–1125. https://doi.org/10.1002/ajim.22504
61. Westaway K, Frank O, Husband A, McClure A, Shute R, Edwards S, Curtis J, Rowett D (2015) Medicines can affect thermoregulation and accentuate the risk of dehydration and heat-related illness during hot weather. J Clin Pharm Ther 40(4):363–367. https://doi.org/10.1111/jcpt.12294
62. Sommet A, Durrieu G, Lapeyre-Mestre M, Montastruc JL, Association of French PharmacoVigilance Centres (2012) A comparative study of adverse drug reactions during two heat waves that occurred in France in 2003 and 2006. Pharmacoepidemiology Drug Safety 21(3):285–288. https://doi.org/10.1002/pds.2307
63. Hausfater P, Megarbane B, Dautheville S, Patzak A, Andronikof M, Santin A, André S, Korchia L, Terbaoui N, Kierzek G, Doumenc B, Leroy C, Riou B (2010) Prognostic factors in non-exertional heatstroke. Intensive Care Med 36(2):272–280. https://doi.org/10.1007/s00134-009-1694-y
64. Victoria State Government (2015) Extreme heat-factsheet for clinicians

65. Tsunematsu N, Yokoyama H, Honjo T, Ichihashi A, Ando H, Shigyo N (2016) Relationship between land use variations and spatiotemporal changes in amounts of thermal infrared energy emitted from urban surfaces in downtown Tokyo on hot summer days. Urban Clim 17:67–79. https://doi.org/10.1016/j.uclim.2016.03.002
66. Toosty NT, Hagishima A, Tanaka KI (2021) Heat health risk assessment analysing heatstroke patients in Fukuoka City, Japan. PLoS ONE 16(6):e0253011. https://doi.org/10.1371/journal.pone.0253011
67. Bai L, Ding G, Gu S, Bi P, Su B, Qin D, Xu G, Liu Q (2014) The effects of summer temperature and heat waves on heat-related illness in a coastal city of China, 2011–2013. Environ Res 132:212–219. https://doi.org/10.1016/j.envres.2014.04.002
68. Gu S, Huang C, Bai L, Chu C, Liu Q (2016) Heat-related illness in China, summer of 2013. Int J Biometeorol 60(1):131–137. https://doi.org/10.1007/s00484-015-1011-0
69. Wang J, Yan Z, Quan X, Feng J (2017) Urban warming in the 2013 summer heat wave in eastern China. Clim Dyn 48:3015–3033. https://doi.org/10.1007/s00382-016-3248-7
70. Climate Central (2021) Hot zones: Urban heat Islands https://medialibrary.climatecentral.org/uploads/general/2021_UHI_Report.pdf
71. Choudhary E, Vaidyanathan A (2014) Heat stress illness hospitalizations-environmental public health tracking program, 20 States, 2001–2010. Morbidity and mortality weekly report. Surveillance Summaries (Washington, DC, 2002), 63(13):1–10. https://www.cdc.gov/mmwr/preview/mmwrhtml/ss6313a1.htm
72. Schramm PJ, Vaidyanathan A, Radhakrishnan L, Gates A, Hartnett K, Breysse P (2021). Heat-related emergency department visits during the Northwestern heat wave-United States, June 2021. MMWR Morb Mortal Wkly Rep. ePub: 16 July 2021. https://doi.org/10.15585/mmwr.mm7029e1
73. Bassil KL, Cole DC, Moineddin R, Lou W, Craig AM, Schwartz B, Rea E (2011) The relationship between temperature and ambulance response calls for heat-related illness in Toronto, Ontario, 2005. J Epidemiol Community Health 65(9):829–831. https://doi.org/10.1136/jech.2009.101485
74. Toloo GS, Yu W, Aitken P, FitzGerald G, Tong S (2014) The impact of heatwaves on emergency department visits in Brisbane, Australia: a time series study. Crit Care 18:R69. https://doi.org/10.1186/cc13826
75. Robine J, Cheung SL, Roy SL, Oyen H (2007) Report on excess mortality in Europe during summer 2003 (EU Community Action Programme for Public Health, Grant Agreement 2005114)
76. Argaud L, Ferry T, Le QH, Marfisi A, Ciorba D, Achache P, Ducluzeau R, Robert D (2007) Short- and long-term outcomes of heatstroke following the 2003 heat wave in Lyon, France. Arch Internal Med 167(20):2177–2183. https://doi.org/10.1001/archinte.167.20.ioi70147
77. Smith S, Elliot AJ, Hajat S, Bone A, Smith GE, Kovats S (2016) Estimating the burden of heat illness in England during the 2013 summer heatwave using syndromic surveillance. J Epidemiol Community Health 70(5):459–465. https://doi.org/10.1136/jech-2015-206079
78. Hartz DA, Golden JS, Sister C, Chuang WC, Brazel AJ (2012) Climate and heat-related emergencies in Chicago, Illinois (2003–2006). Int J Biometeorol 56(1):71–83. https://doi.org/10.1007/s00484-010-0398-x
79. Dematte JE, O'Mara K, Buescher J, Whitney CG, Forsythe S, McNamee T, Adiga RB, Ndukwu IM (1998) Near-fatal heat stroke during the 1995 heat wave in Chicago. Ann Int Med 129(3):173–181. https://doi.org/10.7326/0003-4819-129-3-199808010-00001
80. Semenza JC, McCullough JE, Flanders WD, McGeehin MA, Lumpkin JR (1999) Excess hospital admissions during the July 1995 heat wave in Chicago. Am J Prev Med 16(4):269–277. https://doi.org/10.1016/s0749-3797(99)00025-2

81. Kunkel KE, Changnon SA, Reinke BC, Arritt RW (1996) The July 1995 heat wave in the midwest: a climatic perspective and critical weather factors. Bull Am Meteor Soc 77(7):1507–1518. https://doi.org/10.1175/1520-0477(1996)077%3c1507:TJHWIT%3e2.0.CO;2
82. Michelozzi P, Accetta G, De Sario M, D'Ippoliti D, Marino C, Baccini M, Biggeri A, Anderson HR, Katsouyanni K, Ballester F, Bisanti L, Cadum E, Forsberg B, Forastiere F, Goodman PG, Hojs A, Kirchmayer U, Medina S, Paldy A, Schindler C, Sunyer J, Perucci CA; PHEWE Collaborative Group (2009) High temperature and hospitalizations for cardiovascular and respiratory causes in 12 European cities. Am J Respirat Crit Care Med 179(5):383–389. https://doi.org/10.1164/rccm.200802-217OC
83. Alessandrini E, Zauli Sajani S, Scotto F, Miglio R, Marchesi S, Lauriola P (2011) Emergency ambulance dispatches and apparent temperature: a time series analysis in Emilia-Romagna,Italy. Environ Res 111(8):1192–1200. https://doi.org/10.1016/j.envres.2011.07.005
84. Huang H, Yang H, Deng X, Zeng P, Li Y, Zhang L, Zhu L (2019) Influencing mechanisms of urban heat island on respiratory diseases. Iran J Public Health 48(9):1636–1646. http://ijph.tums.ac.ir
85. Ponjoan A, Blanch J, Alves-Cabratosa L, Martí-Lluch R, Comas-Cufí M, Parramon D, Garcia-Gil MD, Ramos R, Petersen I (2017) Effects of extreme temperatures on cardiovascular emergency hospitalizations in a Mediterranean region: a self-controlled case series study. Environ Health 16:32. https://doi.org/10.1186/s12940-017-0238-0
86. Marti-Soler H, Gubelmann C, Aeschbacher S, Alves L, Bobak M, Bongard V, Clays E, de Gaetano G, Di Castelnuovo A, Elosua R, Ferrieres J, Guessous I, Igland J, Jørgensen T, Nikitin Y, O'Doherty MG, Palmieri L, Ramos R, Simons J, Sulo G, Vanuzzo D, Vila J, Barros H, Borglykke A, Conen D, De Bacquer D, Donfrancesco C, Gaspoz JM, Giampaoli S, Giles GG, Iacoviello L, Kee F, Kubinova R, Malyutina S, Marrugat J, Prescott E, Ruidavets JB, Scragg R, Simons LA, Tamosiunas A, Tell GS, Vollenweider P, Marques-Vidal P (2014) Seasonality of cardiovascular risk factors: an analysis including over 230,000 participants in 15 countries. Heart (Br Cardiac Soc) 100(19):1517–1523. https://doi.org/10.1136/heartjnl-2014-305623
87. Kovats RS, Hajat S, Wilkinson P (2004) Contrasting patterns of mortality and hospital admissions during hot weather and heat waves in Greater London, UK. Occup Environ Med 61(11):893–898. https://doi.org/10.1136/oem.2003.012047
88. Mastrangelo G, Hajat S, Fadda E, Buja A, Fedeli U, Spolaore P (2006) Contrasting patterns of hospital admissions and mortality during heat waves: are deaths from circulatory disease a real excess or an artifact? Med Hypotheses 66(5):1025–1028. https://doi.org/10.1016/j.mehy.2005.09.053
89. Cheng J, Xu Z, Bambrick H, Prescott V, Wang N, Zhang Y, Su H, Tong S, Hu W (2019) Cardiorespiratory effects of heatwaves: a systematic review and meta-analysis of global epidemiological evidence. Environ Res 177:108610. https://doi.org/10.1016/j.envres.2019.108610
90. Halaharvi H, Schramm JP, Vaidyanathan A (2020) Heat exposure and cardiovascular health: a summary for health departments climate and health technical report series climate and health program. Centers for Disease Control and Prevention, July 2020
91. Lu P, Xia G, Zhao Q, Xu R, Li S, Guo Y (2020) Temporal trends of the association between ambient temperature and hospitalisations for cardiovascular diseases in Queensland, Australia from 1995 to 2016: a time-stratified case-crossover study. PLoS Med 17(7):e1003176. https://doi.org/10.1371/journal.pmed.1003176
92. Bao J, Cao Z, Huang C, Lei L, Yang Y, Peng J, Shi X (2021) Modification effect of urban landscape characteristics on the association between heat and stroke morbidity: a small-scale intra-urban study in Shenzhen, China. Sci Total Environ 786(2021):147223. https://doi.org/10.1016/j.scitotenv.2021.147223
93. Sohail H, Kollanus V, Tiittanen P, Schneider A, Lanki T (2020) Heat, heatwaves and cardiorespiratory hospital admissions in Helsinki, Finland. Int J Environ Res Public Health 17(21):7892. https://doi.org/10.3390/ijerph17217892
94. Kang SH, Oh IY, Heo J, Lee H, Kim J, Lim WH, Cho Y, Choi EK, Yi SM, Sang D, Kim H, Youn TJ, Chae IH, Oh S (2016) Heat, heat waves, and out-of-hospital cardiac arrest. Int J Cardiol 221:232–237. https://doi.org/10.1016/j.ijcard.2016.07.071

95. Kranc H, Novack V, Shtein A, Sherman R, Novack L (2021) Extreme temperature and out-of-hospital-cardiac-arrest. Nationwide study in a hot climate country. Environ Health: Glob Access Sci Source 20(1):38. https://doi.org/10.1186/s12940-021-00722-1
96. Doan TN, Wilson D, Rashford S, Bosley E (2021) Ambient temperatures, heatwaves and out-of-hospital cardiac arrest in Brisbane, Australia. Occup Environ Med. oemed-2020-107018. Advance Online Publication. https://doi.org/10.1136/oemed-2020-107018
97. Cianconi P, Betrò S, Janiri L (2020) The impact of climate change on mental health: a systematic descriptive review. Front Psych 11:74. https://doi.org/10.3389/fpsyt.2020.00074
98. Wong LP, Alias H, Aghamohammadi N, Aghazadeh S, Nik Sulaiman NM (2018) Physical, psychological, and social health impact of temperature rise due to urban heat island phenomenon and its associated factors. Biomed Environ Sci: BES 31(7):545–550. https://doi.org/10.3967/bes2018.074
99. Padhy SK, Sarkar S, Panigrahi M, Paul S (2015) Mental health effects of climate change. Ind J Occup Environ Med 19(1):3–7. https://doi.org/10.4103/0019-5278.156997
100. Anderson CA (2001) Heat and violence. Curr Dir Psychol Sci 10(1):33–38. https://doi.org/10.1111/1467-8721.00109
101. Burke M, González F, Baylis P, Heft-Neal S, Baysan C, Basu S, Hsiang S (2018). Higher temperatures increase suicide rates in the United States and Mexico. Nature Clim Change 8:723–729. https://doi.org/10.1038/s41558-018-0222-x
102. Huang H, Deng X, Yang H, Li S, Li M (2020) Spatial evolution of the effects of urban heat island on residents' health Tehnički vjesnik 27(5):1427–1435. https://doi.org/10.17559/TV-20200503211912
103. Noelke C, McGovern M, Corsi DJ, Jimenez MP, Stern A, Wing IS, Berkman L (2016) Increasing ambient temperature reduces emotional well-being. Environ Res 151:124–129. https://doi.org/10.1016/j.envres.2016.06.045
104. Hansen A, Bi P, Nitschke M, Ryan P, Pisaniello D, Tucker G (2008) The effect of heat waves on mental health in a temperate Australian City. Environ Health Perspect 116(10):1369–1375. https://doi.org/10.1289/ehp.11339
105. Thompson R, Hornigold R, Page L, Waite T (2018) Associations between high ambient temperatures and heat waves with mental health outcomes: a systematic review. Public Health 161:171–191. https://doi.org/10.1016/j.puhe.2018.06.008
106. Almendra R, Loureiro A, Silva G, Vasconcelos J, Santana P (2019) Short-term impacts of air temperature on hospitalizations for mental disorders in Lisbon. Sci Total Environ 647:127–133. https://doi.org/10.1016/j.scitotenv.2018.07.337
107. Vida S, Durocher M, Ouarda TB, Gosselin P (2012) Relationship between ambient temperature and humidity and visits to mental health emergency departments in Québec. Psychiat Services (Washington, DC) 63(11):1150–1153. https://doi.org/10.1176/appi.ps.201100485
108. Wang X, Lavigne E, Ouellette-kuntz H, Chen BE (2014) Acute impacts of extreme temperature exposure on emergency room admissions related to mental and behavior disorders in Toronto, Canada. J Affect Disord 155:154–161. https://doi.org/10.1016/j.jad.2013.10.042
109. Basu R, Gavin L, Pearson D, Ebisu K, Malig B (2018) Examining the association between apparent temperature and mental health-related emergency room visits in California. Am J Epidemiol 187(4):726–735. https://doi.org/10.1093/aje/kwx295
110. Lõhmus M (2018) Possible Biological mechanisms linking mental health and heat—A contemplative review. Int J Environ Res Public Health 15(7):1515. https://doi.org/10.3390/ijerph15071515
111. Yi F, Zhou T, Yu L, McCarl B, Wang Y, Jiang F, Wang Y (2021) Outdoor heat stress and cognition: Effects on those over 40 years old in China. Weather Clim Extremes 32:100308. ISSN 2212-0947. https://doi.org/10.1016/j.wace.2021.100308
112. Rifkin DI, Long MW, Perry MJ (2018) Climate change and sleep: a systematic review of the literature and conceptual framework. Sleep Med Rev 42:3–9. https://doi.org/10.1016/j.smrv.2018.07.007
113. Okamoto-Mizuno K, Mizuno K (2012) Effects of thermal environment on sleep and circadian rhythm. J Physiol Anthropol 31(1):14. https://doi.org/10.1186/1880-6805-31-14

114. Min KB, Lee S, Min JY (2021) High and low ambient temperature at night and the prescription of hypnotics. Sleep 44(5). https://doi.org/10.1093/sleep/zsaa262
115. Arifwidodo S, Chandrasiri O, Abdulharis R, Kubota T (2019) Exploring the effects of urban heat island: a case study of two cities in Thailand and Indonesia. APN Sci Bull 9(1). https://doi.org/10.30852/sb.2019.539
116. Zheng G, Li K, Wang Y (2019) The effects of high-temperature weather on human sleep quality and appetite. Int J Environ Res Public Health 16(2):270. https://doi.org/10.3390/ijerph16020270
117. Obradovich N, Migliorini R, Mednick SC, Fowler JH (2017) Nighttime temperature and human sleep loss in a changing climate. Sci Adv 3(5):e1601555. https://doi.org/10.1126/sciadv.1601555
118. Weinreich G, Wessendorf TE, Pundt N, Weinmayr G, Hennig F, Moebus S, Möhlenkamp S, Erbel R, Jöckel KH, Teschler H, Hoffmann B, Heinz Nixdorf Recall Study Group (2015) Association of short-term ozone and temperature with sleep disordered breathing. Eur Respir J 46(5):1361–1369. https://doi.org/10.1183/13993003.02255-2014
119. Nastos PT, Matzarakis A (2008) Human-Biometeorological effects on sleep disturbances in Athens, Greece: a preliminary evaluation. Indoor Built Environ 17(6):535–542. https://doi.org/10.1177/1420326X08097706
120. Kloog I (2019) Air pollution, ambient temperature, green space and preterm birth. Curr Opin Pediatr 31(2):237–243. https://doi.org/10.1097/MOP.0000000000000736
121. Lin Y, Hu W, Xu J, Luo Z, Ye X, Yan C, Liu Z, Tong S (2017) Association between temperature and maternal stress during pregnancy. Environ Res 158:421–430. https://doi.org/10.1016/j.envres.2017.06.034
122. Kim J, Lee A, Rossin-Slater M (2019) What to expect when it gets hotter: the impacts of prenatal exposure to extreme heat on maternal and infant health. In: IZA discussion papers 12685. Institute of Labor Economics (IZA)
123. Ilango SD, Weaver M, Sheridan P, Schwarz L, Clemesha R, Bruckner T, Basu R, Gershunov A, Benmarhnia T (2020) Extreme heat episodes and risk of preterm birth in California, 2005–2013. Environ Int 137:105541. https://doi.org/10.1016/j.envint.2020.105541
124. Spolter F, Kloog I, Dorman M, Novack L, Erez O, Raz R (2020) Prenatal exposure to ambient air temperature and risk of early delivery. Environ Int 142:105824. https://doi.org/10.1016/j.envint.2020.105824
125. Qu Y, Zhang W, Ryan I, Deng X, Liu X, Lin S (2021) Ambient extreme heat exposure in summer and transitional months and emergency department visits and hospital admissions due to pregnancy complications. Sci Total Environ 777:146134. https://doi.org/10.1016/j.scitotenv.2021.146134
126. Dastoorpoor M, Khanjani N, Khodadadi N (2021) (2021) Association between Physiological Equivalent Temperature (PET) with adverse pregnancy outcomes in Ahvaz, southwest of Iran. BMC Pregnancy Childbirth 21:415. https://doi.org/10.1186/s12884-021-03876-5
127. Johnson RJ, Sánchez-Lozada LG, Newman LS, Lanaspa MA, Diaz HF, Lemery J, Rodriguez-Iturbe B, Tolan DR, Butler-Dawson J, Sato Y, Garcia G, Hernando AA, Roncal-Jimenez CA (2019) Climate change and the kidney. Ann Nutr Metab 74(Suppl 3):38–44. https://doi.org/10.1159/000500344
128. De Lorenzo A, Liaño F (2017) High temperatures and nephrology: The climate change problem. Altas temperaturas y nefrología: a propósito del cambio climático. Nefrologia: publicacion oficial de la Sociedad Espanola Nefrologia 37(5):492–500. https://doi.org/10.1016/j.nefro.2016.12.008
129. Remigio RV, Jiang C, Raimann J, Kotanko P, Usvyat L, Maddux FW, Kinney P, Sapkota A (2019) Association of extreme heat events with hospital admission or mortality among patients with end-stage renal disease. JAMA Netw Open 2(8):e198904. https://doi.org/10.1001/jamanetworkopen.2019.8904
130. Kim S, Lee H, Kim J, Lee Y, Kang M, Hijioka Y, Kim H (2019) Temperature as a risk factor of emergency department visits for acute kidney injury: a case-crossover study in Seoul, South Korea. Environ Health 18:1. https://doi.org/10.1186/s12940-019-0491-5

131. Abreu Júnior J, Ferreira Filho SR (2020) Influence of climate on the number of hospitalizations for nephrolithiasis in urban regions in Brazil. Jornal brasileiro de nefrologia : 'orgao oficial de Sociedades Brasileira e Latino-Americana de Nefrologia 42(2):175–181. https://doi.org/10.1590/2175-8239-JBN-2019-0155
132. Dhainaut J, Claessens Y, Ginsburg C, Riou B (2004) Unprecedented heat-related deaths during the 2003 heat wave in Paris: consequences on emergency departments. Crit Care 8:1. https://doi.org/10.1186/cc2404
133. Claessens YE, Taupin P, Kierzek G, Pourriat JL, Baud M, Ginsburg C, Jais JP, Jougla E, Riou B, Dhainaut JF, Landais P (2006) How emergency departments might alert for prehospital heat-related excess mortality? Crit Care (London, England) 10(6):R156. https://doi.org/10.1186/cc5092
134. Johnson H, Kovats S, McGregor G, Stedman J, Gibbs M, Walton H (2005) The impact of the 2003 heat wave on daily mortality in England and Wales and the use of rapid weekly mortality estimates. Euro Surveill 10(7):pii=558. https://doi.org/10.2807/esm.10.07.00558-en
135. Matthies F, Bickler G, Marín NC, Hales S (2008) Heat–health action plans guidance, 1st edn. World Health Organization, 18 Dec 2008
136. Matthies F, Menne B (2009) Prevention and management of health hazards related to heatwaves. Int J Circum Health 68(1):8–12. https://doi.org/10.3402/ijch.v68i1.18293
137. Maxwell J, Blashki G (2016) Teaching about climate change in medical education: an opportunity. J Public Health Res 5(1):673. https://doi.org/10.4081/jphr.2016.673
138. Hackett F, Got T, Kitching GT, MacQueen K, Cohen A (2020) Training Canadian doctors for the health challenges of climate change. Lancet Planetary Health 4(1):e2–e3. https://doi.org/10.1016/S2542-5196(19)30242-6
139. Xie E, de Barros EF, Abelsohn A, Stein AT, Haines A (2018) Challenges and opportunities in planetary health for primary care providers. Lancet Planetary Health 2(5):e185–e187. https://doi.org/10.1016/S2542-5196(18)30055-X
140. Kovats RS, Osborn D (2016) UK climate change risk assessment evidence report: chapter 5, people and the built environment. In: Humphrey K, Thompson D, Johns D, Ayres J, Bates, P, Baylis M, Bell S, Church A, Curtis S, Davies M, Depledge M, Houston D, Vardoulakis S, Reynard N, Watson J, Mavrogianni A, Shrubsole C, Taylor J, Whitman G (eds) Report prepared for the Adaptation Sub-Committee of the Committee on Climate Change, London
141. Ferron C, Trewick D, Le Conte P, Batard ER, Girard L, Potel G, (2006) Canicule de l'été 2003: étude descriptive des décès par coup de chaleur au CHU de Nantes [Heat stroke in hospital patients during the summer 2003 heat wave: a nosocomial disease]. Presse medicale (Paris, France, 1983), 35(2 Pt 1):196–199. https://doi.org/10.1016/s0755-4982(06)74553-5
142. Lenzer B, Rupprecht M, Hoffmann C, Hoffmann P, Liebers U (2020) Health effects of heating, ventilation and air conditioning on hospital patients: a scoping review. BMC Public Health 20:1287. https://doi.org/10.1186/s12889-020-09358-1
143. Gupta R, Howard A, Davies M, Mavrogianni A, Tsoulou I, Oikonomou E, Wilkinson P (2021) Examining the magnitude and perception of summertime overheating in London care homes. Build Serv Eng Res Technol. https://doi.org/10.1177/01436244211013645
144. Carmichael C, Bickler G, Kovats S, Pencheon D, Murray V, West C, Doyle Y (2013) Overheating and hospitals—What do we know? J Hosp Admin 2(1):1–7. https://doi.org/10.5430/jha.v2n1p1
145. Wondmagegn BY, Xiang J, Williams S, Pisaniello D, Bi P (2019) What do we know about the healthcare costs of extreme heat exposure? A comprehensive literature review. Sci Total Environ 657:608–618. https://doi.org/10.1016/j.scitotenv.2018.11.479
146. Merrill CT, Miller M, Steiner C (2008) Hospital stays resulting from excessive heat and cold exposure due to weather conditions in U.S. community hospitals, 2005: statistical brief #55. In: Healthcare cost and utilization project (HCUP) statistical briefs. Agency for Healthcare Research and Quality (US)
147. Schmeltz MT, Petkova EP, Gamble JL (2016) Economic burden of hospitalizations for heat-related illnesses in the United States, 2001–2010. Int J Environ Res Public Health 13(9):894. https://doi.org/10.3390/ijerph13090894

148. Knowlton K, Rotkin-Ellman M, Geballe L, Max W, Solomon GM (2011) Six climate change-related events in the United States accounted for about $14 billion in lost lives and health costs. Health Affairs (Project Hope) 30(11):2167–2176. https://doi.org/10.1377/hlthaff.2011.0229
149. Adélaïde L, Chanel O, Pascal M (2021) Health effects from heat waves in France: an economic evaluation. Eur J Health Econ. https://doi.org/10.1007/s10198-021-01357-2

Chapter 4
Heat and Mental Health in Cities

Nasrin Aghamohammadi, Chng Saun Fong, Nik Daliana Nik Farid, Logaraj Ramakreshnan, and Parshad Agha Mohammadi

Abstract Extreme heat wave events (EHE) are one of the most unprecedented threats posed by climate change. The escalation of global temperature is expected to cause more frequent and prolong heat waves which may have a disastrous impact on the human population and the ecosystem. Currently, more than 50% of the world population lives in urban areas and is expected to increase to 6.4 billion by 2050. Global climate change with the local urban heating phenomenon in cities is expected to impose immense threat in terms of increased surface temperature affecting the urban well-being and livability status. While the different aspects of climate change may affect urban health through direct and indirect pathways, the implications of heat toward mental health are less studied. High ambient temperatures have a range of mental health effects. A study conducted in 19 different countries suggested a significant relationship between heat and psychological conditions where heat stress and consistent exposure to high temperature were found to cause depression and anxiety. Other scholarly studies found an increase in mental and behavioral disorder reports during heat wave period. The strongest evidence was found for increased suicide risk. Six broad mental health outcome categories were identified: suicide and heat; bipolar disorder, mania and depression, and heat; schizophrenia and heat; organic mental health outcomes and heat including dementia; alcohol and substance misuse and heat; and multiple mental health outcomes/mental health service usage

N. Aghamohammadi (✉)
Centre for Epidemiology and Evidence-Based Practice, Department of Social and Preventive Medicine, Faculty of Medicine, University of Malaya, 50603 Kuala Lumpur, Malaysia
e-mail: nasrin@ummc.edu.my

Centre for Energy Sciences, University of Malaya, 50603 Kuala Lumpur, Malaysia

C. S. Fong · L. Ramakreshnan
Institute for Advanced Studies, University of Malaya, 50603 Kuala Lumpur, Malaysia

N. D. N. Farid
Department of Social and Preventive Medicine, Faculty of Medicine, University of Malaya, 50603 Kuala Lumpur, Malaysia

P. A. Mohammadi
Department of Civil Engineering, Faculty of Engineering, Science and Research Branch of Azad University, 1477893855 Tehran, Iran

N. Aghamohammadi and M. Santamouris (eds.), *Urban Overheating: Heat Mitigation and the Impact on Health*, Advances in Sustainability Science and Technology,
https://doi.org/10.1007/978-981-19-4707-0_4

and heat. A review from fifteen studies showed an increased suicide risk with heat (relative risk (RR) 1.014–1.37 per 1 °C, $P < 0.05$; $r = 0.10$–0.64, $P < 0.05$). Increased risks of mental health-related admissions and emergency department visits at higher temperatures were also found. Despite mounting evidence that mental health treatments are essential, especially during EHE, mental health services are found to be scarce in most developing and low-income nations. Raising public awareness of the consequences of EHE on mental health is critical to educate people on how to reduce risks and safeguard mental health. However, understanding the people's attitudes and behaviors on EHE is the first step in developing a more targeted public awareness program. As a result, individual-level research is required to identify vulnerable populations and assist in the development of adaption strategies, such as heat action plans. In summary, evidence for the impact of heat on other mental health outcomes was mixed. Knowledge gaps exist on the impact of high temperatures on many common mental health disorders. Mental health impacts should be incorporated into plans for the public health response to high temperatures.

Keyword Extreme heat event · Mental health · Sustainable cities · Urban health · Urban heat island

1 Global Climate Change, Extreme Heat Events, and Urban Overheating Phenomenon

Climate change is defined as a long-term shift in the average climate conditions of a particular region, such as temperature and rainfall [1]. In the United States (US), the average temperature has increased from 1.3 to 1.9 °F, and most of the increase has occurred since 1970. The current decade has been the nation's warmest ever reported, and temperatures in the US are expected to rise [2]. The average long-term changes on Earth are referred to as global climate change. Some examples of the impact of this change are increasing temperature levels and precipitation changes, as well as the impact of global warming such as rising sea levels, receding mountain glaciers, ice melting at a faster rate than typical in Greenland, Antarctica, and the Arctic, and changes in flower and plant blooming seasons [3].

The global temperature has risen by 0.14 °F (0.08 °C) per decade since 1880, and the rate of warming over the past 40 years is more than twice than that since 1981, i.e., 0.32 °F (0.18 °C) per decade [4]. It is worth noting that some parts of the globe are warming more rapidly than others. The global air temperatures near the Earth's surface have increased by around 2 °F on average over the last century, and the last five years have been the warmest in millennia. The intensity and amount of rainfall during storms such as hurricanes are projected to grow as the Earth's temperature continues to increase. As the climate warms, droughts and heat waves are predicted to become increasingly severe. When the global temperature rises or falls by one or two degrees, it can have a significant influence on human health [5].

The Earth's climate is influenced by various factors. Research has provided evidence that human activities have led the Earth to warm in the last 50–100 years. Anthropogenic activities, including burning fossil fuels, logging forests, and raising livestock, contribute to the warming of the atmosphere, ocean, and land. Change in the reflectance of the Earth's surface (albedo), emissions from burning forests, urban heat island effects, and changes in the natural water cycle are all the outcome of human-driven changes in land use and land cover, such as urbanization [5]. These activities contribute massive amounts of greenhouse gases (such as carbon dioxide, methane, and nitrous oxide) to those already present in the atmosphere, thus amplifying the greenhouse effect and contributing to global warming.

The greenhouse effect is a natural mechanism that causes the Earth's surface to warm. Heat is trapped in the atmosphere by greenhouse gases, which raises air and sea temperatures. When sunlight strikes the Earth's atmosphere, some of it is reflected to space, while the remainder is absorbed and re-radiated by greenhouse gases. Water vapor, carbon dioxide, methane, nitrous oxide, ozone, and some man-made substances such as chlorofluorocarbons (CFCs) are examples of greenhouse gases. The absorbed energy warms the Earth's atmosphere and surface. This mechanism keeps the Earth's temperature roughly 33 °C warmer than it would be, allowing life to flourish on the planet.

Extreme heat wave events (EHE) are one of the most unprecedented threats posed by climate change since they have resulted in more deaths than other climate disasters such as lightning, rain, floods, hurricanes, and tornadoes combined [6]. Heat waves are projected to become more common and long-lasting as global temperatures rise, potentially wreaking havoc on the human population and ecology [7, 8]. Cities occupy approximately 2% of the earth's surface, and in the world today, urban populations are rapidly increasing in size and complexity because more people are leaving rural areas to migrate to cities. More than half of the world's population already lives in cities, and that number is predicted to rise to the estimated 6 billion by the year 2050 [9]. Due to the rising population, cities require large quantities of energy to function properly. In fact, city dwellers consume over 75% of the total energy resources as a result of activities carried out in the urban environment. Part of this energy is dissipated in the form of heat, which is intensified by solar radiation. Due to the complexity of urban structures in cities, this heat is entrapped, accumulated, and slowly dissipated at night. This creates an effect known as urban heat island [UHI], whereby urban environments become warmer compared to their rural counterparts. Climate projections show not only that cities will warm faster than rural areas, but that this UHI intensity is increased at night [10]. Figure 1 depicts the variables that contribute to a city's thermal alteration, resulting in the formulation of a self-contained urban heat "dome."

Elevated temperatures in city centers have been found to exert substantial pressure on microclimate patterns, impacting precipitation patterns, air circulation, climatic disasters, water quality degradation, and air pollution levels. As a result, increasing air temperature in a city produces a slew of environmental concerns, such as heat waves and poor urban air quality [12–14]. UHI also contributes to urban pollution by serving as a catalyst for photochemical reactions in the atmosphere [15].

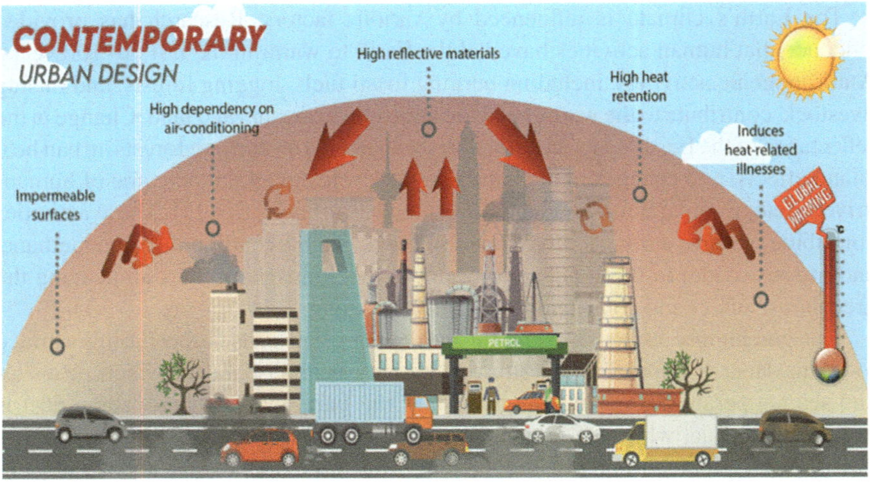

Fig. 1 Urban heat island "dome". *Source* Law and Fong [11]

Various research has been done in the past few decades to determine the variables leading to the genesis of UHI since the first discovery of the phenomena in the 1810s [16]. The occurrence of UHI is a result of inadequate urban design in quickly growing cities. It arose from significant development in order to support a growing population while also attaining national growth. In addition, the study found evidence of a link between urban growth and UHI and several socioeconomic factors, including population, electricity use, and GDP [17, 18]. Therefore, it can be postulated that UHI is driven by the urbanization effect, which is characterized by a large number of densely constructed buildings and urban structures, forming narrow street canyons, insufficient vegetation, and pollution [19–23].

UHI can be caused by a number of factors. In fact, the absorption and retention of urban heat have been linked to the complexity of urban morphology [10, 17, 24–30]. For example, dark surfaces absorb significantly more solar radiation, causing urban concentrations of roads and buildings to heat up more during the day than suburban and rural areas [31]. Materials commonly used in urban areas for pavement and roofs, such as concrete and asphalt, have significantly different thermal bulk properties [including heat capacity and thermal conductivity] and surface radiative properties [albedo and emissivity] than the surficial materials [31]. This changes the energy budget of the city, resulting in higher temperatures in the city than in surrounding rural areas [32]. As the volumetric heat capacity of the urban surface increases, the day–night cycle of the city is delayed. Geometric effects are another source of an UHI phenomena. Tall structures in many urban locations provide various surfaces for sunlight reflection and absorption, improving the efficiency with which cities are heated. Another effect of buildings is the obstruction of wind, which hampers convection cooling and the dissipation of heat and atmospheric pollutants.

Another key factor is the absence of evapotranspiration in urban areas due to lack of greeneries [33]. Typically, there are not many plants in cities as compared to rural areas. Cities lose shade and the evaporative cooling benefits of trees when the amount of vegetation decreases due to land use changes. There are instead sidewalks, streets, parking lots, and high-rise structures, and this would not exist unless humans build them. Cement, asphalt, brick, glass, steel, and dark roofs are common materials used in these structures. Asphalt, steel, and brick come in a variety of dark colors, including black, brown, and gray. A black-colored object absorbs all wavelengths of light energy and converts them to heat, resulting in a heated object. A white object, on the other hand, is better at reflecting wavelengths of light. Consequently, dark objects such as construction materials would absorb more heat from the light. Also, many modern urban building materials have impermeable surfaces. This prevents water from flowing through surfaces such as a brick or patch of cement in the same way that it would through a plant. These surfaces have nothing to cool them down, unless water is moving and evaporating.

The UHI is exacerbated by anthropogenic waste heat such as from vehicles, air conditioning, plants and industries, and other sources [34, 35]. High amounts of pollution in metropolitan areas can further enhance UHI, as many types of pollution alter the atmosphere's radiative characteristics [32]. Since ozone is a greenhouse gas wherein synthesis accelerates as temperature rises, UHI not only elevates urban temperatures, but also increases the ozone concentration [36].

Anthropogenic greenhouse gas emissions are changing our climate system, and more frequent and intense heat-related calamities such as heat waves, droughts, and wildfires have already resulted in higher death and morbidity rates. Increased cardiorespiratory mortality and morbidity are linked to increased ozone levels caused by greater temperatures and air pollutants from burning fossil fuels, presumably by increasing cholesterol levels and systemic inflammation. Other debilitating impacts of urban heat include an increase in the prevalence of kidney disease and mental illness [37].

2 The Implication of Heat Toward Mental Health in Cities Across the Globe

The degradation of the urban environment has a big impact on a city's microclimate [25] and the quality of life of urban communities. Places with poorer development had higher negative impact on the people's mental health. Similarly, the exacerbation of climate change in the process of developing infrastructures and mega cities is expected to have similar influence on the deterioration of mental health. The urban environment is more vulnerable to thermal stress than rural areas due to the UHI phenomena. The UHI phenomenon exposes urban residents, particularly those residing in the city center, to higher and longer heat exposure. There are also considerable concerns about the decline in outdoor thermal comfort level and heat-related diseases. The health and well-being of those who are susceptible and vulnerable, such as children, pregnant women, the elderly, and those who have pre-existing serious health problems, are especially at risk.

Mental health is not merely the absence of formal psychiatric diagnoses, but also the presence of mental wellness [38]. To be in good mental health, one must be able to recognize one's own abilities, manage with everyday challenges, and experience a state of well-being in which one recognizes one's own abilities, works successfully, and contributes to one's community [38]. Mental illness has a huge toll on society. Every year, more than 700,000 individuals die by suicide around the world [39]. Almost one billion people worldwide suffer from a mental condition, which ranges from addiction, to dementia, to schizophrenia [40]. The World Bank predicted that 340 million people will suffer from depression by 2020; and by 2030, depression will be the most disabling disease, as it will have the highest number of days lost due to a disability [41].

According to the Lancet Global Health, poor mental health cost the global economy $2.5 trillion per year in lost productivity and bad health in 2010, with a projected increase to $6 trillion by 2030 [40]. The wider economic cost of mental illness in England was estimated to be around £100 billion per year [42]. Anxiety and depression, the two most frequent mental diseases, have cost the world economy $1 trillion per year in lost productivity [43]. Major depressive disorder caused 49.4 million (33.6–68.7), while disability-adjusted life years (DALYs) and anxiety disorders caused 44.5 million (30.2–62.5) globally in 2020 [43]. In the US, almost one in every five people suffers from a mental illness. According to the Open Minds Market Intelligence Report, the cost of treating and delivering services for mental diseases in the US reached $225 billion in 2019, which is an increment of 25% from 2009. Therapy and prescription drugs, as well as stints in psychiatric or substance misuse rehabilitation institutions, are all included in the fee [44].

Heat-related fatalities are expected to exceed 260,000 per year by 2050 as a consequence of climate change, and high-heat-related work-hour losses in several Asian and African nations may cost billions of dollars in productivity losses [45]. However, heat-related fatalities are mostly avoidable, but interventions for the most vulnerable groups need to be improved, particularly given that UHI is becoming one of the climatic dangers to the impoverished urban population in developing nations [46, 47].

3 The Prevalence of Heat Wave Events and Mental Health Across the Globe

In the twentieth and twenty-first centuries, global temperatures have risen to an alarming state. The early twentieth century was a warming phase, the middle (1940–1969) was a mild cooling period, and since 1970s, there has been a gradual rise in temperature. The warming since 1970s has been seen as a decrease in cold-tail temperature extremes, high-latitude warming, and nighttime and winter warming. The frequency of hot weather during the summer, on the other hand, has been increasing. Although there has been a general increase in the frequency of extremely hot temperatures during the recent warming period, the literature on the spatial pattern is variable, primarily due to sensitivity to time period, meteorological/bio-meteorological variables, daily extrema and/or hot weather definition, statistical methodology, climate data used, and time of year [48].

A heat wave is defined as a period of two or more consecutive days during which the daily lowest apparent temperature (actual temperature adjusted for humidity) in a given city surpasses the 85th percentile of that city's historical July and August temperatures (1981–2010). Heat waves have been occurring more frequently in major cities across the US than they were in the 1960s, with an average of two per year in the 1960s and more than six per year in the 2010s. In 50 large cities, the average heat wave season is 47 days longer than it was in the 1960s. Between 1960 and 2010s, a majority of metropolitan regions witnessed a statistically significant rise in heat wave frequency and the length of the heat wave season [49].

One study looked at variations in the regularity of heat waves in the continental United States using spatially continuous and homogenized observational surface climate data. This includes looking at heat waves based on daytime temperatures only, nighttime temperatures only, and a combination of both. The findings revealed a significant increase in the frequency of heat waves from the mid-1970s to the conclusion of the dataset (2015), which was preceded by a slight drop since the initiation of the dataset (1948) [48].

Heat waves, or heat and hot weather that can last for several days, can have a significant impact on society, including a rise in heat-related deaths. Heat waves are among the most dangerous of natural hazards, but rarely receive adequate attention because their death tolls and destruction are not always immediately obvious. From 1998–2017, more than 166,000 people died due to heat waves, including more than 70,000 who died during the 2003 heat wave in Europe.

Population exposure to heat is increasing due to climate change. Globally, extreme temperature events are observed to be increasing in their frequency, duration, and magnitude. Between 2000 and 2016, the number of people exposed to heat waves increased by around 125 million. The excessive heat that many countries experience has serious mental and physical health consequences. People with pre-existing mental health disorders, for example, are particularly vulnerable. Extreme heat has been linked to a variety of mental health effects in studies over the years, including increased irritation and depressive symptoms, as well as an increase in suicidal thoughts. It can also have an impact on behavior, leading to increased hostility, domestic violence, and the use of alcohol or other narcotics, to cope with stress [50].

A study conducted in 19 different countries found that there is a significant relationship between heat and psychological conditions [51]. In a study conducted in South Australia, it was found that the cases for mental and behavioral disorders had increased by 7.3% during heat wave period [52]. Previous literature review revealed that heat stress and consistent high temperature leads to higher depression and anxiety. As climate change influence the social well-being, this further deteriorate the mental state of human being [53].

Among the studies on climate change and health outcome, weather and heat events were the most to be studied. Seventy-nine percent of the studies were carried out in high and upper-middle-income nations. However, the research to identify the relationship between EHE and mental condition was not clearly explained in certain region such as Africa [54]. Through research, Houghton et al. [55] found that rural area is more vulnerable to EHE although other urban areas are more prone to the events of UHI. In other study, the deteriorating mental and physical health of homeless individual was found to be closely associated with the climatic events indicating their high vulnerability to the EHE [56]. There are also studies conducted to explore the confounding factors affecting individuals to be more vulnerable to EHE. In summary, individuals who consume drug or alcohol, females, individuals with over 65 years old, with lower education level, and in denser urban design were more susceptible to EHE [57].

Studies have shown that people with pre-existing mental health problems are generally more susceptible to heat waves and may even exacerbate their conditions [52]. This is due to the reason that the physical or mental disability linked with their mental illness may impair their capacity to cope with and adjust to temperature changes. According to Bouchama et al. [58], having a pre-existing mental health problem increases the chance of mortality by more than three times during a heat wave, while using psychotropic medications reduces the risk to only nearly two times across all age groups. Therefore, individuals with mental illnesses should be closely monitored to reduce the increased risk of mental disease and given appropriate medical check-ups and prescription medications for their mental health conditions.

Higher temperature was proven to be closely associated with mental-related illness as it was due to human body response to regulate the internal heat stress [59]. Statistics from as early as 1998 showed that during heat waves, the mortality of mental illness patients was doubled [60]. To survive the high temperature, the human body will need to undergo a series of internal regulation influencing one's physical and mental health [61]. From a biological perspective, the elevation of body core temperature lowers the circulation of oxygen transfer efficiency in brain which potentially affects the nervous system in a human body. As a consequence, the blood–brain barrier will increase in its permeability and becomes more susceptible to microorganism and toxic neurochemicals [62]. There are also studies which suggested that the increase in temperature causes death due to cardiovascular and respiratory complication [63].

4 Prevalence of Mental Health Across the Globe

Mental disorders include depression, bipolar disorder, schizophrenia, and other psychoses and dementia [64]. Depression is a widespread mental illness that is also one of the leading causes of disability around the world. Depression affects an estimated 264 million individuals worldwide. Studies found that females are more likely to be affected than males. Sadness, loss of interest or pleasure, feelings of guilt or low self-worth, interrupted sleep or food, weariness, and impaired concentration are all symptoms of depression. People who are depressed may also experience a variety of bodily ailments that have no obvious physical reason. Depression can be chronic or recurrent, limiting people's ability to function at work or school and cope with day-to-day living. In its most extreme form, depression can lead to suicide [64].

Around 45 million people worldwide suffer from bipolar disorder. It usually includes both manic and depressed episodes, with periods of normal mood in between. Manic episodes are characterized by a heightened or irritated mood, excessive activity, quick speech, increased self-esteem, and a reduced desire for sleep. Bipolar disorder can also be diagnosed in those who have manic episodes, but do not have depressed episodes [64].

Schizophrenia is a serious mental illness that affects 20 million individuals on a global scale. Psychoses, including schizophrenia, are characterized by distortions in thinking, perception, emotion, language, sense of self, and behavior. Hallucinations (hearing, seeing, or feeling things that are not there) and delusions are common psychotic experiences (fixed false beliefs or suspicions that are firmly held, even when there is evidence to the contrary). People with the disease may find it difficult to work or study normally [64]. Schizophrenia is most commonly diagnosed in late adolescence or early adulthood. Medicines and emotional assistance are effective treatments. Affected persons can live productive lives and integrate into society with the right therapy and social assistance. Facilitation of assisted living, supported housing, and supported employment can serve as a foundation for people with serious mental illnesses, such as schizophrenia, to achieve a variety of rehabilitation goals, as they often struggle to find or keep a place to live and work [64].

Dementia affects around 50 million people worldwide. Dementia is a chronic or progressive disease in which cognitive function (i.e., the ability to process thoughts) deteriorates beyond what would be expected with normal aging. Memory, cognition, orientation, comprehension, computation, learning capacity, language, and judgment are all affected. Deterioration in emotional regulation, social behavior, or motivation is frequently associated with, and occasionally preceded by, cognitive impairment [64]. Dementia is caused by a range of brain disorders and accidents, including Alzheimer's disease and stroke. While there is currently no cure for dementia or a method to slow its progression, many treatments are in various phases of scientific studies. However, there is much that can be done to help and improve the lives of people with dementia, as well as their caregivers and relatives [64].

In 2019, 20.6% of adults in the US suffered from mental illness (51.5 million people). This equates to one out of every five adults. In terms of major mental illness, 5.2% of adults in the US experienced it in 2019 (13.1 million people). In 2016, 16.5% of children and adolescents aged 6 to 17 had a mental health problem (7.7 million people). In 2019, 3.8% of adults in the US had a co-occurring drug use disorder and mental illness (9.5 million people) [65].

5 Mental Health Outcomes of Climate Change

There is considerable evidence that climate change is affecting the geographic range, seasonality, and transmission of certain diseases, and increasing morbidity and mortality associated with extreme weather events [66]. Epidemiological evidence suggests probable biological pathways that might explain heat wave-related worsening of mental disorder morbidity. The vulnerability of a community group toward heat waves is influenced by different climate conditions, population characteristics, and local adaptation measures at the regional levels [67, 68].

Previous research findings are sufficient to indicate that climate change has a harmful impact on mental health. Heat waves and high temperatures have been linked to an increase in suicide rates and hot weather has been related to an increase in violent crime, which could have a severe influence on mental health. Not only that, high temperatures can also affect sleep quality and the ability to work, resulting in financial loss, which can have a severe influence on mental health as well [42].

Cities are among the most vulnerable to climate change. This is because, increasing storm and rainfall intensity will compound the effects of rising sea levels, which will in turn endanger major coastal cities with increased floods. Buenos Aires, Argentina; Tokyo, Japan; Cairo, Egypt; Shanghai, China; and New York City, United States are all major coastal cities which are more vulnerable to climate change as compared to other cities. Flooding and storms have been related to an increase in the prevalence of depression, post-traumatic stress disorder, and other anxiety disorders. Such weather occurrences can also have significant economic effects, such as if a person's business is harmed, it puts his or her mental health in jeopardy. Other effects of climate change, such as relocation and migration, as well as insufficient access to food and water, can impair mental health both directly and indirectly. They also endanger the health and well-being of the community. In conclusion, every consequence of climate change that might cause physical ill health can also have mental health consequences, given that the two are strongly linked, through numerous mechanisms such as job loss, inability to obtain health treatment, or a sense of loss of control [42].

Other factors that contribute to climate change have been connected to mental health issues (Fig. 2). Poor air quality, for example, caused by the burning of fossil fuels, might have detrimental mental health consequences. Traffic noise can be a mental health stressor, and fossil-fuel vehicles are noisier than electric vehicles and active modes of transportation such as walking or cycling. The UHI effect, which has a negative influence on mental health, can be exacerbated by using fossil fuels for energy in buildings, such as air conditioning [42].

The UHI effect is specific to cities and will grow more prominent as heat waves caused by climate change become more often. The effect is particularly obvious in cities with extensive stretches of urban pavement and concrete structures that absorb radiant heat and, on the other hand, a dearth of reflected or green places to cool. City electricity grids are also more vulnerable to extreme weather events as a result of the

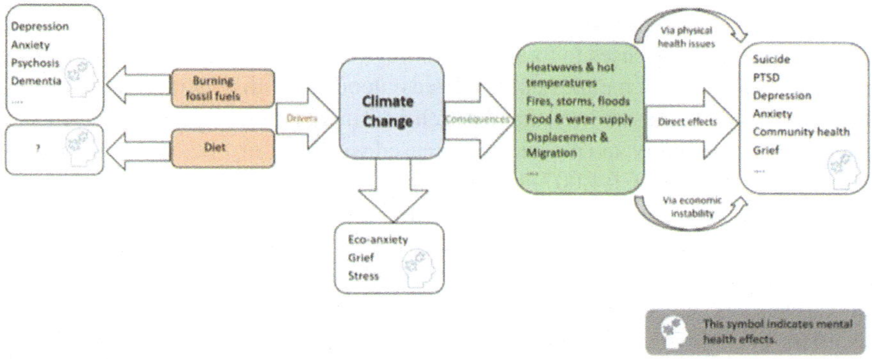

Fig. 2 Factors connected to climate change issues causing mental health issues. *Source* Huebner [42]

UHI effect. During heat waves, power needs for air conditioning spike, resulting in a vicious cycle of increased climate change emissions [69].

A substantial number of persons who are exposed to natural disasters caused by climate change or weather have stress and major mental health implications. As a consequence of climate change, several natural disasters may become more prominent. Post-traumatic stress disorder (PTSD), depression, general anxiety, increased substance uses or abuse, and suicidal thoughts are all reactions to the extreme events that cause life disruption, such as loss of life, resources, social support and social networks, or extensive relocation [70].

Peritraumatic experience has been linked to acute stress during and immediately after a disaster, which is likely to lead to the start of PTSD. Other implications for survivors emerge later, such as a reduction in everyday activities and a loss of "feeling of place." These circumstances may have an impact on mental health concerns and increase them with time. To certain extent, climate change news might even trigger individuals in feeling anxious, agitated, depressed, and powerless [70].

Various forms of psychopathological reactions to these occurrences are triggered by the concrete impact of those changes in life. Extreme weather occurrences such as floods, hurricanes, and wildfires are expected to have acute effects of mental trauma toward unprotected and helpless people. Subacute consequences include powerful emotions experienced by individuals who indirectly observe the effects of climate change and trigger anxiety relating to the survivorship of human and other species, and having a sense of being blocked, disoriented, and apathy. Examples of long-term outcomes involving large-scale social and communal repercussions are represented with erupting into forms of violence, conflict over scarce resources, displacement and forced migration, post-disaster adjustment, and chronic environmental stress [70].

Research has linked high temperatures with a wide range of mental health impacts. A systematic review aimed to summarize the epidemiological evidence and investigate the quantitative effects of high ambient temperatures and heat waves on mental health-related mortality and morbidity outcomes identified 53 high temperatures/heat wave studies that comprised over 1.7 million mental health-related mortality and 1.9 million morbidity cases in total. The findings suggest associations between heat exposure and a range of mental health-related outcomes. With regard to high temperatures, the study findings showed that for each 1 °C increase in temperature, the mental health-related mortality and morbidity increased with a RR of 1.022 (95% CI: 1.015–1.029) and 1.009 (95% CI: 1.007–1.015), respectively. The greatest mortality risk was attributed to substance-related mental disorders (RR, 1.046; 95% CI: 0.991–1.101), followed by organic mental disorders (RR, 1.033; 95% CI: 1.020–1.046). A 1 °C temperature rise was also associated with a significant increase in morbidity such as mood disorders, organic mental disorders, schizophrenia, and neurotic and anxiety disorders [71]. In recent years, the climate-related health hazards have severely increased on a regional scale because the impact from climate change has intensified the frequency and intensity of extreme heat events [70].

6 Associations Between Heat Waves and Mental Disease

Physical and mental health, as well as human well-being and heat waves, appear to be inseparably linked. Heat stress, which is a health outcome from the exposure to heat waves, has been linked to mood disorders, anxiety, and other negative mental health outcomes [72]. People with mental illness were three times more likely than those without mental illness to die as a result of a heat wave exposure [73].

A systematic review investigated the possible link between heat exposure (both high ambient temperatures and heat waves) and mental health-related mortality and morbidity. Five databases (PubMed, Embase, Scopus, Web of Science, and PsycINFO) were used to conduct a comprehensive search of peer-reviewed epidemiological research on heat exposure and mental health outcomes published between January 1990 and November 2020 [71].

The researchers considered studies in the general population that investigated the relationship between high ambient temperatures and/or heat waves and mental health-related mortality and morbidity (e.g., hospital admissions and emergency department visits). Heat exposure was linked to a variety of mental health consequences. In the case of high temperatures, the meta-analysis found that the risk of mental health-related mortality and morbidity increased with an RR of 1.022 (95% CI: 1.015–1.029) and 1.009 (95% CI: 1.007–1.015) for each 1 °C increase in temperature, respectively. Substance-related mental illnesses had the highest mortality risk (RR, 1.046; 95% CI: 0.991–1.101), followed by organic mental disorders (RR, 1.033; 95% CI: 1.020–1.046) [71].

A 1 °C increase in temperature was also linked to an increase in morbidity, including mood disorders, organic mental diseases, schizophrenia, and neurotic and anxiety disorders. The findings also point to indications of risk among populations residing in tropical and subtropical climate zones, as well as persons over the age of 65 [71].

7 Heat Waves and Mental Disease Exacerbation

Exposure to extreme heat can worsen psychiatric symptoms. People with schizophrenia can experience difficulties with body temperature regulation. Changes in temperature can also change symptoms of mood disorders. Some psychiatric medications, including some anti-depressants and anti-psychotics, can affect the way the body regulates temperature. In addition, people with pre-existing psychiatric symptoms are at an increased risk of emergency department visits for heat-related health issues and people with dementia are at increased risk for hospitalization and death, during heat wave events.

The 2003 European heat wave caused more than 70,000 additional deaths in 16 European countries [74]. Much research on the vulnerability analyses of distinct groups in society has been published. The population suffering from various types of mental or behavioral disorders was one of these categories where evidence of heat vulnerability was typically shown to be stronger than in any other [58]. Prior to 2003, it had been known for many years that ambient temperature induces psychotic worsening of core symptoms in people with mental and behavioral disorders. During the 1970s, a large number of patients at the New York State psychiatric hospital died as a result of heat waves [60]. According to a study of mortality data from the New York State psychiatric hospital from 1950 to 1984, psychiatric patients had twice the risk of dying during a heat wave as the general population during this time period. The probability of dying was also shown to be higher in the 1950s, prior to the widespread use of anti-psychotic drugs, than in the 1980s. However, in the 1970s, when very large doses of certain drugs were utilized, mortality was significantly higher. Consequently, both psychiatric disease and the use of anti-psychotic medication were found to increase the chance of dying during heat waves [60].

Following year 2003, an increasing number of epidemiological research works concentrating on the impact of temperature on mental health were conducted. Hansen et al. [52] calculated the impact of heat waves on hospital admissions and deaths due to mental, behavioral, and cognitive disorders in Adelaide, South Australia, from 1993 to 2006. For all mental illnesses, they identified a 7.3% increase in hospital admissions beyond a 26.7 °C ambient temperature threshold [52]. Hospital admission rates for organic illnesses, dementia, mood affective disorders, neurotics, stress-related disorders, somatoform disorders, psychological development disorders, and senility were all significantly affected by higher air temperatures [52]. The study also observed an increase in mortality attributed to mental disorders in the 65- to 74-year-old age group, and in schizophrenia, schizotypal, and delusional disorder patients.

Many other countries have also observed a relationship between heat and hospital admissions or emergency visits. A relationship between higher temperatures and emergency room visits connected with mental and behavioral diseases was documented in Toronto, Canada, with substantial trends for schizophrenia, and mood and neurotic disorders. Studies on schizophrenia [75, 76], depression [77], dementia [78, 79], and mania [78, 79] have all linked ambient temperatures and warm seasons to psychiatric emergencies and hospital admissions caused by mental disorders [80], as well as to Parkinson's [81] and Alzheimer's diseases [82]. These were conducted in Italy [78–83], Spain [79, 81, 82], Taiwan [75], China [84], Vietnam [77, 85], and Denmark [80], among others.

8 Mechanisms Linking Mental Health and Heat

In healthy individuals, the body attempts to maintain a near-constant core temperature of 37 °C, regardless of the external temperature [86]. In a warm setting, cutaneous vasodilation, which increases skin blood flow, is an adaptation reaction that transports metabolic heat to the body surface, thus improving heat transmission to the surrounding environment [87]. Evaporation of perspiration from the skin's surface is also vital for heat dissipation. The sweat response, on the other hand, is reliant on cutaneous vasodilation, since blood plasma is required as a precursor fluid for sweat synthesis [88]. Increased blood flow in the epidermis is accompanied by vasoconstraction (blood flow reduction) in the visceral organs in humans [87]. An increase in cardiac output is required to compensate for the lower blood flow in visceral tissues, ensuring adequate oxygen delivery to inner organs [86].

When variation in perspiration and skin blood flow does not result in significant heat dissipation, body core temperature rises constantly, resulting in hyperthermia. Heat stroke, the most serious complication of heat exposure, is a life-threatening condition marked by a core temperature of over 40 °C and organ failure, including the central nervous system. However, indications such as weariness, weakness, disorientation, and diminished awareness are witnessed before heat stroke develops [89]. Heat exhaustion from prolonged exposure can increase the risk of heat-related disease.

An increase in external temperature can damage the proper functioning of the central nervous system. The ideal temperature for optimal performance is around 22 °C. The external temperature has a variety of effects on the chance of developing or maintaining mental problems. Temperature stress, for example, might impair psychophysiological activities by modifying biochemical levels such as altering serotonin and dopamine production or interrupting thermoregulation equilibrium [62].

The underlying physiological mechanisms of heat exposure on mental health are complex. The plausible etiological pathways may be explained in part by the pathophysiological effects on the neurological system. High temperature has been reported to impact the levels and balance of the neuro-transmitters serotonin and dopamine in the brain, which have roles in mood regulation, cognitive function, and complex task performance. Additionally, when high temperatures are sustained for several days or weeks, individuals' physiological and behavioral adaptation strategies can be challenged. As a result, irritability and psychological distress (including risky behaviors such as substance abuse and alcohol consumption), aggression, violence, and suicides become more common [62].

Higher suicide rates in the spring and early summer have been claimed to be due to seasonal fluctuations in serotonin, a neurotransmitter that can influence impulsivity and aggression, possibly leading to suicide. L-tryptophan is a serotonin precursor, and low L-tryptophan levels in the blood have been linked to serious depressive illnesses [90]. This indicates a plausible influence of ambient temperature toward the control of L-tryptophan levels which directly links heat with mental health outcomes.

9 Mental Health Effects of Urban Heat Island in Global Cities

Climate change's drivers and repercussions can both cause or worsen mental health illnesses, posing a threat to overall emotional well-being or also known collectively as mental health. There is a well-established link between environmental exposures and mental health. These environmental exposures, such as green space, noise, air pollution, weather conditions, and housing conditions, may either cause or protect mental diseases, promoting stress reduction and mental healing. Although some environmental elements were investigated and given significant attention, others, such as heat in urban settings, have received relatively less attention [91–94]. As a result, a more holistic knowledge of the link between heat and mental health is necessary to comprehend the societal cost of poor mental health and to prepare effectively for community adaptation and resilience to extreme heat events in future.

The urban population has expanded by more than 100 times in the last two centuries. More than half of the world's population now resides in cities, with predictions that this number will increase to 70% by year 2050. Increased ambient heat, high temperatures, frequent and extended hot spells, and thermal stress are already being exacerbated by poor choice of urban fabrics such as dark-colored building surfaces, roads, pavements, car emissions, and decreasing urban green spaces [95]. For example, in the twentieth century, the mean temperature in six large cities in Japan, including Tokyo and Nagoya, had risen by 2–3 °C, while the globally averaged temperature had risen by 0.6 °C. The impact from UHI phenomena in large cities in Japan is significant when compared with the trend of global climate change [96].

According to empirical research, heat stress is increasing in tropical cities, as a result of UHIs and poor unplanned urbanization, especially in developing countries. Heat stress has a significant negative impact on the population's mortality and morbidity. In fact, mental health and well-being are two of the health consequences of heat stress identified in Thailand and Indonesia. Though not as severe as other countries, these countries have seen an increase in the occurrence of heat-related ailments [97]. A case study of two cities in Thailand and Indonesia explored the effects of UHI on the household energy consumption and perceived health effects in Bangkok, Thailand, and Bandung, Indonesia. The perceived health effect of UHI was examined using heat stress variables and health outcomes. The data suggest that respondents with heat stress problems have lower life satisfaction, lower energy levels, and experience more frequent emotional problems. These findings are similar to the results in previous studies, i.e., heat stress significantly reduces health outcomes and well-being. One study by Lan et al. [98] found that people working in hot environments had a lower motivation to work and experienced negative moods during work. Another study also found that there is an effect of increasing temperature on mortality in Bangkok, Thailand. This study found that heat stress not only affects their working life, but also interferes with other aspects of daily life such as sleep quality, daily travel, and intention to exercise [99].

UHI impacts were also shown to be prominent in various Malaysian cities, including Kuala Lumpur and Putrajaya. The influence of rising temperatures on the city community in Kuala Lumpur, particularly on psychological and social health, was investigated. In Greater Kuala Lumpur, Malaysia, a study was done to investigate the knowledge, attitudes, preventative activities, and health impact of temperature rise related with UHI. The survey was completed by 558 people in total between October 2016 and May 2017. Nearly all the participants reported that they suffered from heat fatigue. Findings show that the urban temperature can alter the rate of psychological distresses, e.g., anxiety, depression, and aggressive behaviors. Anxiety (79%) was the most reported psychological health impact, followed by depression (74.4%) and aggressive behavior (74.4%). According to this study, a substantial percentage of study participants suffered from a variety of negative health effects. From another similar study to provide evidence regarding the health effects of UHI, a community of people working in Greater Kuala Lumpur were recruited. Among all psychological health experiences queried, depression (64.7%) was reported highest, while 33% experienced anxiety [100]. This high incidence of negative health effects demonstrates the significant impact that UHI-related temperature rise has on Malaysian city dwellers [101].

A study was conducted in Isfahan to investigate the effects of UHI on overall citizen health. Isfahan is the third-largest metropolis in Iran. The city has grown substantially in population and industrialization over the recent decades. The occurrence and spatial pattern of the UHI in Isfahan in 2013–2015 was earlier investigated by Ahmadi and Dadashi Roudbari [102]. Their results show that there is a sharp temperature difference among the various regions of the city. This extreme heat variance in different regions of the city is postulated to cause disturbances in the citizen health parameters [103]. In the study, the land surface temperature (LST) estimation and UHI monitoring of heat and cold island regions in Isfahan were determined using Landsat-8 satellite data. The identified heat map leads to a benchmark that can quantify the different general health sub-scales (physical health, anxiety and sleep, social function, and depression) of the citizens in the heat and cold island regions. The GHQ-28 was used to analyze the effects of UHIs on citizen health status over a three-year period (2016–2018) in the Isfahan metropolitan area. The results of statistical analysis showed that there was significant difference between the two groups of citizens in terms of the following: (1) social function, (2) depression, and (3) anxiety and sleep ($P < 0.05$). In other words, the citizens experiencing UHIs have shown more severe responses in terms of social function, depression, and anxiety and sleep [104].

In a study of the consequences of heat waves in Australia cities, it was found that the most at-risk groups include the elderly, children, pregnant women, patients with chronic diseases, people with physical and mental disabilities, and low-income communities [52]. Extreme heat was suggested to be a significant concern to the health and well-being of the mentally ill. In Adelaide, South Australia, researchers focused on mental, behavioral, and cognitive impairments that could be induced or exacerbated by heat waves, potentially predisposing people to heat-related morbidity and mortality. The influence of heat waves on hospital admissions and mortalities ascribed to mental, behavioral, and cognitive disorders was assessed using health outcome data from Adelaide, South Australia, from 1993 to 2006. The study discovered a link between ambient temperature and hospital admissions for mental and behavioral illnesses above a threshold of 26.7 °C. The hospital admissions increased by 7.3% during heat waves, compared to non-heat-wave periods. Organic illnesses such as symptomatic mental disorders, dementia, mood (affective) disorders, neurotics use, stress-related disorders, somatoform disorders, psychological development disorders, and senility were among the illnesses for which admissions increased. During heat waves, deaths from mental and behavioral problems rose in the 65- to 74-year-old age group, as well as in people with schizophrenia, schizotypal, and delusional illnesses. Dementia deaths have increased among people under the age of 65 [52].

In another case study, the association between ambient temperature and suicide in multiple cities in three East Asian countries was evaluated [90]. A time-stratified case-crossover method was used to explore the relationship between temperature and suicide, adjusting for potential time-varying confounders and time-invariant individual characteristics. Sex- and age-specific associations of temperature with suicide were estimated, as well as the interactions between temperature and these variables. A random-effects meta-analysis was used to estimate country-specific pooled associations of temperature with suicide. An increase in temperature corresponding to half of the city-specific standard deviation was positively associated with suicide in most cities, although average suicide rates varied substantially. Pooled country-level effect estimates were 7.8% (95% CI: 5.0, 10.8%) for a 2.3 °C increase in ambient temperature in Taiwan; 6.8% (95% CI: 5.4, 8.2%) for a 4.7 °C increase in Korea; and 4.5% (95% CI: 3.3, 5.7%) for a 4.2 °C increase in Japan. The association between temperature and suicide was significant even after adjusting for sunshine duration. The association between sunshine and suicide was found to be insignificant. The associations were greater among men than women in 12 of the 15 cities, although not significantly so. There was little evidence of a consistent pattern of associations with age. In general, associations were strongest with temperature on the same day or the previous day, with little evidence of associations with temperature over longer lags (up to 5 days) [90].

Luan et al. [105] estimated the association between ambient temperature and suicide in 31 capital cities in China during 2008–2013. Multivariate meta-analysis was used to pool the city-specific estimates to explore the overall relative risk in China. The study found that high temperature had a significant impact on suicide death. The country-level relative risk of high temperature on suicide was 1.37 (95%

CI, 0.96–2.57), and the RR was higher in the male and age <65-year-old group than that in the female and age ≥65-year-old group [105].

A study in a subtropical city in Hong Kong investigated the short-term association between temperatures (mean annual temperature over 21 °C) and mental disorder hospitalizations. This study found a positive temperature-mental-disorder admission association in a warm subtropical region (cumulative relative risk at 28 °C vs. 19.4 °C (interquartile range, lag 0–2 days) = 1.09 (95% confidence interval 1.03, 1.15), and the association was most prominent among the elderly. Transient mental disorders and episodic mood disorders demonstrated a robust positive association with temperature, while drug-related mental disorders showed a positive association with temperatures above 20 °C [106].

In a case study conducted in Toronto, Canada, health data from 2002 to 2010 was utilized to investigate the effects of severe ambient temperature on hospital emergency room visits (ER) related to mental and behavioral problems. The researchers discovered a substantial link between ER visits for mental and behavioral illnesses (MBD), and a mean daily temperature of 28 °C. For exposure to hot temperatures, the link was strongest after a period of 0–4 days. Over a 7-day period, a 29% (RR = 1.29, 95% CI 1.09–1.53) increase in MBD ER visits was seen after exposure to high ambient temperature (99th percentile vs. 50th percentile). Schizophrenia, mood disorders, and neurotic disorders were linked in the same manner. Interestingly, no significant associations with cold temperatures were reported [107].

In another case study involving Shanghai, China, the short-term effects of daily mean temperature on hospital admissions for mental illnesses were studied for a period from January 2008 to December 2015 using daily hospital admission data for mental illnesses and daily meteorological and ambient pollution data in Shanghai [108]. During the study period, there are 93,971 mental illness cases admitted to hospitals. There was a significant positive correlation between temperature over predefined threshold (24.6 °C) and mental illness hospital admission visits at a lag of 0–1 days with a reference of median temperature (18.3 °C). Over the lag 0–1 days, the relative risks of severe hot temperatures (33.1 °C, 99% percentile) were 1.266 times higher than the median temperature (95% confidence intervals: 1.074–1.493). There was no influence of cold weather on hospital admissions for mental illnesses [108].

In a study conducted in Hefei, China, the association between ambient temperature and hospital admissions for schizophrenia was evaluated [109]. Anhui Mental Health Centre and Hefei Bureau of Meteorology provided the daily statistics on hospital admissions for schizophrenia and meteorological data for the warm season (May–October), from 2005 to 2014. Findings from this research highlighted that high temperatures represent significant hazards for schizophrenia, with a significant link discovered between ambient temperature and schizophrenia admissions [109]. This research ascertains the impact of heat on mental health outcomes even in subtropical climates.

In Adelaide, South Australia, health burden is closely associated with hot weather during summer and heat waves. A study was conducted by Williams et al. [110] to examine the heat thresholds and temperature relationships for mortality and morbidity outcomes in Adelaide. Most health outcomes showed a positive relationship with daily temperatures over predefined thresholds. When adjusted for air pollutants, a 10 °C increase in maximum temperature was associated with a 4.9% increase in daily ambulance callouts (IRR 1.049; 95% CI 1.027–1.072), and a 3.4% increase in mental health-related hospital admissions (IRR 1.034; 95% CI 1.009–1.059), for the all-age population. Daily temperatures were also associated with all-cause and mental health-related ED presentations [110].

For a case study in South Korea, Lee et al. [111] investigated the burden of mental disorder caused by high temperatures in six main cities (Seoul, Incheon, Daejeon, Daegu, Busan, and Gwangju) in South Korea using data on daily temperature and emergency admissions (EA) for mental disorders over an 11-year period between 2003 and 2013. Within a period of 0–4 days of high temperature exposure, the strongest link between mental disorders and high temperature was observed. Extreme heat was shown to be the cause of 14.6% of EA for mental disorder, with the elderly being particularly vulnerable (19.1%). Anxiety, dementia, schizophrenia, and depression are few of the mental illnesses that have been linked to high temperatures. Extremely hot temperatures were the cause for 31.6% of all EA for anxiety [111].

In a study conducted in Quebec, Canada, Vida et al. [112] investigated the data on the number of emergency room visits as a function of temperature and humidity in three geographic locations of Quebec. Findings indicate a higher mean temperature and humidity was associated with increased usage of emergency rooms for mental and psychological disorders [112].

10 Strengthening the Public and Institutional Preparedness and Response to Heat-Related Mental Health Events Across the Globe

The drastic climate change followed by urbanization across the globe caused not only the increase in ambient temperature, but also the epidemic of non-communicable diseases (NCDs) such as obesity, cardiovascular disease, and mental health disorder [113]. In order to address these issues, interventions in society and the environment serve as the most preventive measures [114]. However, mitigating the consequences of climate change and adapting to current or anticipated effects entails a variety of potential measures to assist in decreased risks, susceptibility, and increased resilience capability that requires multi-sectoral approaches and initiatives. In addition, these actions are expected to be successful on a variety of geographical and temporal dimensions, if it is planned in advance, or driven by new planning rules, market demand, or even societal pressure. Stigma on mental health should also be addressed to reduce the negative impact of climate on mental health [115].

If adequate emergency prevention, preparedness, response, and recovery measures are implemented in a sustainable and timely manner, the magnitude of mental health problems caused by heat can be reduced. While many efforts can be done to curb heat-related mental health, health professionals should assist in creating adaptation and mitigation measures that will help individuals better adapt to the changing environment which will help reduce emotional and psychological impacts. Preventive measures against extreme temperatures should be taken to protect public health, such as planning for public heat stress level monitoring, conduct more holistic outdoor thermal comfort studies, and providing sustainable countermeasures to combat the negative impact of UHI.

11 Conclusion

In line with climate change and extreme heat events, global mental health services should be improved especially in middle and lower-income countries. Scientific findings from various regions around the world have elucidated the significant relationship of heat events on mental health. In particular, the vulnerable group of urban dwellers with high exposure to urban heat along with the worsening of extreme heat events and global climate change urges a need for immediate action. Coordinated efforts from a multi-lateral collaboration are needed to mobilize resources for both research and practical activities across all levels. More studies are needed to investigate and suggest effective measures to be taken to reduce the negative impact of extreme heat events on mental health. One of the effective approaches could be linked to a better re-design of urban area. The urban development program should also look into the development of network of sensors to address the issue with data scarcity. Concurrently, a better network of sensors will also be beneficial for the development of early warning system to enhance public resilience against extreme heat event threats in the near future. From various studies, the identified group that are vulnerable to heat waves consists of the elderly, socially isolated, chronically ill, homeless, mentally ill, people with cognitive abnormalities, and anyone taking drugs that impact cognition and thermoregulation, or have photosensitive side effects. The reduced vulnerability of community will contribute to a better mental health state while simultaneously improved lifestyle and livelihood.

References

1. United Nations (2021) What is climate change? [Internet]. [cited 2022 Mar 19]. Available from: https://www.un.org/en/climatechange/what-is-climate-change
2. U. S. Global Change Research Program (2014) Climate change impacts in the United States climate trends and regional impacts
3. National Geographic Society (2019) Climate change [Internet]. [cited 2022 Mar 19]. Available from: https://www.nationalgeographic.org/encyclopedia/climate-change/

4. Shaftel H, Jackson R, Callery S, Bailey D (2020) Global warming versus climate change. Resources—climate change: vital signs of the planet [Internet]. NASA—California Institute of Technology. [cited 2022 Mar 19]. Available from: https://climate.nasa.gov/resources/global-warming-vs-climate-change/
5. NASA (2022) Evidence. Facts—climate change: vital signs of the planet [Internet]. [cited 2022 Mar 19]. Available from: https://climate.nasa.gov/evidence/
6. Klinenberg E (2015) Heat wave: a social autopsy of disaster in Chicago. University of Chicago Press, 305
7. Ng E, Ren C (2018) China's adaptation to climate & urban climatic changes: a critical review. Urban Clim 1(23):352–372
8. Viceto C, Pereira SC, Rocha A (2019) Climate change projections of extreme temperatures for the Iberian Peninsula. Atmosphere [Basel] [Internet]. 2019 Apr 29 [cited 2020 Mar 10];10[5]:229. Available from: https://www.mdpi.com/2073-4433/10/5/229
9. Department of Economic and Social Affairs PDUN (2014) World urbanization prospects. Demogr Res [Internet]. [cited 2017 Nov 17];12:32. Available from: https://esa.un.org/unpd/wup/Publications/Files/WUP2014-Report.pdf
10. Gago EJ, Roldan J, Pacheco-Torres R, Ordóñez J (2013) The city and urban heat islands: a review of strategies to mitigate adverse effects. Renew Sustain Energy Rev 25:749–758
11. Law J, Fong CS (2020) Urban heat mitigation in tropical city [Internet]. [cited 2022 Mar 19]. Available from: https://rouqilaw0908.wixsite.com/portfolio/competition
12. Ibrahim I (2014) UHI and temporal study of urban growth: a study of the Klang valley area. Asian Conf Sustain Energy Environ [Internet]. [cited 2018 Jan 18]; Available from: https://www.researchgate.net/profile/Illyani_Ibrahim/publication/280722157_UHI_and_Temporal_Study_of_Urban_Growth_A_Study_of_the_Klang_Valley_area/links/562f299208ae518e3483c4f5.pdf
13. Kondoh K (2007) The role of urban communities in responding to global climate change and the urban heat island effect [Internet]. Ph.D. [cited 2020 Mar 16], p 167. Available from: http://proquest.umi.com/pqdweb?did=1414132981&Fmt=7&clientId=18803&RQT=309&VName=PQD
14. Yang L, Niyogi D, Tewari M, Aliaga D, Chen F, Tian FQ et al (2016) Contrasting impacts of urban forms on the future thermal environment: example of Beijing metropolitan area. Environ Res Lett 11(3):10
15. Kardinal Jusuf S, Wong NH, Hagen E, Anggoro R, Hong Y (2007) The influence of land use on the urban heat island in Singapore. Habitat Int 31(2):232–242
16. Roth M, Chow WTL (2012) A historical review and assessment of urban heat island research in Singapore. Singap J Trop Geogr [Internet]. Nov 1 [cited 2018 Jan 6]; 33(3):381–397. Available from: https://doi.org/10.1111/sjtg.12003
17. Ibrahim I (2017) Measuring socio-economic indicators for climatic analysis. Plan Malaysia [Internet]. [cited 2020 Mar 16]; 15(1):289–294. Available from: http://planningmalaysia.org/index.php/pmj/article/view/245
18. Zhang K, Wang R, Shen C, Da L (2010) Temporal and spatial characteristics of the urban heat island during rapid urbanization in Shanghai, China. Environ Monit Assess 169(1–4):101–112
19. Elsayed ISM (2012) A Study on the urban heat island of the City of Kuala Lumpur, Malaysia. JKAU Met, Env Arid L Agric Sci. 23(2):121–134
20. Nuruzzaman M (2015) Urban heat island: causes, effects and mitigation measures—a review. Int J Environ Monit Anal [Internet]. 2015 Mar 10 [cited 2019 Jul 9]; 3(2):67. Available from: http://www.sciencepublishinggroup.com/journal/paperinfo.aspx?journalid=162&doi=10.11648/j.ijema.20150302.15
21. Rizwan AM, Dennis LYC, Liu C (2008) A review on the generation, determination and mitigation of urban heat island. J Environ Sci 20(1):120–128
22. Santamouris M (2015) Analyzing the heat island magnitude and characteristics in one hundred Asian and Australian cities and regions. Sci Total Environ [Internet]. 2015 Apr [cited 2019 Jul 9]; 512–513:582–598. Available from: https://linkinghub.elsevier.com/retrieve/pii/S0048969715000753

23. Shaharuddin A, Noorazuan M, Yaakob M (2009) Of MY-GMJ, 2009 U. Fenomena Pulau Haba Bandar dan isu alam sekitar di Bandaraya Kuala Lumpur. Geogr Malaysian J Soc Sp 5(3):57–67
24. Benrazavi RS, Dola K, Ujang N, Sadat Benrazavi N (2016) Effect of pavement materials on surface temperatures in tropical environment. Sustain Cities Soc [Internet]. [cited 2017 Dec 14]; 22:94–103. Available from: http://www.sciencedirect.com/science/article/pii/S2210670716300117
25. Chen L, Ng E, An X, Ren C, Lee M, Wang U, et al (2012) Sky view factor analysis of street canyons and its implications for daytime intra-urban air temperature differentials in high-rise, high-density urban areas of Hong Kong: a GIS-based simulation approach. Int J Climatol [Internet]. 2012 Jan 1 [cited 2018 Jan 16]; 32(1):121–136. Available from: https://rmets.onlinelibrary.wiley.com/doi/full/10.1002/joc.2243
26. Ibrahim I, Samah AA (2011) Preliminary study of urban heat island: measurement of ambient temperature and relative humidity in relation to landcover in Kuala Lumpur. In: Proceedings—2011 19th international conference on geoinformatics, geoinformatics 2011 [Internet]. 2011 [cited 2020 Mar 16]. Available from: https://ieeexplore.ieee.org/abstract/document/5981068/
27. Ibrahim I, Samah AA, Fauzi R (2012) Land Surface Temperature and Biophysical Factors in Urban Planning. World Acad Sci Eng Technol [Internet]. [cited 2020 Mar 18]; 6:1518–1523. Available from: http://www.academia.edu/download/58010434/Land_Surface_Temperature_and_Biophysical20181216-3067-mr4pyz.pdf
28. Nakano A (2015) Urban weather generator user interface development: towards a usable tool for integrating urban heat island effect within urban design process. [cited 2018 Jan 8]. Available from: https://dspace.mit.edu/handle/1721.1/99251
29. Shahmohamadi P, Che-Ani AI, Maulud KNA, Tawil NM, Abdullah NAG (2011) The impact of anthropogenic heat on formation of urban heat island and energy consumption balance. Urban Stud Res. 2(2011):1–9
30. Zhao L, Lee X, Smith RB, Oleson K. Strong contributions of local background climate to urban heat islands. Nature [Internet]. 2014 Jul 9 [cited 2019 Jul 9]; 511(7508):216–219. Available from: http://www.ncbi.nlm.nih.gov/pubmed/25008529
31. Solecki WD, Rosenzweig C, Parshall L, Pope G, Clark M, Cox J et al (2011) Mitigation of the heat island effect in urban New Jersey. https://doi.org/10.1016/j.hazards200412002 [Internet]. [cited 2022 Mar 19]; 6(1):39–49. Available from: https://www.tandfonline.com/doi/abs/10.1016/j.hazards.200412002
32. OKE T (1982) The energetic basic of the urban heat island. Q J R Meteorol Soc [Internet]. 1982 Jan 1 [cited 2019 Jul 8]; 108(455):1–24. Available from: http://doi.wiley.com/10.1002/qj.49710845502
33. Kumar R, Mishra V, Buzan J, Kumar R, Shindell D, Huber M (2017) Dominant control of agriculture and irrigation on urban heat island in India. Sci Rep 71 [Internet]. 2017 Oct 25 [cited 2022 Mar 19]; 7(1):1–10. Available from: https://www.nature.com/articles/s41598-017-14213-2
34. Sailor DJ (2011) A review of methods for estimating anthropogenic heat and moisture emissions in the urban environment. Int J Climatol [Internet]. 2011 Feb 1 [cited 2022 Mar 19]; 31(2):189–99. Available from: https://onlinelibrary.wiley.com/doi/full/10.1002/joc.2106
35. Chen F, Kusaka H, Bornstein R, Ching J, Grimmond CSB, Grossman-Clarke S et al (2011) The integrated WRF/urban modelling system: development, evaluation, and applications to urban environmental problems. Int J Climatol [Internet]. 2011 Feb 1 [cited 2018 Jan 2]; 31(2):273–288. Available from: http://doi.wiley.com/10.1002/joc.2158
36. Velders G (1997) Effect of greenhouse gas emissions on stratospheric ozone depletion.. PBL Netherlands Environmental Assessment Agency [Internet]. [cited 2022 Mar 19]. Available from: https://www.pbl.nl/en/publications/Effect_of_greenhouse_gas_emissions_on_stratospheric_ozone_depletion
37. Tong S, Prior J, McGregor G, Shi X, Kinney P (2021) Urban heat: an increasing threat to global health. BMJ [Internet]. 2021 Oct 25 [cited 2022 Mar 19]; 375. Available from: https://pubmed.ncbi.nlm.nih.gov/34697023/

38. World Health Organization (2004) Promoting mental health concepts, emerging evidence, practice summary report. A report of the World Health Organization
39. WHO (2021) Suicide [Internet]. [cited 2022 Mar 19]. Available from: https://www.who.int/news-room/fact-sheets/detail/suicide
40. The Lancet Global Health (2020) Mental health matters. Lancet Glob Heal [Internet]. 2020 Nov 1 [cited 2022 Mar 19]; 8(11):e1352. Available from: http://www.thelancet.com/article/S2214109X20304320/fulltext
41. Vigo D, Thornicroft G, Atun R (2016) Estimating the true global burden of mental illness. The Lancet Psychiatry [Internet]. 2016 Feb 1 [cited 2022 Mar 19]; 3(2):171–178. Available from: https://pubmed.ncbi.nlm.nih.gov/26851330/
42. Huebner G (2021) Climate change and mental health. The Bartlett—UCL—University College London [Internet]. [cited 2022 Mar 19]. Available from: https://www.ucl.ac.uk/bartlett/news/2021/jul/climate-change-and-mental-health
43. Santomauro DF, Mantilla Herrera AM, Shadid J, Zheng P, Ashbaugh C, Pigott DM et al (2021) Global prevalence and burden of depressive and anxiety disorders in 204 countries and territories in 2020 due to the COVID-19 pandemic. Lancet [Internet]. 2021 Nov 6 [cited 2022 Mar 19]; 398(10312):1700–1712. Available from: http://www.thelancet.com/article/S0140673621021437/fulltext
44. Leonhardt M (2021) Cost and accessibility of mental health care in America [Internet]. CNBC. [cited 2022 Mar 19]. Available from: https://www.cnbc.com/2021/05/10/cost-and-accessibility-of-mental-health-care-in-america.html
45. Global Clean Cooling Landscape Assessment (2018) Invest in a cooler world [Internet]. [cited 2019 Sep 23]. Available from: https://www.clean-cooling.ac.uk/
46. O'Neill MS, Ebi KL (2009) Temperature extremes and health: impacts of climate variability and change in the United States. J Occup Environ Med [Internet]. 2009 Jan [cited 2020 Mar 11]; 51(1):13–25. Available from: http://journals.lww.com/00043764-200901000-00004
47. Wolfe RM, Pauleau A (2003) Death in heat waves. Br Med J 327:1228–1229
48. Oswald EM (2018) An analysis of the prevalence of heat waves in the United States between 1948 and 2015. J Appl Meteorol Climatol [Internet]. 2018 Jul 1 [cited 2022 Mar 19]; 57(7):1535–1549. Available from: https://journals.ametsoc.org/view/journals/apme/57/7/jamc-d-17-0274.1.xml
49. NOAA (2019) USGCRP Indicator Details | GlobalChange.gov [Internet]. 2019 [cited 2022 Mar 19]. Available from: https://www.globalchange.gov/browse/indicators/us-heat-waves
50. American Psychiatric Association [APA] (2021) Extreme heat contributes to Worsening mental health, especially among vulnerable populations [Internet]. [cited 2022 Mar 19]. Available from: https://www.psychiatry.org/newsroom/news-releases/extreme-heat-contributes-to-worsening-mental-health-especially-among-vulnerable-populations
51. Thompson R, Hornigold R, Page L, Waite T (2018) Associations between high ambient temperatures and heat waves with mental health outcomes: a systematic review. Public Health [Internet]. 2018 Aug 1 [cited 2022 Mar 19]; 161:171–191. Available from: https://pubmed.ncbi.nlm.nih.gov/30007545/
52. Hansen A, Bi P, Nitschke M, Ryan P, Pisaniello D, Tucker G (2008) The effect of heat waves on mental health in a temperate Australian city. Environ Health Perspect [Internet]. [cited 2022 Mar 19]; 116(10):1369–1375. Available from: https://pubmed.ncbi.nlm.nih.gov/18941580/
53. Palinkas LA, Wong M (2020) Global climate change and mental health. Curr Opin Psychol [Internet]. 2020 Apr 1 [cited 2022 Mar 19]; 32:12–16. Available from: https://pubmed.ncbi.nlm.nih.gov/31349129/
54. Berrang-Ford L, Sietsma AJ, Callaghan M, Minx JC, Scheelbeek PFD, Haddaway NR et al (2021) Systematic mapping of global research on climate and health: a machine learning review. Lancet Planet Heal [Internet]. 2021 Aug 1 [cited 2022 Mar 19]; 5(8):e514–e525. Available from: http://www.thelancet.com/article/S2542519621001790/fulltext
55. Houghton A, Austin J, Beerman A, Horton C (2017) An approach to developing local climate change environmental public health indicators in a rural district. J Environ Public Health 2017

56. Bezgrebelna M, McKenzie K, Wells S, Ravindran A, Kral M, Christensen J et al (2021) Climate change, weather, housing precarity, and homelessness: a systematic review of reviews. Int J Environ Res Public Heal 2021, 18:5812 [Internet]. 2021 May 28 [cited 2022 Mar 19]; 18(11)]5812. Available from: https://www.mdpi.com/1660-4601/18/11/5812/htm
57. Ellena M, Breil M, Soriani S (2020) The heat-health nexus in the urban context: a systematic literature review exploring the socio-economic vulnerabilities and built environment characteristics. Urban Clim 1(34):100676
58. Bouchama A, Dehbi M, Mohamed G, Matthies F, Shoukri M, Menne B (2007) Prognostic factors in heat wave-related deaths: a meta-analysis [Internet], Archives of Internal Medicine. Arch Intern Med [cited 2022 Mar 19]. 167:2170–2176. Available from: https://pubmed.ncbi.nlm.nih.gov/17698676/
59. Cusack L, de Crespigny C, Athanasos P (2011) Heatwaves and their impact on people with alcohol, drug and mental health conditions: a discussion paper on clinical practice considerations. J Adv Nurs [Internet]. 2011 Apr [cited 2022 Mar 19]; 67(4):915–922. Available from: https://pubmed.ncbi.nlm.nih.gov/21214621/
60. Bark N (1998) Deaths of psychiatric patients during heat waves. Psychiatr Serv [Internet]. [cited 2022 Mar 19]; 49(8):1088–1090. Available from: https://pubmed.ncbi.nlm.nih.gov/9712220/
61. Ebi KL, Capon A, Berry P, Broderick C, de Dear R, Havenith G et al (2021) Hot weather and heat extremes: health risks. Lancet [London, England] [Internet]. 2021 Aug 21 [cited 2022 Mar 19]; 398(10301):698–708. Available from: https://pubmed.ncbi.nlm.nih.gov/34419205/
62. Lõhmus M (2018) Possible biological mechanisms linking mental health and heat—a contemplative review. Int J Environ Res Public Health [Internet]. 2018 Jul 18 [cited 2022 Mar 19]; 15(7). Available from: /pmc/articles/PMC6068666/
63. Chan EYY, Ho JY, Hung HHY, Liu S, Lam HCY (2019) Health impact of climate change in cities of middle-income countries: the case of China. Br Med Bull [Internet]. 2019 Jun 19 [cited 2022 Mar 19]; 130(1):5–24. Available from: https://pubmed.ncbi.nlm.nih.gov/31070715/
64. WHO (2019) Mental disorders [Internet]. [cited 2022 Mar 19]. Available from: https://www.who.int/news-room/fact-sheets/detail/mental-disorders
65. National Alliance on Mental Illness (2022) Mental health by the numbers. NAMI: National Alliance on Mental Illness [Internet]. [cited 2022 Mar 19]. Available from: https://www.nami.org/mhstats
66. Ebi KL, Ogden NH, Semenza JC, Woodward A (2017) Detecting and attributing health burdens to climate change. Environ Health Perspect [Internet]. 2017 Aug 1 [cited 2022 Mar 19]; 125(8). Available from: https://pubmed.ncbi.nlm.nih.gov/28796635/
67. de' Donato FK, Leone M, Scortichini M, De Sario M, Katsouyanni K, Lanki T et al (2015) Changes in the effect of heat on mortality in the last 20 years in Nine European cities. Results from the PHASE project. Int J Environ Res Public Health [Internet]. 2015 Dec 8 [cited 2022 Mar 19]; 12(12):15567. Available from: /pmc/articles/PMC4690942/
68. Åström C (2017) Health effects of heatwaves Short and long term predictions. [cited 2022 Mar 19]. Available from: http://umu.diva-portal.org/
69. World Health Organization (2021) Global health observatory. Choice Rev Online [Internet]. 2021 [cited 2021 Feb 12]; 51(11):51-6210. Available from: https://www.who.int/data/gho
70. Cianconi P, Betrò S, Janiri L (2020) The impact of climate change on mental health: a systematic descriptive review. Front Psychiatry [Internet]. [cited 2020 Oct 22]; 11. Available from: https://www.ncbi.nlm.nih.gov/pmc/articles/PMC7068211/
71. Liu J, Varghese BM, Hansen A, Xiang J, Zhang Y, Dear K et al (2021) Is there an association between hot weather and poor mental health outcomes? A systematic review and meta-analysis. Environ Int [Internet]. 2021 Aug 1 [cited 2022 Mar 19]; 153. Available from: https://pubmed.ncbi.nlm.nih.gov/33799230/
72. Padhy SK, Sarkar S, Panigrahi M, Paul S (2015) Mental health effects of climate change. Indian J Occup Environ Med [Internet]. 2015 Jan 1 [cited 2022 Mar 19]; 19(1):3–7. Available from: https://pubmed.ncbi.nlm.nih.gov/26023264/

73. Clayton S, Cunsolo A, Derr V, Doherty T, Kotcher J, Silka L et al (2017) Mental health and our changing climate. Am Acad Pediatr [Internet]. 2017 [cited 2022 Mar 19]; +&é&é(March):13–28. Available from: https://www.apa.org/news/press/releases/2017/03/mental-health-climate.pdf
74. Robine JM, Cheung SLK, Le Roy S, Van Oyen H, Griffiths C, Michel JP et al (2008) Death toll exceeded 70,000 in Europe during the summer of 2003. C R Biol 331(2):171–178
75. Sung TI, Chen MJ, Lin CY, Lung SC, Su HJ (2011) Relationship between mean daily ambient temperature range and hospital admissions for schizophrenia: results from a national cohort of psychiatric inpatients. Sci Total Environ [Internet]. 2011 Dec 1 [cited 2022 Mar 19]; 410–411:41–46. Available from: https://pubmed.ncbi.nlm.nih.gov/22018962/
76. Zhao D, Zhang X, Xie M, Cheng J, Zhang H, Wang S et al (2016) Is greater temperature change within a day associated with increased emergency admissions for schizophrenia? Sci Total Environ [Internet]. 2016 Oct 1 [cited 2022 Mar 19]; 566–567:1545–1551. Available from: https://pubmed.ncbi.nlm.nih.gov/27320736/
77. Trang PM, Rocklöv J, Giang KB, Nilsson M (2016) Seasonality of hospital admissions for mental disorders in Hanoi, Vietnam. Glob Health Action [Internet]. 2016 Dec 1 [cited 2022 Mar 19]; 9(1). Available from: https://pubmed.ncbi.nlm.nih.gov/27566716/
78. Conti S, Masocco M, Meli P, Minelli G, Palummeri E, Solimini R et al (2007) General and specific mortality among the elderly during the 2003 heat wave in Genoa [Italy]. Environ Res [Internet]. 2007 Feb [cited 2022 Mar 19]; 103(2)267–274. Available from: https://pubmed.ncbi.nlm.nih.gov/16890219/
79. Linares C, Culqui D, Carmona R, Ortiz C, Díaz J (2017) Short-term association between environmental factors and hospital admissions due to dementia in Madrid. Environ Res [Internet]. 2017 Jan 1 [cited 2022 Mar 19]; 152:214–220. Available from: https://pubmed.ncbi.nlm.nih.gov/27810678/
80. Medici CR, Vestergaard CH, Hadzi-Pavlovic D, Munk-Jørgensen P, Parker G (2016) Seasonal variations in hospital admissions for mania: examining for associations with weather variables over time. J Affect Disord [Internet]. 2016 Nov 15 [cited 2022 Mar 19]; 205:81–86. Available from: https://pubmed.ncbi.nlm.nih.gov/27423064/
81. Linares C, Martinez-Martin P, Rodríguez-Blázquez C, Forjaz MJ, Carmona R, Díaz J (2016). Effect of heat waves on morbidity and mortality due to Parkinson's disease in Madrid: a time-series analysis. [cited 2022 Mar 19]. Available from: https://doi.org/10.1016/j.envint.2016.01.017
82. Culqui DR, Linares C, Ortiz C, Carmona R, Díaz J (2017) Association between environmental factors and emergency hospital admissions due to Alzheimer's disease in Madrid. Sci Total Environ [Internet]. 2017 Aug 15 [cited 2022 Mar 19]; 592:451–457. Available from: https://pubmed.ncbi.nlm.nih.gov/28342386/
83. Cervellin G, Comelli I, Lippi G, Comelli D, Rastelli G, Ossola P et al (2014) The number of emergency department visits for psychiatric emergencies is strongly associated with mean temperature and humidity variations. Results of a nine year survey. Emerg Care J [Internet]. 2014 Jul 21 [cited 2022 Mar 19]; 10(1):2271. Available from: https://www.pagepressjournals.org/index.php/ecj/article/view/2271
84. Zhao L, Zhou X, Li L, He S, Chen R (2016) Study on outdoor thermal comfort on a campus in a subtropical urban area in summer [Internet]. Sustain Cities Soc [cited 2018 Nov 14]. 22:164–170. Available from: https://linkinghub.elsevier.com/retrieve/pii/S2210670716300221
85. Trang PM, Rocklöv J, Giang KB, Kullgren G, Nilsson M (2016) Heatwaves and hospital admissions for mental disorders in Northern Vietnam. PLoS One [Internet]. 2016 May 1 [cited 2022 Mar 19]; 11(5):e0155609. Available from: https://journals.plos.org/plosone/article?id=10.1371/journal.pone.0155609
86. Kenny GP, Yardley J, Brown C, Sigal RJ, Jay O (2010) Heat stress in older individuals and patients with common chronic diseases. C Can Med Assoc J [Internet]. 2010 Jul 13 [cited 2022 Mar 19]; 182(10):1053. Available from: /pmc/articles/PMC2900329/
87. Morrison SF, Nakamura K (2011) Central neural pathways for thermoregulation. Front Biosci [Internet]. 2011 Jan 1 [cited 2022 Mar 19]; 16(1):74. Available from: /pmc/articles/PMC3051412/

88. Smith CJ, Johnson JM (2016) Responses to hyperthermia. Optimizing heat dissipation by convection and evaporation: neural control of skin blood flow and sweating in humans. Auton Neurosci Basic Clin [Internet]. [cited 2022 Mar 19]; 196:25–36. Available from: http://libres.uncg.edu/ir/asu/, https://doi.org/10.1016/j.autneu.2016.01.002. Publisher version of record available at: https://www.autonomicneuroscience.com/article/S1566-0702
89. Martin-Latry K, Goumy MP, Latry P, Gabinski C, Bégaud B, Faure I et al (2007) Psychotropic drugs use and risk of heat-related hospitalisation. Eur Psychiatry [Internet]. [cited 2022 Mar 19]; 22(6):335–338. Available from: https://pubmed.ncbi.nlm.nih.gov/17513091/
90. Kim Y, Kim H, Honda Y, Guo YL, Chen BY, Woo JM et al (2016) Suicide and Ambient temperature in East Asian countries: a time-stratified case-crossover analysis. Environ Health Perspect [Internet]. 2016 Jan 1 [cited 2022 Mar 19]; 124(1):75–80. Available from: https://pubmed.ncbi.nlm.nih.gov/26069051/
91. Helbich M (2018) Mental health and environmental exposures: an editorial. Int J Environ Res Public Health [Internet]. 2018 Oct 10 [cited 2022 Mar 19]; 15(10)] Available from: /pmc/articles/PMC6210156/
92. Fong CS, Aghamohammadi N, Ramakreshnan L, Sulaiman NM (2020) evaluation of secondary school student's outdoor thermal comfort during peak urban heating hours in Greater Kuala Lumpur. J Heal Transl Med [Internet]. [cited 2020 Aug 28]; 23:3–11. Available from: https://jummec.um.edu.my/article/view/25806
93. Aghamohammadi N, Fong CS, Idrus MHM, Ramakreshnan L, Sulaiman NM (2021) Environmental heat-related health symptoms among community in a tropical city. Sci Total Environ 15(782):146611
94. Fong CS, Aghamohammadi N, Ramakreshnan L, Sulaiman NM, Mohammadi P (2019) Holistic recommendations for future outdoor thermal comfort assessment in tropical Southeast Asia: a critical appraisal [Internet]. Sustain Cities Soc 46:101428. Available from: https://linkinghub.elsevier.com/retrieve/pii/S2210670718319577
95. Yenneti K, Ding L, Prasad D, Ulpiani G, Paolini R, Haddad S et al (2020) Urban overheating and cooling potential in Australia: an evidence-based review. Climate 2020 8:126 [Internet]. 2020 Nov 4 [cited 2022 Mar 19]; 811):126. Available from: https://www.mdpi.com/2225-1154/8/11/126/htm
96. Goto T, Mitamura T, Yoshino H, Tamura A, Inomata E (2007) Long-term field survey on thermal adaptation in office buildings in Japan. Build Environ [Internet]. 2007 Dec [cited 2018 Aug 20]; 42(12):3944–3954. Available from: http://linkinghub.elsevier.com/retrieve/pii/S0360132306003659
97. Arifwidodo SD, Chandrasiri O, Abdulharis R, Kubota T (2019) Exploring the effects of urban heat island: A case study of two cities in Thailand and Indonesia. APN Sci Bull [Internet]. 2019 Jun 3 [cited 2022 Mar 19]; 9(1). Available from: https://www.apn-gcr.org/bulletin/?p=539
98. Lan L, Lian Z, Pan L (2010) The effects of air temperature on office workers' well-being, workload and productivity-evaluated with subjective ratings. Appl Ergon [Internet]. [cited 2022 Mar 19]; 42(1):29–36. Available from: https://pubmed.ncbi.nlm.nih.gov/20478555/
99. Guo Y, Punnasiri K, Tong S (2012) Effects of temperature on mortality in Chiang Mai city, Thailand: a time series study. Environ Heal A Glob Access Sci Source [Internet]. 2012 Jul 9 [cited 2022 Mar 19]; 11(1):1–9. Available from: https://ehjournal.biomedcentral.com/articles/10.1186/1476-069X-11-36
100. Wong LP, Alias H, Aghamohammadi N, Aghazadeh S, Nik Sulaiman NM (2017) Urban heat island experience, control measures and health impact: a survey among working community in the city of Kuala Lumpur. Sustain Cities Soc [Internet]. 2017 Nov 1 [cited 2017 Nov 27]; 35:660–668. Available from: https://www.sciencedirect.com/science/article/pii/S2210670717307904
101. Wong LP, Alias H, Aghamohammadi N, Aghazadeh S, Sulaiman NMN (2018) Physical, psychological, and social health impact of temperature rise due to urban heat island phenomenon and its associated factors. Biomed Environ Sci 31:545–550

102. Ahmadi M, Dadashi Roudbari AA (2017) The identification of urban thermal islands based on an environmental approach, case study: Isfahan Province [Internet]. 2017 [cited 2022 Mar 19]. Available from: https://www.sid.ir/en/journal/ViewPaper.aspx?id=566712
103. Mirzaei PA, Haghighat F (2010) Approaches to study urban heat island—abilities and limitations. Build Environ [Internet]. 2010 Oct 1 [cited 2018 Nov 28]; 45(10):2192–2201. Available from: https://www.sciencedirect.com/science/article/pii/S0360132310001083
104. Mirzaei M, Verrelst J, Arbabi M, Shaklabadi Z, Lotfizadeh M (2020) Urban heat island monitoring and impacts on citizen's general health status in Isfahan Metropolis: a remote sensing and field survey approach. Remote Sens 2020, Vol 12, Page 1350 [Internet]. 2020 Apr 24 [cited 2022 Mar 19]; 12(8):1350. Available from: https://www.mdpi.com/2072-4292/12/8/1350/htm
105. Luan G, Yin P, Wang L, Zhou M (2019) Associations between ambient high temperatures and suicide mortality: a multi-city time-series study in China. Environ Sci Pollut Res 26(20):20377–20385
106. Chan EYY, Lam HCY, So SHW, Goggins WB, Ho JY, Liu S et al (2018) Association between Ambient temperatures and mental disorder hospitalizations in a subtropical city: a time-series study of Hong Kong Special Administrative Region. Int J Environ Res Public Health [Internet]. 2018 Apr 14 [cited 2022 Mar 19]; 15(4). Available from: https://pubmed.ncbi.nlm.nih.gov/29662001/
107. Wang X, Lavigne E, Ouellette-Kuntz H, Chen BE (2014) Acute impacts of extreme temperature exposure on emergency room admissions related to mental and behavior disorders in Toronto, Canada. J Affect Disord [Internet]. 2014 Feb [cited 2022 Mar 19]; 155(1):154–161. Available from: https://pubmed.ncbi.nlm.nih.gov/24332428/
108. Peng Z, Wang Q, Kan H, Chen R, Wang W (2017) Effects of ambient temperature on daily hospital admissions for mental disorders in Shanghai, China: a time-series analysis. Sci Total Environ [Internet]. 2017 Jul 15 [cited 2022 Mar 19]; 590–591:281–286. Available from: https://pubmed.ncbi.nlm.nih.gov/28274603/
109. Wang S, Zhang X, Xie M, Zhao D, Zhang H, Zhang Y et al (2018) Effect of increasing temperature on daily hospital admissions for schizophrenia in Hefei, China: a time-series analysis. Public Health [Internet]. 2018 Jun 1 [cited 2022 Mar 19]; 159:70–77. Available from: https://pubmed.ncbi.nlm.nih.gov/29567010/
110. Williams S, Nitschke M, Sullivan T, Tucker GR, Weinstein P, Pisaniello DL et al (2012) Heat and health in Adelaide, South Australia: assessment of heat thresholds and temperature relationships. Sci Total Environ [Internet]. 2012 Jan 1 [cited 2022 Mar 19]; 414:126–133. Available from: https://pubmed.ncbi.nlm.nih.gov/22169392/
111. Lee S, Lee H, Myung W, Kim EJ, Kim H (2018) Mental disease-related emergency admissions attributable to hot temperatures. Sci Total Environ [Internet]. 2018 Mar 1 [cited 2022 Mar 19]; 616–617:688–694. Available from: https://pubmed.ncbi.nlm.nih.gov/29126638/
112. Vida S, Durocher M, Ouarda TBMJ, Gosselin P (2012) Relationship between ambient temperature and humidity and visits to mental health emergency departments in Québec. Psychiatr Serv [Internet]. 2012 Nov 1 [cited 2022 Mar 19]; 63(11):1150–1153. Available from: https://pubmed.ncbi.nlm.nih.gov/23117515/
113. van den Bosch M, Ode Sang (2017) Urban natural environments as nature-based solutions for improved public health—a systematic review of reviews. Environ Res [Internet]. 2017 [cited 2022 Mar 19]; 158:373–384. Available from: https://pubmed.ncbi.nlm.nih.gov/28686952/
114. WHO-Regional Committee for Europe (2012) EUR/RC61/12 Action plan for implemetation of the European strategy for the prevention and control of noncommunicable diseases 2012–2016. WHO Rep [Internet]. [cited 2022 Mar 19]. Available from: http://www.euro.who.int/en/publications/abstracts/action-plan-for-implementation-of-the-european-strategy-for-the-prevention-and-control-of-noncommunicable-diseases-20122016
115. Berry HL, Bowen K, Kjellstrom T (2010). Climate change and mental health: a causal pathways framework. Int J Public Health [Internet]. 2010 [cited 2022 Mar 19]; 55(2):123–132. Available from: https://pubmed.ncbi.nlm.nih.gov/20033251/

Chapter 5
Human Adaptation to Higher Ambient Temperature

Ilaria Pigliautile, Roberta Jacoby Cureau, and Anna Laura Pisello

Abstract The on-going climate change is leading to a progressively warmer world where extreme weather events, including heat waves (HWs), are going to be more frequent and intense. While limiting the anthropogenic contribution to the changing climate (through mitigation strategies), it is fundamental to put in place effective adaptation strategies for reducing overheating risks and enhancing communities' resiliency. In fact, the increasing exposure to higher ambient temperatures has negative effects on human health and the overheating risk is associated with individuals and communities' heat exposure and vulnerabilities. Hyperlocal environmental monitoring coupled to socioeconomical assessment is needed to point out the most vulnerable areas in a city, thus providing references on where and how to implement the proper adaptation strategy. In the body of this chapter, several strategies related to human adaptation to higher ambient temperature are presented by grouping those strategies according to their scale of application and impact. The proposed spatial scales are human, building, and city scales. Human adaptation can be physiological (related to body thermoregulatory systems) or behavioural (actions taken to increase thermal tolerance). Building and city adaptation are related to greenery (e.g. green roofs and façades and green urban infrastructure) and the development and implementation of new materials. Each strategy should be evaluated in terms of its local benefits, but to guarantee sustained societal functioning in this progressively hotter climate, these adaptation actions need to be implemented now.

Keywords Hyperlocal monitoring · Cross-scale adaptation · Human-centred solutions · Exposure · Vulnerability · Overheating risk

I. Pigliautile · R. J. Cureau · A. L. Pisello (✉)
CIRIAF Interuniversity Research Centre On Pollution and Environment Mauro Felli, University of Perugia, via g. Duranti 67, 06125 Perugia, Italy
e-mail: anna.pisello@unipg.it

I. Pigliautile · A. L. Pisello
Department of Engineering, University of Perugia, via g. Duranti 67, 06125 Perugia, Italy

N. Aghamohammadi and M. Santamouris (eds.), *Urban Overheating: Heat Mitigation and the Impact on Health*, Advances in Sustainability Science and Technology,
https://doi.org/10.1007/978-981-19-4707-0_5

1 Introduction

Climate change is currently one of the main concerns worldwide, with considerable impacts also at local and regional scales [21]. One of the consequences of climate change is the significant increase in frequency, intensity, and duration of extreme weather events, including heat waves [57]. People living in urban environments are even more vulnerable to heat extreme events because of the Urban Heat Island (UHI) phenomenon and the population density [3]. Indeed, the rapid urbanization and anthropogenic activities in cities contribute to increase the ambient temperature in considerable levels [70]. It is known that thermal properties of materials used in cities, urban canyons radiative distribution, urban greenhouse effect, and diminished evaporative surfaces and turbulent transfer are some of the factors causing urban overheating [70].

The higher air temperature in cities leads to a rise in energy consumption for cooling buildings and peak electricity demand [74]. It was also stated that urban overheating has negative impacts on air quality [35]. In addition, it harmfully affects human health by increasing heat-related mortality and morbidity [75]. However, these impacts are not the same for all. Some groups, like elderly people, subjects with pre-existing medical conditions, low-income communities, ethnic minorities, and socially isolated people, have a higher heat-related vulnerability [67].

Efficient reactions to the challenges involving the growing heat-related vulnerability are needed [73], and adaptation is one of the most common strategies adopted to deal with this [38]. Cirrincione et al. [8] mention that adaptation measures are important to make cities more sustainable and resilient.

According to the Intergovernmental Panel on Climate Change (IPCC, the United Nations body for assessing the science related to climate change), adaptation is the procedure of adjustments to actual or expected climate and its effects [25]. Some examples of adaptation include the intrinsic human acclimatization process, increasing the amount of greenery areas in the cities, and the development of new materials to be used in the built environment.

The IPCC [26] describes two types of adaptation: autonomous and planned. Autonomous adaptation is a reaction to the current climate and represents an additional change in an existing system. Planned adaptation is proactive; that is, it is designed to eliminate or reduce a problem before it happens and is characterized by an adjustment or a transformation of a system.

Rohat et al. [67] showed that the efficacy of an adaptation measure depends on "(i) the aspect of heat risk it targets, (ii) the type of socioeconomic development, (iii) the level of climate change (for strategies targeting the heat hazard), (iv) the communities it targets, and (v) the location where it is implemented". They also complement that the efficiency of planned adaptation is supplemented by autonomous adaptation. Therefore, both types of strategies need to be considered.

Despite the type of adaptation, these strategies need to be implemented. Santamouris [72] highlighted the importance of adaptation at both urban and building scales to effectively face the future growth of cooling energy needs and the increase of climatic vulnerability. Then, it is essential to evaluate the benefits of adaptation measures on minimizing the negative overheating effects, such as reducing emerging demand for cooling purposes, improving thermal comfort indoors and outdoors, and decreasing heat-related health risks.

This chapter explores the role of human adaptation triggered by higher ambient temperatures, especially in urban environments. Risks for humans and vulnerabilities associated with overheating are presented, highlighting the role of environmental and socioeconomical factors. Environmental data collection methods and the most common indexes for the human heat stress assessment are briefly pointed out as fundamentals for an accurate overheating risk mapping and thus an efficient adaptation strategies' planning. Finally, examples of adaptation in different scales are reported.

2 Human Overheating Risk and Vulnerabilities in the Framework of Climate Change

The IPCC recognized that human-induced warming reached approximately 1 °C above pre-industrial levels in 2017. This value is assessed accounting for the combination of surface air and sea surface temperatures averaged over the globe and over a 30-year period. Therefore, greater warming levels have already been experienced in many regions and during specific seasons. More specifically, land regions are facing higher rise in temperature than ocean regions that shows a slower warming rate. Within land regions, cities further experience the UHI phenomenon due to their specific metabolism and morphology. The construction materials in urban environments (e.g. concrete and asphalt) usually absorb heat more than natural surfaces, and the high density of buildings hinders air circulation, which causes an increase in air temperature in comparison with the surrounding less-urbanized areas [39].

Any further rise in temperature, beyond 1 °C, would likely be less than 0.5 °C on the next three decades by reducing all the anthropogenic emissions to zero immediately. In this view, the 1.5 °C emission pathways are defined as those scenarios providing high chance in limiting global warming below 1.5 °C or returning to 1.5 °C after an overshoot (of different magnitude according to the specific scenario) by 2100, at current knowledge state. The consistent 1.5 °C pathways propose to eliminate CO_2 emissions promptly and to reduce widely other Green House Gases (GHG) emissions, which can be accomplished through transformations in energy, land, urban and infrastructure (including transportations and buildings), and industrial systems. The industry sector, for example, could reach the target reductions by electrification, hydrogen, sustainable bio-based feedstock, product substitution, and carbon capture, utilization, and storage [27]. Different 1.5 °C pathways are associated with different projected impacts at a global average warming.

As observed and projected by the scientific community, climate change is associated with a global warming trend which additionally turns into weather extremes that are foreseen to be more frequent and intense in the coming future. Among these extreme events, heat waves (HW) are particularly dangerous for people due to the rapid increase in their occurrences and intensity, their synergistic interaction with the UHI, and the high mortality rate associated [18, 46, 66]. The World Meteorological Organization (WMO) recommended HW definition as follows [83]: *a period of marked unusual hot weather over a region persisting for at least three consecutive days during the warm period of the year based on local climatological conditions, with thermal conditions recorded above given threshold.* Such a general definition underlines the impossibility of providing a unique common definition worldwide. Different thresholds for the observed temperature and different durations of these temperatures have been used in different studies according to site-specific climate conditions and accounting for physical, social, and cultural adaptation capabilities. McCarthy et al. [44] re-adapt the WMO definition of HW for different regions of the UK by setting appropriate thresholds in accordance with the UK climate. Pascal et al. [56] test multiple methodologies for defining temperature thresholds in different cities in France, further accounting for the estimated excess in mortality associated with different HW temperature thresholds. Robinson [66] highlights that the basic HW definition implicitly accounts for two fundamental aspects named as the "physiological" (related to thermoregulation capability of the human body) and "sociological" (centred on local adaptation to climate) aspects. In this framework, the main driver for the HW definition results in being its impact on human health and that is also the reason behind the need of developing effective warning systems [52, 55]. The WMO [84] has also produced guidance on the development of heat-health warning systems, HHWS, whose aim is *to alert decision-makers and the general public to impending dangerous hot weather*. The following sub-section presents what are the heat-related health risks for humans, accounting for diversified population vulnerabilities and what has been done so far in literature for their recognition and quantification.

2.1 Overheating Risks for Humans

A prolonged heat exposure has negative impact on human health as it increases morbidity (e.g. heat exhaustion and heat stroke) and mortality (death) [5, 39]. Indeed, HWs are a major cause of weather-related deaths [42]. Prolonged exposition to extremely high temperatures impacts human health due to heat-related illnesses with syndromes that vary from less severe (heat syncope), to severe and lethal (heat stroke) [78]. Extreme heat also worsens the effects of chronic diseases (including respiratory diseases, cardiovascular diseases, and kidney problems) while it has been observed that high temperatures and air pollution lead to synergistic negative health effects [11]. High temperatures negatively impact also emotional and psychological health [50]. Extreme weather events, in general, affect mental health in several ways [51]. Negative health outcomes of urban climate extremes vary according to individuals' vulnerabilities meaning that some groups are more susceptible to these risks. More specifically, the health risk due to overheating results from the combination of both individuals' exposure and specific vulnerabilities. Characteristics that affect the individual vulnerabilities include age, pre-existing medical conditions, economic conditions (including energy poverty and poor housing), level of instruction, accessibility to urban services and facilities (such as the Public Leisure Spaces, PLSs), and health education and awareness, all affecting the adaptive capacity of people to environmental stressors. Greater heat-related vulnerability is associated with the elderly and very young, people with pre-existing medical conditions, low-income communities, people without access to air conditioning, ethnic minorities, and socially isolated individuals [67]. Climate change projections also show that Central and South America, Africa, Middle East, Pakistan, India, Bangladesh, eastern China, south-east Asia, and northern Australia are the regions at greater future heat-related risks [14, 48].

Mallen et al. [42] affirm that the vulnerability depends on the exposure, sensitivity, and adaptive capacity of a person. Regarding HWs, the exposure is related to the event intensity and its spatial distribution, the sensitivity is associated with how much the increased exposure to these events will affect people physically; and the adaptive capacity is the capability to diminish or deal with the increased personal exposure. Some specialists proposed vulnerability indexes, that is, a metric that combines indicators considered to be representative of this issue used to describe the vulnerability of a population [25]. Alonso and Renard [1], for example, presented a proposal for vulnerability indexes that included physiological (response capacity of the organism) and socioeconomical aspects. These indexes are important to identify the locations and groups of people more heat-vulnerable to plan pointed interventions and strategies that will be more effective [6, 42]. Another methodology for assessing population overheating vulnerability in urban contexts is provided by Macintyre et al. [39] by combining multiple factors associated with heat-health outcomes, mainly spatial distribution of the UHI intensity, dwelling types, and socioeconomic factors. The selection of proper indexes, highlighting human overheating risks, and of effective methods for mapping such risks is going to be the benchmarks for developing decision-makers' supporting tools, driving adaptation and mitigation strategies' implementation.

Even limiting global warming to 1.5 °C, adaptation actions still are fundamental to face the effects of climate change [54]. Adaptation to the changing climate accounts for any kind of action aimed at managing the impacts of climate change by lowering both vulnerability and exposure to its detrimental effects for individuals, communities, ecosystems, or the same built environment (e.g. deterioration of cultural heritage sites, of infrastructures, etc.). In addition to limit climate change-induced risks, to adapt to changing climate also means to exploit any potential benefits. Along with the adaptation capabilities in face of the on-going climate change, mitigation actions are urgent to limit the anthropogenic contribution to global warming. Indeed, mitigation actions are all those strategies implemented at different scales (international, national, local/community, and individual level) for directly limiting GHG emissions. While to adapt to changing climate is needed since we are already facing relevant effects, we cannot indefinitely adapt, and thus, mitigation is urgent to avoid worsening of the same effects. Adaptation and mitigation actions can interact with each other resulting in synergies or trade-offs, and this is recognized in several sectors, such as agricultural, infrastructure planning and construction, and tourism [28, 30]. This is the reason why policies and measure efficacy could be significantly enhanced by accounting for these types of actions jointly. Furthermore, planned adaptation is an important strategy to decrease the negative effects of climate change on health within the most vulnerable groups [67].

A quantitative description of the achievable benefits due to the implementation of adaptation or mitigation actions allows to present and compare more or less effective strategies to policy and decision-makers. The following sub-section specifically focusses on those indexes more commonly used to quantify human health outcomes of overheating risks.

2.2 Assessment of Human Heat Stress

The most common methodologies for assessing heat stress refer to (i) single meteorological variables, (ii) simple biometeorological indexes, or (iii) outputs from numerical human heat-budget models.

The physical environmental component of the heat accounts for the interactions among solar radiation, atmospheric temperature and moisture, and wind speed and direction. These data are generally retrieved by fix weather stations that may be located in not-highly populated areas, or in proximity of airports, sites that cannot be assumed as representative of the real conditions a citizen is exposed to. The weather data collection site has to be considered during the heat load analysis and forecast when the study output is the human health outcome. Recent studies present alternative methodologies for retrieving environmental data in areas more representative of daily citizen exposure (e.g. within the Urban Canopy). Among those proposals, Pigliautile and Pisello [60] developed a miniaturized weather station that can be worn as a common bike helmet. This system allowed to map microclimate variations within different neighbourhood of the same city [59]. Different microclimatic conditions correspond to different heat exposure within the monitored urban environment. The heat exposure is a main component in overheating heat assessment: the higher the spatial resolution of the monitoring, the higher the accuracy in risk mapping. For the heat exposure assessment, it is worth noting that people use to spend the majority of their life in indoors. In this view, the assessment of indoor climate for a range of dwellings typologies is needed. Important factors to be considered in the indoor thermal load estimation include building thermal mass and orientation, as well as force or passive ventilation.

Simplified biometeorological indexes combine more than one environmental parameter for describing the heat exchange between the human body and the surroundings. Many indexes have been developed throughout years, and reasons beyond the choice of a specific index could concern the availability of required data. All those indexes aim at determining an apparent temperature that humans perceive given a specific combination of two or more environmental parameters. Some biometeorological indexes are presented along with their computation equation in Table 1.

Other indexes further account for human body characteristics. These are the outcomes of simplified heat-budget models describing heat exchanges occurring between the body and the environment. In addition to the internal heat production (i.e. metabolic rate) and personal power expenditure (i.e. mechanical power) which are related to the individual performed activities, the heat stored in the body

Table 1 Some of the most common biometeorological indexes

Index	Acronym	Equation
Heat Index	HI	$\mathrm{HI} = -42.379 + 2.04901523(T_f) + 10.14333127(\mathrm{RH})$ $- 0.22475541(T_f)(\mathrm{RH}) - (6.83783 \times 10^{-3})(T_f^2)$ $- (5.481717 \times 10^{-2})(\mathrm{RH}^2) + (1.22874 \times 10^{-3})(T_f^2)(\mathrm{RH})$ $+ (8.5282 \times 10^{-4})(T_f)(\mathrm{RH}^2) - (1.99 \times 10^{-6})(T_f^2)(\mathrm{RH}^2)$ where T_f is the air temperature [°F] and RH is the relative humidity (expressed as a whole number)
Humidex	Humidex	$\mathrm{Humidex} = (\text{air temperature}) + h$ $h = (0.5555)(e - 10.0)$ $e = 6.11 \times \exp^{\left(5417.7530 \times \left(\left(\frac{1}{273.16}\right) - \left(\frac{1}{\text{dewpoint}}\right)\right)\right)}$
Net effective temperature	NET	$\mathrm{NET} = 37 - \dfrac{37 - T}{0.68 - 0.0014(\mathrm{RH}) + 1/(1.76 + 1.4v^{0.75})}$ $- 0.29T(1 - 0.01(\mathrm{RH}))$ where T is the air temperature [°C], v is the wind speed [m/s], and RH is the relative humidity [%]
Wet-bulb globe temperature	WBGT	$\mathrm{WBGT} = 0.567 \times T_a + 0.393 \times e + 3.94$ where T_a is the air temperature (°C) and e is the water-vapour pressure (hPa)
Apparent temperature	AT	$\mathrm{AT} = T_a + 0.348 \times e - 0.70 \times \mathrm{ws}$ $+ 0.70 \times \dfrac{Q}{\mathrm{ws} + 10} - 4.25$ $\mathrm{AT} = T_a + 0.33 \times e - 0.70 \times \mathrm{ws} - 4.00$ where T_a is the dry-bulb temperature (°C), e is the water-vapour pressure (hPa), ws is the wind speed (m/s) at an elevation of 10 m, and Q is the net radiation absorbed per unit area of body surface (W/m^2)
Excess heat index	EHI	$\mathrm{EHI} = \frac{T_i + T_{i+1} + T_{i+2}}{3} - T_{95}$ where T_{95} is the 95th percentile of daily temperature (T_i) for a climate reference period, calculated using all days of the year

results from the balance of sensible and latent heat fluxes at the skin surface plus the respiratory component. Among the most used heat-budget-based indices, there are:

- Standard Effective Temperature (SET*), the equivalent temperature of an isothermal environment at 50% of humidity where the subject presents the same heat stress (i.e. skin temperature) and thermoregulatory strain (i.e. skin wettedness) as the actual environment; for the SET* computation, the human physiology model is the Pierce "two-node" model [17]. The same index is further adapted for the outdoors (OUT_SET*) [58],
- Predicted Mean Vote (PMV), perceived thermal sensation computed assuming that it depends on the physiological strain imposed to the human body by the surrounding environment;
- Physiological Equivalent Temperature (PET), the equivalent temperature of an isothermal reference environment characterized by a water-vapour pressure of 12 hPa and a light air of 0.1 m/s where the core and the skin temperature of a reference person are the same of the conditions to be assessed;
- Universal Thermal Climate Index (UTCI) [29], the equivalent temperature associated with a strain index representing the synergistic behaviour of core temperature, skin wettedness, blood flow, and sweat rate. This index resulted from the multidisciplinary cooperation of scientists from 22 countries.

3 The Scales of Human Adaptation

Despite the classification of adaptation strategies in autonomous or planned, they can also be differentiated among their different spatial scales [67]. Then, in this section, adaptation actions triggered by overheating will be explored on different scales, namely human, building, and city scales.

3.1 Human Scale

Bringing the general definition of adaptation to the human scale, it is the individual's ability to adjust to current or expected climate conditions [38]. In this context, human-scale adaptation can be physiological or behavioural. Both are autonomous adaptation types [70]; the previous associated with the energy balance of the body and its thermoregulatory systems, while the latter mainly related to individual exposure limitation.

Physiological adaptation is related to human acclimatization. According to IPCC [26], it is defined as the adjustment of functional or morphological characteristics that allow the body to adapt to new climatic conditions. These adjustments can occur once or seasonally and enable human performance maintenance under different environmental conditions [25]. It could take several years to get completely acclimatized to unfamiliar thermal environment (long-term adaptation). This process results in a lower rise in core temperature and a lower increase in heart rate at a given heat load. The main physiological responses to heat adaptation consist of "an earlier onset and higher rate of skin blood flow; an increased sweat rate with earlier onset and more dilute concentration of sweat; decreased electrolyte loss and greater resistance to dehydration; a decrease in basal metabolic rate and heart rate; a decrease in perceived exertion; and a decrease in oxygen consumption at a given activity level or metabolic rate" [10]. Physiological adaptation is one of the main reasons why same temperatures generate different health outcomes in different population. Indeed, these impacts are more related to the heat wave event duration, the local climate, and thus the acclimatization status of the population than to the magnitude of experienced temperature.

However, the physiological adaptation capability of a human body varies throughout life. Children [86] and elderly people [45] are more sensitive to heat than adults due to their different physiological characteristics. Elderly have reduced thermoregulatory responses (e.g. sweating rate, cardiovascular function, etc.) compared to young adults. Chronic diseases and the use of some medications can also limit acclimatization [23]. Moreover, it is still unknown to what extent physiological adaptation can mitigate the heat effects on health [24]. For this reason, it is impossible that human autonomous adaptation relies only on the physiological aspect.

Behavioural adaptation is related to changes in actions that people take to adapt [22]. Some examples of behavioural adaptations related to higher ambient temperature are wearing lighter clothes, reducing physical activity, going to a colder place outdoors (e.g. parks), opening windows, closing blinds or shades, and using air conditioning systems [69]. Behaviours usually complement the physiological adaptation to enlarge human thermal tolerance [23]. Furthermore, social mechanisms recognized in specific behavioural factors are the result of a long-term community adaptation. Adapting behaviours more specifically associated with cultural backgrounds of a community include the time spent in outdoors or local-specific working schedule (e.g. the siesta in southern Europe). As for the adaptation physiological capability, specific categories can be recognized as more vulnerable to overheating risk due to lack in their behavioural adaptation opportunities. For instance, either age extremes or cognitively impaired people generally rely upon others for an adequate liquid intake, and thus, these population are more prone to dehydration. Dehydration slows sweating rate and causes cardiac outputs becoming a common cause of hyperthermia and death [20].

3.2 Building Scale

Urban overheating will increase the air temperature and hours of discomfort in indoor environments, especially the non-conditioned ones [21]. Whenever active systems are at disposal, it will cause growth in cooling energy consumption of buildings and it will raise the peak in electricity demand [70]. These effects are highly correlated to individual building characteristics [43]. Therefore, the climate change adaptation process at the building scale concerns all those solutions thought to limit the consequences of overheating indoors. Their benefits are mostly evaluated in terms of thermal comfort improvements, indoor air temperature or superficial temperature reduction, and cooling loads decrease.

Many solutions of adaptation at building scale involve passive strategies to decrease the use of cooling systems while limiting the overheating risks for occupants. As the thermal comfort indoors is affected by the thermo-physical properties of the building envelope [47], building adaptation has a close relation to the development and implementation of new materials and techniques for the building envelope. Under hot conditions, the envelope needs to have low solar absorptance [79]. In this context, the application of cool materials is a good example of adaptation strategy at the building scale. These materials have properties that reduce solar energy gains and increase the longwave radiative heat dissipation leading to lower surface temperatures [34]. Cool clay tiles [62], cool roof membranes, and cool façade painting [61] are options that have already been demonstrated to contribute to reducing the indoor air temperature in the hot season. Figure 1 presents an example of the use of cool membranes and cool painting.

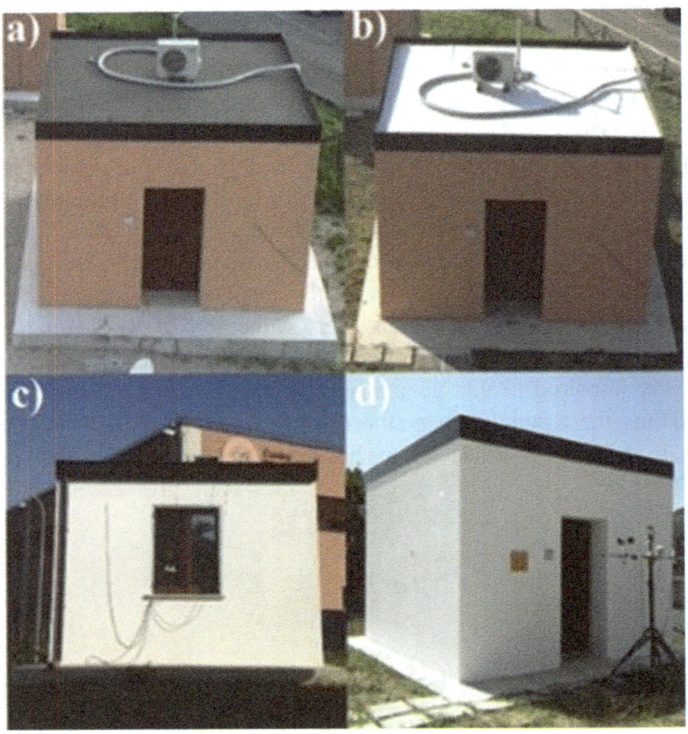

Fig. 1 Original prototype building (**a**) and implementation of a cool membrane (**b**) and cool painting (**c**, **d**) [61]

Fig. 2 Photoluminescent paint (**a**), grit (**b**), and tiles (**c**) [7]

Cool roof solutions can be improved by thermochromic roof as proposed by Fabiani et al. [16]. Thermochromic materials can alter their optical properties when the temperature reaches a determined value, and the results of the study showed that thermochromic roof is still capable of significantly reducing cooling loads while limiting winter penalties that are typical of cool finishing solutions, e.g. not raising the heating load during winter [34].

In the same context, photoluminescent materials are presented as an innovative solution to be implemented in the built environment with cooling potential. Photoluminescent materials absorb energy and re-emit it as visible light instead of long-wave radiation [32]. Figure 2 presents some examples of photoluminescent materials. Chiatti et al. [7] presented a characterization of photoluminescent paints concerning their thermal emissivity and solar reflectance and found that they have good performance as cool materials. Rosso et al. [68] presented a study of applying photoluminescent paint as an external envelope finishing layer. They demonstrated that this solution can reduce indoor temperature because of its high solar reflectance.

Additionally, incorporating different phase change materials (PCMs) in building elements is a topic that has been gaining attention in terms of improving building thermal performance. PCMs absorb and release a large amount of latent heat during their phase transition [37]. This technology is useful in cold climates, but other investigations also revealed a high potential in reducing indoor temperature and peak cooling loads [4, 64, 77].

Green roofs [8] and façades [88] are also adaptation strategies that have proven to be very effective in keeping the internal temperature lower in hot conditions. The main effect of these green elements on enhancing indoor thermal comfort is by reducing the heat transfer to the building [8].

There are also adaptation strategies related to glazing. In hot conditions, glazing solutions should have a low solar heat gain coefficient, which according to [13] is the most important factor for reducing cooling loads. Examples of glasses that have been shown to maintain lower indoor air temperatures are double glazing and low-e (low emissive) coating [76].

Another passive adaptation option is shading, which avoids or reduces the amount of incoming solar radiation and that can be achieved, among others, through building's geometrical elements, vegetation, and blinds [79]. Some studies have already shown the main benefits associated with the shading systems in reducing indoor or superficial temperature [15, 41]. Shading strategies are often related to glazing because it was already demonstrated that shading could compensate for the low thermal performance of some glasses [36].

Several investigations on these passive strategies have already shown their benefits for improving the indoor environment under overheating conditions. However, the effectiveness of human adaptation at building scale may depend on the building typology and architecture and on the local climate. For this reason, simulation tools are important in choosing the best strategy since the design stage in order by testing one or combinations of different solutions and assessing their efficiency. Furthermore, given the projections in climate change, the simulation tool allows to assess the overheating adaptation efficacy of specific solutions even under future climate conditions [21].

Finally, it is worth noting that sometimes the passive solutions are not sufficient to maintain adequate thermal comfort levels, and in these cases, cooling systems are needed and their usage still belongs to human overheating adaptation solutions. The use of more efficient heat exchangers and compressors, more advanced fan motors, and appropriate control systems are the main factors on which the energy efficiency of cooling systems depends [72]. However, more efficient systems must be combined with actions to mitigate urban overheating and the implementation of buildings' passive solutions to compensate for the expressive increase in cooling energy demand that will be faced in the next years due to climate changes [72].

3.3 City Scale

Urban overheating is provoking a significant increase in energy consumption for cooling and peak energy demand, in addition to its negative effects on people's vulnerability level and the raised heat-related mortality and morbidity [70]. To minimize these consequences, efficient urban adaptation strategies need to be developed and implemented [73].

The urban green infrastructure has already proven to be a good alternative to mitigate urban heat [21, 53, 85]. The urban green infrastructure is the combination of planned and unplanned green areas and includes, for example, native vegetation, parks, street trees, green roofs, and green walls [53]. It cools the urban environments through evapotranspiration and shading, and the climatic factors, design, and plant selection affect its cooling efficiency [63]. Regarding parks, their size influence not only their cooling potential inside the park but also the outside effects [85].

In the same context, some studies have been investigating the effects of waterbodies to minimize urban heat. The main process by which water contributes to reducing urban heat is evapotranspiration [82]. Its cooling potential is influenced by the waterbody area and shape index and the land-use characteristics in the surrounding [89]. Recently, the combining effects of green infrastructure and waterbodies are also being studied, evaluating their mutual environmental results in counteracting urban overheating [9, 19, 87] and attesting their social and health benefits [31].

From the material point of view, academia and industry are studying and developing different solutions capable of reducing the negative impacts of urban overheating [34]. They can be evaluated considering their capacity to reduce the air temperature and the surface temperature.

Doulos et al. [12] compared several materials in terms of their thermal properties. They concluded that "cold" materials are the ones with a smooth and light-coloured surface made of marble, mosaic, and stone, while "warm" materials have a rough and dark-coloured surface made of pebble, pave stone, and asphalt. Other studies also found that increasing the surfaces (pavements, roofs, and/or walls) albedo can considerably reduce the air temperature [49, 80].

In fact, many adaptation measures developed for buildings (Sect. 3.2) have presented benefits also in the outdoor environment, such as cool roofs [40], thermochromic roofs [16], and photoluminescent materials [7]. If these materials are implemented on a large scale in buildings, they could help on minimizing the overheating effects in the urban environments besides contributing to keeping the indoor temperature lower.

Some adaptation strategies are also developed for being implemented in pavements. Permeable pavements, for example, can minimize urban overheating effects through evaporative cooling [81]. Santamouris [71] showed that white pavements, which have a higher albedo than dark surfaces, have a significantly lower surface temperature than asphalt. Kousis et al. [32] proposed a solution for concrete pavements with photoluminescent aggregates (Fig. 2b) to improve their reflectivity against surface overheating. They concluded that this pavement maintained lower superficial temperatures (up to 3.3 °C) than the reference (cool concrete).

The heterogeneity of the urban environment gives rise to variations on microclimatic conditions on a micro-scale (hyperlocal), and these differences should be investigated so that personalized and site-specific interventions can be proposed for each situation [59]. Then, the adaptation process on a city scale should involve microclimatic monitoring. The most used techniques used to map urban microclimate are remote sensing using satellite data [2], fixed weather station network [65], mobile transects equipped on dedicated vehicles [33], and, more recently, wearable devices able to monitor physical parameters from a pedestrian point of view [59, 60]. Each technique has its pros and cons, but, ideally, they should be combined to get a granular perspective of the microclimate on a large scale. Regardless of the method adopted, urban microclimate monitoring is important when choosing the best adaptation strategy for each specific case.

4 Conclusions

The climate change is recognized as an on-going process, mostly influenced by anthropogenic activities. The observed global warming suggests that we must put in place effective mitigation strategies, to slow down the process by reducing the anthropogenic impact, along with adaptation strategies. Humans need to enhance their capabilities of adapting to the changing climate and to enhance their resilience to extreme weather events (including heat waves) that are going to become more frequent and intense in the coming future. In this framework, overheating risk is a major concern for the global population that mainly live in urbanized areas, places already affected by the Urban Heat Island phenomenon. A hyperlocal assessment of the overheating risk is needed for planning effective adaptation and mitigation strategies, capable of addressing site-specific criticalities and local communities' vulnerabilities.

Moving from an overview of the current and the expected overheating risks, specifically focussing on the synergistic effect of HWs and UHI in terms of human health detriment, adaptation strategies have been presented, grouped according to their scale of application and impact in three categories, as follows: human-, building-, and city-scale adaptation strategies.

Adaptation at human scale involves individuals' capabilities of adapting to the changing climate by reducing their vulnerability and limiting their exposure to heat. Two main mechanisms are identified: physiological and behavioural adjustment. The human body has a thermoregulatory system that allows to keep almost constant its own core temperature reducing the heat stress. The physiological adjustment is observed on both the short and the long term. The latter guarantees the human body performance under different environmental conditions: it could take years to get acclimatized to unfamiliar thermal environment. Nevertheless, physiological adaptation can be compromised due to ageing or health status and cannot mitigate the heat-health outcomes completely. Behavioural adjustment procedures are generally put in place to limit the individual exposure to heat, being generally adapted to the cultural background and thus also related to local climate conditions (e.g. spending the hottest hours indoors, drinking water, adjust clothing insulation, etc.).

Adaptation strategies at the building scale mainly aim at enhancing indoor thermal conditions. Active and passive solutions are both considered adaptation practices but in the view of limiting anthropogenic contribution to climate change (mitigation actions), passive solutions are preferable and the main purpose of enhancing indoor thermal comfort must be combined to the aim of minimizing building energy consumptions. The most effective solutions for the building envelope belong to the broad category of "cool materials" characterized by specific radiative and optical properties which limit the warming-up process during day-time and the re-emitting energy component during night-time (e.g. cool paintings, thermochromic, photoluminescent, PCMs, etc.). Other constructive elements that contribute to reducing building internal heat loads include green roofs and green facades. A main component of the greenery mitigation potential concerns the evapotranspiration and so the increment of the latent component within the element energy budget with respect to the sensitive one.

Green and blue infrastructures are among the most effective overheating adaptation solutions at the urban scale. Cities are particularly critical areas due to a general lack of greenery and permeable surfaces and their complex morphology that limits the wind field and thus advection cooling process. A proper design of public spaces and parks, accounting for cool and permeable pavements and green and blue elements, could provide overheating mitigation not limited to the area of intervention but extended to its surroundings. Furthermore, urban microclimate would benefit from the large-scale application of the building-scale adaptation solutions.

In a warming world, it is fundamental to enhance humans' adaptation capability, especially given the well-stated relation between urban overheating and heat-related mortality and morbidity. Coordinated adaptation strategies in all scales are urgent to cope with the climate change harmful impacts by reducing the associated risks and vulnerability. Each strategy should be evaluated in terms of its local benefits, even if the particularities and complexities of urban environments hinder this process. However, to avoid worse consequences and guarantee sustained societal functioning in this progressively hotter climate, these adaptation actions need to be implemented now.

5 Acknowledgments

The Authors acknowledgments are due to the European Union's Horizon 2020 program under grant agreement no. 765057 (SAFERUP) and no. 890345 (NRG2peers) supporting their research. The research is further supported by the European Union, through the ERC HELIOS project (grant agreement no. 101041255). Ilaria Pigliautile would like to thank the Italian funding programme Fondo Sociale Europeo REACT EU – Programma Operativo Nazionale Ricerca e Innovazione 2014-2020 (D.M. n.1062 del 10 agosto 2021) for supporting her research through the RED-TO-GREEN project.

References

1. Alonso L, Renard F (2020) A comparative study of the physiological and socio-economic vulnerabilities to heat waves of the population of the Metropolis of Lyon (France) in a climate change context. Int J Environ Res Public Health 17:1004. https://doi.org/10.3390/ijerph17031004
2. Avdan U, Jovanovska G (2016) Algorithm for automated mapping of land surface temperature using LANDSAT 8 satellite data. J Sens 2016:1480307. https://doi.org/10.1155/2016/1480307

3. Boeckmann M, Rohn I (2014) Is planned adaptation to heat reducing heat-related mortality and illness? A systematic review. BMC Public Health 14. https://doi.org/10.1186/1471-2458-14-1112
4. Bohórquez-Órdenes J, Tapia-Calderón A, Vasco DA et al (2021) Methodology to reduce cooling energy consumption by incorporating PCM envelopes: a case study of a dwelling in Chile. Build Environ 206:108373. https://doi.org/10.1016/J.BUILDENV.2021.108373
5. Campbell S, Remenyi TA, White CJ, Johnston FH (2018) Heatwave and health impact research: a global review. Health Place 53:210–218. https://doi.org/10.1016/J.HEALTHPLACE.2018.08.017
6. Cheng W, Li D, Liu Z, Brown RD (2021) Approaches for identifying heat-vulnerable populations and locations: a systematic review. Sci Total Environ 799:149417. https://doi.org/10.1016/J.SCITOTENV.2021.149417
7. Chiatti C, Fabiani C, Cotana F, Pisello AL (2021) Exploring the potential of photoluminescence for urban passive cooling and lighting applications: a new approach towards materials' optimization. Energy 231:120815. https://doi.org/10.1016/J.ENERGY.2021.120815
8. Cirrincione L, Marvuglia A, Scaccianoce G (2021) Assessing the effectiveness of green roofs in enhancing the energy and indoor comfort resilience of urban buildings to climate change: methodology proposal and application. Build Environ 205. https://doi.org/10.1016/J.BUILDENV.2021.108198
9. Cruz JA, Blanco AC, Garcia JJ et al (2021) Evaluation of the cooling effect of green and blue spaces on urban microclimate through numerical simulation: a case study of Iloilo River Esplanade, Philippines. Sustain Cities Soc 74. https://doi.org/10.1016/J.SCS.2021.103184
10. De Freitas CR, Grigorieva EA (2015) Role of acclimatization in weather-related human mortality during the transition seasons of autumn and spring in a thermally extreme mid-latitude continental climate. Int J Environ Res Public Heal 12
11. De Sario M, Katsouyanni K, Michelozzi P (2013) Climate change, extreme weather events, air pollution and respiratory health in Europe. Eur Respir J 42:826 LP–843. https://doi.org/10.1183/09031936.00074712
12. Doulos L, Santamouris M, Livada I (2004) Passive cooling of outdoor urban spaces. The role of materials. Sol Energy 77:231–249. https://doi.org/10.1016/J.SOLENER.2004.04.005
13. Dutta A, Samanta A (2018) Reducing cooling load of buildings in the tropical climate through window glazing: a model to model comparison. J Build Eng 15:318–327. https://doi.org/10.1016/J.JOBE.2017.12.005
14. Eun-Soon I, PJ S, EEA B (2021) Deadly heat waves projected in the densely populated agricultural regions of South Asia. Sci Adv 3:e1603322. https://doi.org/10.1126/sciadv.1603322
15. Evangelisti L, Guattari C, Asdrubali F, de Lieto Vollaro R (2020) An experimental investigation of the thermal performance of a building solar shading device. J Build Eng 28. https://doi.org/10.1016/J.JOBE.2019.101089
16. Fabiani C, Pisello AL, Bou-Zeid E et al (2019) Adaptive measures for mitigating urban heat islands: the potential of thermochromic materials to control roofing energy balance. Appl Energy 247:155–170. https://doi.org/10.1016/J.APENERGY.2019.04.020
17. Gagge AP, Stolwijk J, Nishi Y (1971) Effective temperature scale based on a simple model of human physiological regulatory response. ASHRAE Trans 77:247–263
18. Gasparrini A, Armstrong B (2011) The impact of heat waves on mortality. Epidemiology 22:68–73. https://doi.org/10.1097/EDE.0b013e3181fdcd99
19. Gunawardena KR, Wells MJ, Kershaw T (2017) Utilising green and bluespace to mitigate urban heat island intensity. Sci Total Environ 584–585:1040–1055. https://doi.org/10.1016/J.SCITOTENV.2017.01.158
20. Guyton AC, Hall JE (2000) Textbook of medical physiology, 10th edn. W.B. Saunders Company, Philadelphia
21. Haddad S, Barker A, Yang J et al (2020) On the potential of building adaptation measures to counterbalance the impact of climatic change in the tropics. Energy Build 229. https://doi.org/10.1016/J.ENBUILD.2020.110494

22. Hamilton M, Fischer AP, Guikema SD, Keppel-Aleks G (2018) Behavioral adaptation to climate change in wildfire-prone forests. WIREs Clim Chang 9:e553. https://doi.org/10.1002/wcc.553
23. Hanna EG, Tait PW (2015) Limitations to thermoregulation and acclimatization challenge human adaptation to global warming. Int J Environ Res Public Heal 12. https://doi.org/10.3390/ijerph120708034
24. Hondula DM, Balling RC, Vanos JK, Georgescu M (2015) Rising temperatures, human health, and the role of adaptation. Clim Chang Hum Heal 1:144–154. https://doi.org/10.1007/s40641-015-0016-4
25. IPCC (2014a) Climate Change 2014: impacts, adaptation, and vulnerability. Part B: regional aspects. In Contribution of Working Group II to the fifth assessment report of the intergovernmental panel on climate change. Cambridge University Press, Cambridge
26. IPCC (2014b) Climate Change 2014: impacts, adaptation, and vulnerability. Part A: global and sectoral aspects. In: Contribution of Working Group II to the fifth assessment report of the intergovernmental panel on climate change. Cambridge University Press, Cambridge
27. IPCC (2018) Summary for policymakers. In: Global warming of 1.5 °C. An IPCC special report on the impacts of global warming of 1.5 °C above pre-industrial levels and related global greenhouse gas emission pathways, in the context of strengthening the global response to the threat of climate change. Cambridge University Press, Cambridge, p 24
28. Jarvis A, Lau C, Cook Si et al (2011) An integrated adaptation and mitigation framework for developing agricultural research: synergies and trade-offs. Exp Agric 47:185–203. https://doi.org/10.1017/S0014479711000123
29. Jendritzky G, de Dear R, Havenith G (2012) UTCI—why another thermal index? Int J Biometeorol 56:421–428. https://doi.org/10.1007/s00484-011-0513-7
30. Kengoum F, Tiani AM (2013) Adaptation and mitigation policies in Cameroon: pathways of synergy 52
31. Kimic K, Ostrysz K (2021) Assessment of blue and green infrastructure solutions in shaping urban public spaces—spatial and functional, environmental, and social aspects. Sustainability 13
32. Kousis I, Fabiani C, Gobbi L, Pisello AL (2020) Phosphorescent-based pavements for counteracting urban overheating—a proof of concept. Sol Energy 202:540–552. https://doi.org/10.1016/J.SOLENER.2020.03.092
33. Kousis I, Pigliautile I, Pisello AL (2021) Intra-urban microclimate investigation in urban heat island through a novel mobile monitoring system. Sci Rep 11:9732. https://doi.org/10.1038/s41598-021-88344-y
34. Kousis I, Pisello AL (2020) For the mitigation of urban heat island and urban noise island: two simultaneous sides of urban discomfort. Environ Res Lett 15. https://doi.org/10.1088/1748-9326/abaa0d
35. Lai LW, Cheng WL (2009) Air quality influenced by urban heat island coupled with synoptic weather patterns. Sci Total Environ 407:2724–2733. https://doi.org/10.1016/J.SCITOTENV.2008.12.002
36. Lau AKK, Salleh E, Lim CH, Sulaiman MY (2016) Potential of shading devices and glazing configurations on cooling energy savings for high-rise office buildings in hot-humid climates: the case of Malaysia. Int J Sustain Built Environ 5:387–399. https://doi.org/10.1016/J.IJSBE.2016.04.004
37. Lei J, Yang J, Yang EH (2016) Energy performance of building envelopes integrated with phase change materials for cooling load reduction in tropical Singapore. Appl Energy 162:207–217. https://doi.org/10.1016/J.APENERGY.2015.10.031
38. Ma ATH, Wong GKL, Cheung LTO et al (2021) Climate change perception and adaptation of residents in Hong Kong. J Clean Prod 288. https://doi.org/10.1016/J.JCLEPRO.2020.125123
39. Macintyre HL, Heaviside C, Taylor J et al (2018) Assessing urban population vulnerability and environmental risks across an urban area during heatwaves—implications for health protection. Sci Total Environ 610–611:678–690. https://doi.org/10.1016/J.SCITOTENV.2017.08.062
40. Macintyre HL, Heaviside C (2019) Potential benefits of cool roofs in reducing heat-related mortality during heatwaves in a European city. Environment International 127:430–441 https://doi.org/10.1016/j.envint.2019.02.065

41. Magri Elouadjeri S, Boussoualim A, Ait Haddou H (2021) Evaluating the effect of external horizontal fixed shading devices' geometry on internal air temperature, daylighting and energy demand in hot dry climate. Case study of Ghardaïa, Algeria. Build 11
42. Mallen E, Stone B, Lanza K (2019) A methodological assessment of extreme heat mortality modeling and heat vulnerability mapping in Dallas, Texas. Urban Clim 30:100528. https://doi.org/10.1016/J.UCLIM.2019.100528
43. Mavrogianni A, Wilkinson P, Davies M et al (2012) Building characteristics as determinants of propensity to high indoor summer temperatures in London dwellings. Build Environ 55:117–130. https://doi.org/10.1016/J.BUILDENV.2011.12.003
44. McCarthy M, Armstrong L, Armstrong N (2019) A new heatwave definition for the UK. Weather 74:382–387. https://doi.org/10.1002/wea.3629
45. Meade RD, Akerman AP, Notley SR et al (2020) Physiological factors characterizing heat-vulnerable older adults: a narrative review. Environ Int 144. https://doi.org/10.1016/J.ENVINT.2020.105909
46. Meehl GA, Tebaldi C (2004) More intense, more frequent, and longer lasting heat waves in the 21st century. Science (80–) 305:994–997. https://doi.org/10.1126/science.1098704
47. Mirrahimi S, Mohamed MF, Haw LC et al (2016) The effect of building envelope on the thermal comfort and energy saving for high-rise buildings in hot–humid climate. Renew Sustain Energy Rev 53:1508–1519. https://doi.org/10.1016/J.RSER.2015.09.055
48. Mora C, Dousset B, Caldwell IR et al (2017) Global risk of deadly heat. Nat Clim Chang 7:501–506. https://doi.org/10.1038/nclimate3322
49. Morini E, Touchaei AG, Castellani B et al (2016) The impact of Albedo increase to mitigate the urban heat Island in Terni (Italy) using the WRF model. Sustain 8
50. Mullins JT, White C (2019) Temperature and mental health: evidence from the spectrum of mental health outcomes. J Health Econ 68:102240. https://doi.org/10.1016/J.JHEALECO.2019.102240
51. Nilsson M (2020) Climate change and its impact on mental health
52. Nori-Sarma A, Benmarhnia T, Rajiva A et al (2019) Advancing our understanding of heat wave criteria and associated health impacts to improve heat wave alerts in developing country settings. Int J Environ Res Public Heal 16
53. Norton BA, Coutts AM, Livesley SJ et al (2015) Planning for cooler cities: a framework to prioritise green infrastructure to mitigate high temperatures in urban landscapes. Landsc Urban Plan 134:127–138. https://doi.org/10.1016/J.LANDURBPLAN.2014.10.018
54. Papadimitriou L, Holman IP, Dunford R, Harrison PA (2019) Trade-offs are unavoidable in multi-objective adaptation even in a post-Paris Agreement world. Sci Total Environ 696:134027. https://doi.org/10.1016/J.SCITOTENV.2019.134027
55. Park J, Kim J (2018) Defining heatwave thresholds using an inductive machine learning approach. PLoS ONE 13:e0206872–e0206872. https://doi.org/10.1371/journal.pone.0206872
56. Pascal M, Wagner V, Le Tertre A et al (2013) Definition of temperature thresholds: the example of the French heat wave warning system. Int J Biometeorol 57:21–29. https://doi.org/10.1007/s00484-012-0530-1
57. Perkins SE (2015) A review on the scientific understanding of heatwaves—their measurement, driving mechanisms, and changes at the global scale. Atmos Res 164–165:242–267. https://doi.org/10.1016/J.ATMOSRES.2015.05.014
58. Pickup J, de Dear R (2000) An outdoor thermal comfort index (Out_SET*)—part 1—the model and its assumptions. In: de Dear R, Kalma J, Oke T, Auliciems A (eds) Biometeorology and urban climatology at the turn of the millennium. World Meteorological Organization, Geneva, pp 279–283
59. Pigliautile I, D'Eramo S, Pisello AL (2021) Intra-urban microclimate mapping for citizens' wellbeing: novel wearable sensing techniques and automatized data-processing. J Clean Prod 279. https://doi.org/10.1016/J.JCLEPRO.2020.123748
60. Pigliautile I, Pisello AL (2018) A new wearable monitoring system for investigating pedestrians' environmental conditions: development of the experimental tool and start-up findings. Sci Total Environ 630:690–706. https://doi.org/10.1016/j.scitotenv.2018.02.208

61. Pisello AL, Castaldo VL, Piselli C et al (2017) Thermal performance of coupled cool roof and cool façade: experimental monitoring and analytical optimization procedure. Energy Build 157:35–52. https://doi.org/10.1016/J.ENBUILD.2017.04.054
62. Pisello AL, Cotana F (2014) The thermal effect of an innovative cool roof on residential buildings in Italy: results from two years of continuous monitoring. Energy Build 69:154–164. https://doi.org/10.1016/J.ENBUILD.2013.10.031
63. Priya UK, Senthil R (2021) A review of the impact of the green landscape interventions on the urban microclimate of tropical areas. Build Environ 205. https://doi.org/10.1016/J.BUILDENV.2021.108190
64. Ramakrishnan S, Wang X, Sanjayan J, Wilson J (2017) Thermal performance of buildings integrated with phase change materials to reduce heat stress risks during extreme heatwave events. Appl Energy 194:410–421. https://doi.org/10.1016/J.APENERGY.2016.04.084
65. Rizvi SH, Alam K, Iqbal MJ (2019) Spatio-temporal variations in urban heat island and its interaction with heat wave. J Atmos Solar-Terr Phys 185:50–57. https://doi.org/10.1016/J.JASTP.2019.02.001
66. Robinson PJ (2001) On the definition of a heat wave. J Appl Meteorol 40:762–775. https://doi.org/10.1175/1520-0450(2001)040%3c0762:OTDOAH%3e2.0.CO;2
67. Rohat G, Wilhelmi O, Flacke J et al (2021) Assessing urban heat-related adaptation strategies under multiple futures for a major U.S. city. Clim Change 164. https://doi.org/10.1007/s10584-021-02990-9
68. Rosso F, Fabiani C, Chiatti C, Pisello AL (2019) Cool, photoluminescent paints towards energy consumption reductions in the built environment. J Phys Conf Ser 1343:012198. https://doi.org/10.1088/1742-6596/1343/1/012198
69. Sampson NR, Gronlund CJ, Buxton MA et al (2013) Staying cool in a changing climate: reaching vulnerable populations during heat events. Glob Environ Chang 23:475–484. https://doi.org/10.1016/J.GLOENVCHA.2012.12.011
70. Santamouris M (2020) Recent progress on urban overheating and heat island research. Integrated assessment of the energy, environmental, vulnerability and health impact. Synergies with the global climate change. Energy Build 207:109482. https://doi.org/10.1016/J.ENBUILD.2019.109482
71. Santamouris M (2013) Energy and climate in the urban built environment. Routledge, London
72. Santamouris M (2016) Cooling the buildings—past, present and future. Energy Build 128:617–638. https://doi.org/10.1016/J.ENBUILD.2016.07.034
73. Santamouris M, Osmond P (2020) Increasing green infrastructure in cities: impact on ambient temperature, air quality and heat-related mortality and morbidity. Buildings 10. https://doi.org/10.3390/buildings10120233
74. Santamouris M, Cartalis C, Synnefa A, Kolokotsa D (2015) On the impact of urban heat island and global warming on the power demand and electricity consumption of buildings—A review. Energy and Buildings 98:119–124 https://doi.org/10.1016/j.enbuild.2014.09.052
75. Schinasi LH, Benmarhnia T, De Roos AJ (2018) Modification of the association between high ambient temperature and health by urban microclimate indicators: a systematic review and meta-analysis. Environ Res 161:168–180. https://doi.org/10.1016/J.ENVRES.2017.11.004
76. Somasundaram S, Chong A, Wei Z, Thangavelu SR (2020) Energy saving potential of low-e coating based retrofit double glazing for tropical climate. Energy Build 206. https://doi.org/10.1016/j.enbuild.2019.109570
77. Souayfane F, Fardoun F, Biwole PH (2016) Phase change materials (PCM) for cooling applications in buildings: a review. Energy Build 129:396–431. https://doi.org/10.1016/J.ENBUILD.2016.04.006
78. Székely M, Carletto L, Garami A (2015) The pathophysiology of heat exposure. Temp (Austin, Tex) 2:452. https://doi.org/10.1080/23328940.2015.1051207
79. Triana MA, De Vecchi R, Lamberts R (2019) Building design for hot and humid climate in a changing world. In: Enteria N, Awbi H, Santamouris M (eds) Building in hot and humid regions. Springer, pp 59–73

80. Tsoka S, Theodosiou T, Tsikaloudaki K, Flourentzou F (2018) Modeling the performance of cool pavements and the effect of their aging on outdoor surface and air temperatures. Sustain Cities Soc 42:276–288. https://doi.org/10.1016/J.SCS.2018.07.016
81. Wang J, Meng Q, Tan K et al (2018) Experimental investigation on the influence of evaporative cooling of permeable pavements on outdoor thermal environment. Build Environ 140:184–193. https://doi.org/10.1016/J.BUILDENV.2018.05.033
82. Wang Y, Ouyang W (2021) Investigating the heterogeneity of water cooling effect for cooler cities. Sustain Cities Soc 75. https://doi.org/10.1016/J.SCS.2021.103281
83. WMO (2015a) Guidelines on the definition and monitoring of extreme weather and climate events
84. WMO (2015b) Heatwaves and health: guidance on warning-system development. Geneva
85. Wong NH, Tan CL, Kolokotsa DD, Takebayashi H (2021) Greenery as a mitigation and adaptation strategy to urban heat. Nat Rev Earth Environ 2:166–181. https://doi.org/10.1038/s43017-020-00129-5
86. Xu Z, Etzel RA, Su H et al (2012) Impact of ambient temperature on children's health: a systematic review. Environ Res 117:120–131. https://doi.org/10.1016/J.ENVRES.2012.07.002
87. Yu Z, Yang G, Zuo S et al (2020) Critical review on the cooling effect of urban blue-green space: a threshold-size perspective. Urban Urban Green 49. https://doi.org/10.1016/J.UFUG.2020.126630
88. Zhang L, Deng Z, Liang L et al (2019) Thermal behavior of a vertical green facade and its impact on the indoor and outdoor thermal environment. Energy Build 204. https://doi.org/10.1016/J.ENBUILD.2019.109502
89. Zhou X, Zhang S, Zhu D (2021) Impact of urban water networks on microclimate and PM2.5 distribution in downtown areas: a case study of Wuhan. Build Environ 203. https://doi.org/10.1016/J.BUILDENV.2021.108073

Chapter 6
Models and Forecasts on the Future Heat-Related Mortality Under Climate Change

John A. Paravantis

Abstract This chapter summarizes a review of selected literature on models and forecasts of heat-related mortality in the context of climate change. Climate change is a global health threat, with urbanization, the urban heat island effect, and heatwaves exacerbating the problem. Heatwaves are consecutive days with unusually high temperatures above certain thresholds. Temperature and humidity measures are combined to formulate various heatwave definitions. The apparent and wet bulb temperature as well as the Humidex index have been used. Heat-related health consequences include increased morbidity and mortality, particularly among people over the age of 65. There is a typically nonlinear relationship between temperature and mortality. Population acclimatization may help to mitigate some of the effects of heat mortality. Missing data are a common concern in heat mortality research with shorter or longer time series and panel data, and they are occasionally hindcasted. The majority of studies have a geographic focus and make use of weather and socioeconomic data. Atmospheric and general circulation models are frequently used, with meteorological simulations and synoptic classification downscaled to regional scale on occasion. Additionally, multiple models are averaged and Monte Carlo simulations are used. Agent-based models have been used in individual-based approaches. Descriptive, case study, time series, spatial, and synoptic weather analyses have been used, with confounders such as trends, seasonal cycles, humidity, and air pollution considered. In the published literature, *t*-tests, smoothing spline regressions, moving averages, trend analyses, multiple linear regression, quantile regression, distributed lags, autoregressive models, logistic and Poisson regression, Poisson and negative binomial count regression, Cox proportional hazard models, meta-regressions, generalized estimating equations, and cluster analysis have been used. Effect modifiers, dummy variables, and variable normalization are frequently employed. To analyze and develop heat mortality projections, multiple futures with shared socioeconomic pathways and representative concentration pathways have been used. A gene expression algorithm, random forest regression, and the Boruta method for automatic identification have been used, along with consideration of artificial neural networks. Quantitative and qualitative data are used to inform scenarios. As global warming

J. A. Paravantis (✉)
University of Piraeus, Piraeus, Greece
e-mail: jparav@unipi.gr

N. Aghamohammadi and M. Santamouris (eds.), *Urban Overheating: Heat Mitigation and the Impact on Health*, Advances in Sustainability Science and Technology,
https://doi.org/10.1007/978-981-19-4707-0_6

worsens, much heat-related mortality will be concentrated in capital cities with larger heat islands. Building information may not be required, but air-conditioning does influence the outcome. Social isolation is a contributing factor, and long-term inpatients in social care facilities are particularly vulnerable. Green roofs and vegetation are examples of published nature-based solutions. The impact of deforestation is a contributing factor. Uncertainties include regional disparities and future demographic shifts. Because research has shown that mitigation efforts are effective, it has been proposed to develop city-specific heatwave definitions and targeted public health interventions.

Keywords Heat mortality · Heatwaves · Simulation · Modeling · Scenarios

1 Introduction

Climate change may be the greatest global health threat of the twenty-first century [26, 30, 33]. Over the last few decades, there has been a significant increase in temperature. Karwat and Franzke [15] found statistically significant increases of all percentiles of the maximum temperature in 46 cities, and all percentiles of the minimum temperature in 44 cities (out of a total of 48 cities). This underscores that heat stress and the health risks associated with it have increased significantly globally over the last few decades.

Global warming has significantly altered the environment in recent decades, increasing the likelihood and severity of meteorological and climatic disasters [31]. Correspondingly, there is growing concern about the effect of rising temperatures on human health [36]. Heatwave duration and intensity in particular are projected to grow as annual temperatures rise. A review of literature projections has revealed that climate change would result in a significant increase in heat mortality [14].

This chapter reviews selected literature on models and forecasts of heat-related mortality under climate change. Section 2 (Heat and mortality) discusses heatwaves and the connection of heat with mortality. Section 3 (models and forecasts) reviews data, models, and forecasts of heat mortality (in connection with climate change). Section 4 (Discussion) discusses selected issues presented in the previous sections. Finally, Sect. 5 (Conclusions) presents conclusions and some recommendations.

2 Heat and Mortality

Heat is a known public health hazard [37], and heat mortality is predicted to rise as a result of anthropogenic climate change [15], with the elderly being more susceptible to heat stress.

City residents are exposed to additional heat stress, as the urban heat island (UHI) phenomenon causes cities to be warmer than their rural surroundings. Paravantis et al. [30] stressed that the adverse effects of global warming are exacerbated by the UHI effect, affecting residents in ways that go beyond mild thermal discomfort [25]. Intensively developed urban areas are characterized by lower albedo, lower vegetative cover, lower moisture, and increased anthropogenic heating, differing from non-urban areas [32].

Heat-related health risks for urban residents are expected to worsen in the twenty-first century, given the impact of increased urbanization on the intensity of the UHI and the fact that heatwaves are expected to become more frequent as a result of climate change [30]. The long timescale over which climate change will occur; the diversity of potential health impacts; and interactions between demographic changes, socioeconomic development, technological innovation, and environmental drivers are all factors that constitute significant challenges [14].

Karwat and Franzke [15] reviewed the literature and concluded that warmer areas in the northern hemisphere are shifting toward even warmer conditions. For example, the climate in London in 2050 is likely to be similar to that of Barcelona (Spain), whereas Madrid's (Spain) future climate may be more similar to that of Marrakesh (Morocco). The concern is that, while Barcelona has the infrastructure to deal with extreme heat, London may not. Increases in the maximum, minimum, and bias-corrected temperature values were found in all locations, implying an increase in the risk of heat stress.

Climate change and exacerbating urban heat have made heat mortality a growing public health concern. Because of the UHI effect, poor urban design and planning, and the interaction of heat and air pollution, urban areas are particularly vulnerable to heat [14]. In many cities, heatwaves are the leading cause of weather-related fatalities. The UHI is amplified during heatwaves, especially during the night.

Heat-related stress is most commonly manifested as cardiovascular and respiratory disorders, as well as heatstroke [31]. Paravantis et al. [30] noted that heatwaves amplify heat-related mortality particularly among people over the age of 65 [7, 8, 11]. Climate modeling predicts that future heatwaves will be more frequent, more intense, and longer [14, 37]. This will result in a significant increase in the risks of heat stress, cardiovascular, pulmonary, and renal disorders [15]. As the world population ages and urbanization accelerates, heatwaves are likely to pose a greater threat to human health [37].

Excess heat exposure has been shown to decrease productivity as well as increase all-cause mortality, heat-related illnesses, and occupational injuries [36]. Extreme temperature frequency and recurrence are proxy indicators of thermal stress [15]. In addition to the maximum and minimum temperatures, the wet bulb temperature is an important parameter for defining climate risk in terms of heat stress, illness, and death [15]. There is evidence that, without artificial cooling, most physical labor becomes unsafe at wet bulb temperatures above 32 °C. The theoretical limit of human survival in the shade for more than a few hours is when the wet bulb temperature exceeds 35 °C, which is likely the upper limit of human tolerance to heat stress. Also, if the temperature rises above 20 °C, the human body is less able to relax at night.

2.1 Heatwaves

2.1.1 Definition

The UK Met Office and the World Meteorological Organization have defined a heatwave as a period of "*marked unusual hot weather (max, min, and daily average) over a region persisting at least two consecutive days during the hot period of the year based on local climatological conditions, with thermal conditions recorded above given thresholds*" [10]. As an example, in a study near Loughborough (English Midlands) there was a two-day heatwave during which the external dry-bulb temperature exceeded 35 °C [10].

Heatwaves are defined differently in different countries [31]. In general, heatwaves are thought to arise when a region's normal climate pattern is disrupted significantly. Different definitions of a heatwave combine the intensity and duration of heat [37]. Prior studies have defined a heatwave as a heat episode lasting one to seven days. The definition of a heatwave used by Dubey et al. [9] was adapted from the WMO Expert Team on Climate Change Detection Indices and defines a heatwave as the number of days above the 90th percentile value for consecutive six days or more for the base period (1958–2005).

According to Paravantis et al. [30], heatwaves are typically defined as periods of consecutive days with maximum temperatures above the 90th percentile of the 1961–1990 (being considered a base) period. Heatwaves are frequently defined as at least two-day high-temperature periods [37] that are typically accompanied by high humidity [2].

Zhang et al. [38] defined heatwaves as periods of a minimum duration of three days with a daily maximum temperature greater than the 92.5th percentile of the temperature of the warm season (without clarifying the exact temperature measure). Kollanus et al. [18] defined heatwaves as periods with a daily average temperature exceeding the 90th percentile of that from May to August in 2000 to 2014 for at least five days. Those authors analyzed risk separately for heatwaves of short (four to five days) and long (at least 10 days) duration.

Heatwaves in South Korea are usually generated by a considerable increase in temperature during periods of stagnant air pressure, which are widespread throughout the country [31]. A heatwave warning is issued in South Korea when the daily maximum temperature is expected to exceed 33 °C for at least two consecutive days.

It is difficult to agree on a definition for a heatwave, also because each region's population adapts and acclimates differently [37]. Those authors found heatwave definitions to differ in three respects: temperature metric, intensity, and duration. Developing a local or regional heatwave definition is a simpler task than defining heatwaves globally. Such efforts have been undertaken in Australia and Chinese megacities.

In the literature, the maximum, mean, apparent temperatures (or heat index) have been used to define a heatwave [37]. Previous studies have used a relative or an absolute threshold to define a heatwave. Some studies have combined the maximum with the minimum (apparent) temperature or the maximum with the mean temperature, in which temperature indicator best predicts mortality is still debatable. It has been suggested that the criteria for selecting the best temperature measure should be based on practical considerations such as missing data values.

Abadie and Polanco-Martínez [1] considered heatwaves to have the following stochastic characteristics: number of heatwaves in a given year; duration of heatwaves in days; and intensity, measured by their exceedance of a critical temperature of 38 °C (for the city of Madrid). These three characteristics (number, duration, and intensity) were considered to have expected values that change over time.

Lo et al. [21] used extreme heatwaves as a reference for the development of an automatic event-tracking method for global heatwaves that can help the search for compound events with a quantitative approach. The proposed method searched long-term historical datasets for extremely high-temperature areas automatically, instead of relying on indexes and local thresholds. The weather systems and climatic conditions that corresponded to the growing, maturation, and decaying stages of events were identified. Two distinct heatwave events in Scandinavia and France in 2003 were used for the validation of this method, while all historical global heatwave events from 1979 to 2018 were tracked.

2.1.2 Health Impacts

The health impacts of heatwaves have been researched extensively, with most studies examining mortality rather than morbidity. It has become obvious that heatwaves are especially dangerous for the elderly and those who live in cities [37]. It has also been proposed that the additional effect of heatwave mortality increases with heatwave duration [37].

The risk of dying in a heatwave increases with age; socio-economic disadvantages; social isolation; geographical remoteness; the presence of disabilities (physical or mental); some prescribed medications; and the absence or non-use of air conditioning or other building heat protection [6]. The pattern of damage caused by heatwaves (even in the same weather) might vary based on factors like population vulnerability [31]. Particularly vulnerable to heatwaves are elderly people with limited mobility, who have limited means to seek cooling and create cooler indoor environments [23]. However, there is a need for more quantitative information on factors affecting sensitivity to the adverse health effects, particularly in countries with cool summer temperatures [18].

Coates et al. [6] conducted a statistical analysis of fatalities associated with heatwaves in Australia from 2001 to 2018. Almost two-thirds (63%) of heatwave fatalities occurred during two severe heatwave years (2009 and 2014), while the record was dominated by male fatalities. At least, 473 heat-related deaths were reported, 354 of which occurred during heatwave conditions; of these, 244 were within buildings. Most indoor heatwave fatalities occurred in older housing stock. Those authors suggested that heatwaves pose a greater threat to Australian mortality than any other natural hazard. It is further suggested that those more likely to die in heatwave events are the elderly, young children, people with existing medical conditions, the isolated, and people who experience social and financial disadvantages.

2.2 Temperature and Mortality

Exposure to extreme heat has been linked to morbidity and mortality [14]. The majority of excess mortality during heatwaves is due to cardiovascular and respiratory issues [15]. Heat stroke, fatigue, cramps, fainting, and edema are also among the illnesses that may be caused by the heat [31].

Under the dual effects of climate change and urban heat islands (UHI), non-optimum temperature-related mortality burdens have been characterized as complex and uncertain [39]. Honda et al. [13] found a minor proportion of excess deaths related to heat, but predicted that climate change will increase the burden of heat mortality. To calculate the net impact of global warming on fatalities, both the cold and heat-related effects on fatalities must be compared. At the time of writing, Huang et al. [14] noted that a precise estimate of years of life lost due to high temperatures was still unknown. More research is also required to determine how the proportion of heat-related and cold-related mortality will change as climate and socioeconomic conditions change. Park et al. [31] have argued that estimating risk using morbidity (rather than just mortality) may be more effective.

When the temperature rises above a certain threshold, the risk of disease and death rises dramatically both on the day of exposure and for a few days afterward [37]. To determine the heatwave threshold temperature, researchers must first identify the temperature inflection point (which may be equal to an uncommon percentile) at which mortality begins to rise significantly [37]. Exposure of the population to high ambient temperatures increases mortality [34] rapidly above the threshold temperature value, which appears to be higher in Mediterranean countries (29.4 °C) than northern and continental European areas (23.3 °C) [3].

Attempts have been published to define a minimum mortality temperature; introduce a climate damage function; establish climate risk indexes; build disaster databases; create interactive risk explorers; and publish climate signal maps [15]. Fatigue is likely to appear at around 26 °C after prolonged exposure to heat or during physical activity. Even low humidity values can result in dangerously high wet bulb temperatures as the air temperature rises. When the temperature is 36 °C (with a relative humidity of 30%), the risk of heat exhaustion or heat stroke is 10% higher than when the temperature is 26 °C.

Karwat and Franzke [15] defined critical days as days with a wet bulb temperature greater than 26 °C (in order to quantify heat stress risk). From 1950 to 2005, all cities studied by those authors experienced an increase in the number of critical days, with Spain, France, and Germany having the greatest increases. Asian regions were shown to be more vulnerable to climate change, with heat-related mortality accounting for 0.6% of all deaths in some locations, compared to 0.1–0.4% in some high-income countries [13]. Since different IPCC regional climate model (RCM) scenarios produce different results, the precise risk increase is considered unknown [15]. The majority of physical activities are expected to become unsafe under the RCM8.5 emission scenario. By 2050, the risk of heat mortality is projected to reach dangerous levels.

The relationship between heat and mortality has been investigated in a number of European cities and countries, including the estimation of a heat wave vulnerability index for London (UK) [15]. Those authors linked temperature parameters to mortality risk using monthly mortality data from selected European cities on a nationwide scale. Nastos and Matzarakis [29] studied the relationship between daily mortality and hot and cold weather in Athens, Greece (1992–2001). Daily maximum and minimum temperatures as well as the daily values of two indexes (of physiological thermal stress) were considered. Significant correlations were found between daily temperature, thermal indexes, and mortality, with a one-day lag also having a significant effect on hot weather mortality.

Heat exposure is commonly measured using maximum and mean temperatures. According to Huang et al. [14], it is important to consider which temperature measure is the best predictor of mortality. In the literature, composite indexes have been used to investigate the combined effects of ambient temperature, humidity, and other meteorological variables. Martens [22] used meta-analysis to derive a pooled estimate of the effect of temperature on mortality in an earlier study. Huang et al. [14] discovered that there is little agreement among the various approaches used to model the temperature-mortality relationship, with greater accuracy on days with the highest excess mortality. Because of socioeconomic development, demographic change, and population acclimatization, temperature-mortality relationships in the same city can be very different between the 1960s and the 2000s.

A significant factor of climate risk is the composite effect of maximum and minimum temperature and relative humidity [15]. Those authors used the maximum and minimum temperatures as well as the relative humidity to calculate the wet bulb temperature (which is related to both comfort and health). Humidity values were not corrected for bias in air temperatures. The wet bulb temperature is an important proxy for the human body's ability to withstand heat, and it is especially useful during prolonged heatwaves. High wet bulb temperatures are associated with discomfort and the risk of the body succumbing to heat death. Karwat and Franzke [15] looked into heat stress by examining an empirically modeled wet bulb temperature in historical climate simulations as well as future climate projections.

Honda et al. [13] attempted to calculate the relationship between ambient temperature and mortality in order to estimate the implications of climate change in the future. Their goal was to find the optimum temperature at which the risk of death is the lowest, so that excess mortality could be calculated. The optimum temperature has been observed to be higher for warmer areas. The applicability of the optimum temperature estimation method to different regions of the world was considered. It was found that the optimum temperature may be estimated as the 80–85th percentile of the daily maximum temperature (although other researchers have used the daily mean temperature instead). Since the range of the daily maximum temperature is wide in northern and narrow in southern regions, the temperature rise in tropical locations is expected to be less pronounced than in temperate and high latitude areas.

Many studies have quantified the effects of cold and heat separately, assuming a linear response below and above a threshold temperature [14, 30]. The heat threshold is the temperature at which the harmful effects of heat begin to occur, whereas the heat slope is the magnitude of this effect. Both the heat threshold and the heat slope have shown significant geographic variation. Heat thresholds are higher in warmer climates, indicating some acclimatization.

Temperature and mortality have a V-shaped relationship in many regions [13, 30], although the optimal temperature indicators may differ between cities and age groups [37]. The V-shaped association was absent in three southern US cities (Dallas/Fort Worth, Houston, and San Diego). It was assumed that air conditioning was widely available in those sweltering cities, and that residents had become accustomed to the heat.

With the exception of tropical places, where the annual temperature variation is very small, it has been argued that setting the optimum temperature at about the 84th percentile of the daily maximum temperature is generalizable internationally [13]. The 84th percentile was found to be about equal to the summer mean temperature. Almost every city with a V-shaped association had an optimum temperature that was similar to the summer average temperature in that city.

According to Baccini et al. [3], the exposure–response curves that depict the relationship between maximum and minimum apparent temperature and log mortality are J-shaped (or U-shaped), with the lowest mortality occurring at a range of moderate temperature values and gradually increasing as temperatures increase or decrease. The maximum apparent temperature had a slightly stronger effect on Mediterranean cities than on northern continental cities. Heat effects were lower in northern continental cities, as measured by percent changes in mortality associated with a 1 °C increase in maximum apparent temperature above a city-specific threshold.

Xu et al. [37] examined the epidemiological evidence regarding the effect of heatwaves on mortality, compared various published heatwave definitions, and pooled the mortality risks associated with different heatwave definitions. Those authors found heatwave mortality to be associated with higher temperature thresholds in 108 US cities, London, Milan, Budapest, and Brisbane (Australia). Heatwave intensity had no effect on heatwave mortality in communities in the Guangdong province (China) or Houston (USA).

Heterogeneity is likely to exist in the heatwave impacts on mortality across different regions [37]. The duration of extreme heat exposure which increases mortality significantly varies across regions. Despite community heterogeneity, heatwave intensity, rather than duration, is likely to be the primary cause of heatwave mortality. Studies comparing the health effects of heatwaves using various definitions have revealed that even a small change in the definition appears to have an effect on the estimated health impacts. As a result, caution should be exercised when comparing the results of studies using various heatwave definitions.

Xu et al. [37] also found disparities in heatwave mortality across studies that used six different heatwave definitions, but the majority of studies found a significant increase in mortality during heatwaves. No single temperature indicator outperformed others in predicting heatwave mortality. A very high threshold temperature may understate heatwave mortality, while a too low threshold temperature may cause premature heatwave early warning, wasting health resources. It has been suggested that local heatwave thresholds should be established, based on the relationship between heatwaves and local population health.

Kollanus et al. [18] observed an increase in mortality for those aged 65–74 years, for cerebrovascular diseases, chronic lower respiratory diseases, and mental and behavioral disorders, with a borderline statistically significant involvement of diseases of the nervous system. The effects were consistently higher for longer heatwaves. In terms of harvesting (short-term mortality displacement), heat exposure only hastens some heat-related deaths in frail populations.

As far as other factors are concerned, Wolff et al. [36] were among the first to model the effects of deforestation and climate change on all-cause mortality and unsafe working conditions from increases in heat exposure in populations in low-latitude, industrializing countries. Spatially explicit data on tree cover change, land surface temperatures, and population distribution were used. It was discovered that heat exposure from deforestation increased all-cause mortality and decreased safe work hours. Heat-related mortality is especially dangerous for the elderly, very young children, and those suffering from chronic diseases.

2.3 Adaptation

Population adaptation (or acclimatization) is another issue pertinent to heat mortality that is often not taken into consideration (e.g., [13]).

Literature sources have predicted that global death rates due to heat would increase by 100–1000% in the 2090s compared to the 1990s, with China, India, and Europe bearing a disproportionate share of the burden due to their high population density [14]. Population acclimatization could reduce this heat mortality by 20–25% (in Los Angeles and New York City) and up to 40% in south-central Canada. Population acclimatization in Lisbon (Portugal) could reduce heat-related deaths by 15% in the current decade (2020s), rising to 40% in the 2050s. Acclimatization can be accounted for by projecting current temperature-mortality relationships into the future. It has been proposed that complete acclimatization to 1 °C takes three decades. When projecting future heat mortality, those authors suggested performing a sensitivity analysis using various approaches to modeling population acclimatization.

While complete acclimatization is unlikely, literature estimates show that acclimatization to an extra 2 °C in the maximum temperature would reduce heat mortality by half (50%) [14]. Abadie and Polanco-Martínez [1] projected a substantial increase in the expected duration of heatwaves, especially conspicuous from the 2040s onwards. A sensitivity adaptation analysis was performed, researching the mortality caused by a mean excess of 3.6 °C over the 38 °C temperature threshold under RCP8.5 for 2100 for Madrid (Spain). Mortality projections increased sharply after 2070 under the RCP8.5 scenario. It was assumed that acclimatization may absorb the impact of half the expected excess temperature, in which case the effects of heatwave mortality may be about 42% lower.

There is little epidemiological evidence to support acclimatization [14], so the literature has suggested that people may adapt to extreme heat by increasing their use of air conditioning, changing their behavior patterns, improving building design, and improving urban planning. Acclimatization capacity will be reduced in vulnerable populations, such as patients with advanced heart disease.

Examining analogous cities, i.e., cities whose current climate approximates the target city's projected future climate, is one method for incorporating acclimatization [14]. Such estimates are acceptable as long as the target and analogous city have similar social, economic, and demographic characteristics. Those authors used New York City as an example, with acclimatization represented by a temperature-modality relationship derived from Washington (DC) and Atlanta (Georgia), both of which had mean summer temperatures for 1973–1994 that were within about 0.6 °C (1°F) of the temperatures projected for NYC in the 2050s [17].

3 Models and Forecasts

3.1 Data and Variables

Temperature, mortality, and socioeconomic data are usually collected and used in heat mortality studies. These are mostly (short or long) time series or panel (longitudinal) observational data.

Park et al. [31] write that data collection technology has advanced at a fast pace. The use of big data and artificial intelligence (AI) for data analysis has grown increasingly important. Documents, images, and signals are examples of big datasets, and their size has been growing at an exponential rate. Statistical approaches, visualization, and AI technologies may be used to get meaningful results from big data.

Oftentimes, the literature is searched for data in relevant publication, with the possible intent of carrying out a meta-analysis. Huang et al. [14] conducted a literature search of journal articles published between January 1980 and July 2010 in PubMed, Scopus, ScienceDirect, and Web of Science, using the following keywords: heat, temperature, mortality, death, climate change, projection, and scenario. Books, reports, and conference abstracts were not considered. Only quantitative empirical studies were taken into account, with reviews and qualitative studies being excluded. Fourteen studies were discovered.

Missing data is a usual concern. Daily mortality data prior to 1990 are frequently unavailable in many cities [14]. Sensor failures may result in missing values in weather data [31]. Park et al. [31] suggested that unrepresentative data, low data accuracy, and irregularities in data frequency limit predictions. Missing values in historical data are often replaced by hindcasting. Many empirical studies in the field suffer from short periods of analysis. For example, data for the missing years 1950–1989 in Germany were reconstructed using a linear relationship between monthly deaths and mean maximum temperature [15].

Another source of systematic bias of (regional) climate models is due to their limited spatial resolution and simplified representation of the climate system. There are several methods for correcting bias. Karwat and Franzke [15] discovered that the delta change method was more appropriate in their case. Honda et al. [13] combined long-term with area-specific data to improve precision.

To give an idea, data of five heatwaves (February 2000 to October 2014) of the Chinese Longitudinal Healthy Longevity Survey (CLHLS) were used by Zhang et al. [38]. Dubey et al. [9] examined the current and future heatwave risks for India. Gridded temperature data (from 395 stations of the India Meteorological Department) and socioeconomic data (from the Census of India 2011, and the Socioeconomic Data and Application Center) during the April to June season were combined for the construction of scenarios, including a baseline scenario (1958–2005) and a projected scenario (2006–2099). Contributing demographic variables included gender, age, education, caste (which is relevant in the case of India), occupation, etc.

Most studies are focused geographically, with much research centered on American, Asian, and individual European cities like London and Paris [15]. The geographic focus of many studies is driven by practical considerations, but it makes sense because disaster damage is linked to geographic factors such as surface relative humidity, wind speed, population density, economic position, and vulnerable occupation groups [31]. Karwat and Franzke [15] used data for Germany (1990–2018) as well as France and Spain (1950–2018) and related deaths per month to the mean maximum temperature. Their goal was to quantify the risk of heat stress and mortality in 48 cities in Germany, France, and Spain. Honda et al. [13] mentioned a study of 47 Japanese prefectures. Park et al. [31] studied the 17 administrative divisions of South Korea, with a temporal resolution selected to match the disease monitoring system. Morais et al. [27] considered daily temperatures and mortality data from the municipality of Lisbon (Portugal) from 1980 to 2016. Mortality data were subdivided by age and cause of death. Kollanus et al. [18] investigated the mortality risk of heatwaves in Finland, depending on age, sex, cause, and place of death.

Regarding long-term trends, Karwat and Franzke [15] used daily gridded meteorological observations from January 1950 to December 2018, with a focus on the summer months (June to August). The daily maximum and minimum temperature were used to evaluate the output of a regional climate model. Sailor et al. [32] combined daily mortality data from the National Center for Health Statistics with historical weather data covering 50 years.

The impact and risk to society posed by weather and climate extremes are not necessarily captured accurately by the analysis of a single variable, but depend upon many possible combinations [20]. Karwat and Franzke [15] concluded that increased mortality was caused by a variety of factors, including socioeconomic factors and an increase in seasonal temperatures. Sensitivity analysis may need to be carried out to test for potential confounding by factors such as air pollution [18].

Huang et al. [14] cited literature sources on factors that influence the risk of heat mortality, such as age, social isolation, low income, low education, poor housing, a lack of access to air-conditioning, and a lack of healthcare services. Park et al. [31] divided the variables into static and dynamic. Government agencies collect static variables in advance and updated them quarterly or yearly. Dynamic variables (such as the floating population and the weather) change over time and are often gathered hourly or monthly. The following socioeconomic and demographic statistics were used as static variables: per capita income, insurance premium per person, resident population, and the number of vulnerable occupation groupings (i.e., agricultural, manufacturing, and construction workers). Insurance premiums, income, and vulnerable demographic groupings could all be used to infer the vulnerable population. The following meteorological data were included in the dynamic variables: maximum and minimum weekly temperatures, mean and median weekly temperatures, variance of weekly temperatures, mean weekly humidity, and mean weekly wind speed. Weekly mean temperature, vulnerable occupational groups, weekly median temperature, floating population, and weekly maximum temperature were determined to have the greatest impact. Those authors reasoned that the floating population (which is greater than the population of registered residents, reflects real-time data, and changes over time) would be a more helpful variable (for forecasting) than the aggregate population.

Indexes are used occasionally. Huang et al. [14] used the apparent temperature and humidity index (Humidex) and calculated a temperature index as the offset of the optimum temperature from the daily maximum temperature. The Excess Heat Factor (EHF, a product of two indexes) and the Generalized Accumulated Thermal Overload (GATO IV) indexes were used by Morais et al. [27]. Both are heat-health warning systems (HHWS) designed to alert the public and decision-makers about the danger of high temperatures. Issues of sensitivity and specificity of the models were also considered. Morais et al. [27] found that, for the daily circulatory and respiratory mortality of individuals of 65 years of age or older, the GATO IV was the only statistically significant index capable of predicting the impact of heatwave mortality. Zhang et al. [38] calculated a normalized difference vegetation index (NDVI) to assess the individual long-term exposure to heatwaves and the role of green flora.

Before feeding their variables into a random forest regression model, Park et al. [31] standardized them. Karwat and Franzke [15] normalized deaths for all cities for 100,000 inhabitants, assuming (for all cities and countries) a constant population size. Dummy variables were used for the maximum temperature thresholds of 33, 35, and 37 °C (with no significant effect).

3.2 Models and Simulations

Citing the IPCC and the World Health Organization (WHO), Huang et al. [14] have argued that the potential impact of climate change on heat mortality has been a growing public concern, and that various methods to model and project heat-related mortality have been used. These include atmospheric models (with nested horizontal and vertical grids), models of weather services, and general circulation models (GCMs).

Sailor et al. [32] investigated heat-related illness and mortality, using an atmospheric model from the National Center for Atmospheric Research. The region studied was the city of Philadelphia, in Pennsylvania (USA). The study area had a boundary of approximately 1200 by 1000 km and was modeled using three nested grids with grid spacing of 2, 6, and 18 km. There was also a vertical grid with 30 levels. Three urban categories were used (with characteristics taken from the literature): urban core; commercial and industrial; and residential.

Sailor et al. [32] selected a number of two to five day episodes from 1997 to 2001 to investigate heat mortality. The study focused on identifying and evaluating certain oppressive air masses that caused significant increases in mortality, particularly during the summer. The researchers wanted to see which meteorological (maximum temperature, dew point temperature, and wind speed) and non-meteorological (consecutive days of oppressive air masses, and timing within the season) variables were associated with the highest mortality. For each episode, meteorological simulations were run. A control run was carried out for the historical land cover, and a second run was carried out for the altered land cover, which was associated with UHI reduction methods and resulted in a 0.1 increase in urban albedo.

The combination of temperature and humidity, which can be measured by the wet bulb temperature, causes heat stress [15]. The apparent temperature and the German Weather Service's Klima-Michel model (a heat budget model) are two further approaches. Humidity is an important physiological element for human health, and hence, it is necessary to include it in studies on the health effects of heat. High humidity has an effect on the body's thermoregulatory processes, such as sweating and evaporation. Those authors cite a published investigation of the threshold between lethal and non-lethal heat episodes that found wet bulb temperatures above 37 °C to cause hyperthermia.

Abadie and Polanco-Martínez [1] pointed out that estimates from stochastic models for characterizing heatwaves and their epidemiological impacts on human health can vary from one climate model to another, so relying on a single climate model may be problematic. In their research, the parameters of 21 temperature models were calibrated with nonlinear least squares, and the results of these models were averaged. Monte Carlo simulations were used for the projection of epidemiological risk.

GCMs have a solid physical foundation and produce credible projections of future climate change. Huang et al. [14] proposed collaborating with climate researchers in the field of environmental health. An index was defined to translate the relative risk to the excess number of deaths [13]. Data from cities in Korea, Taiwan, the USA, and Europe were used to test robustness. Heat-related relative risk is lower in younger people, according to the study. The distribution of the daily maximum temperature was required to estimate risk. One of the GCMs given by the WHO Global Burden of Diseases project (BCM2), with a grid resolution of one degree, was used to forecast future climate.

Dubey et al. [9] used the ROM regional earth system model, which coupled with the Regional Atmosphere Model (REMO); the Max Planck Institute Ocean Model (MPIOM); the Hydrological Discharge model (HD); and the global marine biogeochemistry Hamburg Ocean Carbon Cycle model (HAMOCC). Heatwave hazard, vulnerability, and (resulting) risk maps for India were calculated and drawn. Districts with high heatwave hazard and high socioeconomic vulnerability had high risk. The ROM model captured critical districts, and the study identified the south-eastern coast and Indo-Gangetic plains and some populous districts with metropolitan regions (Mumbai, Delhi, and Kolkata) as the most hazardous, vulnerable, and risk-prone regions under the current scenario.

Marvuglia et al. [23] used a microsimulation modeling approach. Together with cellular automata and agent-based models (ABMs), microsimulation models are individual-based models. A spatial microsimulation ABM was implemented to simulate the temperature impacts of green roofs installations in cities and their capacity to attenuate the effects of heatwave episodes [23]. The model, implemented using the Netlogo platform (version 6.0.4, https://ccl.northwestern.edu/netlogo/), considered the elderly citizens as agents in a city area and simulated the heatwave-related health impacts, which are measured in mortality likelihood. The model simulated a generalized 1.5–3 °C indoor temperature reduction range induced by green roofs (based on inferences from green roof literature) in four different European cities: Szeged (Hungary), Alcalá de Henares (Spain), Metropolitan City of Milan (Italy), and Çankaya municipality (Turkey). Population and land data (including 29 types of land) were used to allocate the elderly (defined as people of at least 65 years of age) to areas of a city. Temperature projections to the years 2030, 2050, 2070, and 2090 were based on the IPCC RCP8.5 scenario. A total of 360 model runs were executed for each city, including a no-heatwave baseline scenario. The results of the simulations showed substantial variation in the impacts of green roofs on heatwave-induced mortality across the cities.

Model parameters, structural uncertainties, and processes in the climate system that are impossible to resolve due to computational constraints, can all-cause uncertainty in climate change estimates [14], which may be investigated with sensitivity analysis.

3.3 Statistical Methods

Statistical methods reviewed in this section include descriptive approaches (*t*-tests); smoothing/cubic splines; moving averages; trend analysis; multiple linear regression (with dummy variables, e.g., for temperature thresholds); nonlinear distributed lag models; quantile regression; conditional logistic and Poisson regression; quasi-Poisson distribution; spatial synoptic classification (SSC) approaches; and multivariate analysis techniques (cluster analysis). Carry-on (lag) effects and mortality displacement (harvesting) are frequently accounted for.

The following study designs have been used to investigate the effects of temperature on mortality: descriptive, case control, case only, case crossover, time series, spatial, and synoptic (weather) analysis [14]. Time series and case crossover designs are thought to be more effective for studying the relationship between temperature and mortality in multiple locations over time. Trends, seasonal cycles, humidity, and air pollution are all potential confounders that must be managed. Temperature-mortality relationships are typically nonlinear.

A statistical approach to quantifying the similarity of a city's climate to that of another location was one example of a more intuitive way to communicate the societal impacts of extreme temperature risks [15]. Many other statistical temperature analyses do not take the urban heat island effect into account because it is not a direct climate effect. Since those authors were interested in the direct effect of heat on humans in their study, the urban heat island effect should not be ignored because it contributes to heat stress in humans. Nonetheless, most climate models do not account for urban surfaces (due to cities accounting for a small fraction of the earth's surface), which likely understated the extreme temperatures in urban areas.

Using a cubic smoothing spline (with six degrees of freedom), optimum temperature estimates were obtained for 47 prefectures in Japan [13]. Some data points were in the very hot region, especially above 35 °C. In most cases, the optimum temperature was around the 84th percentile of the daily maximum temperature. The 83.6th percentile was found to be the mean of all 47 prefectures, and it was decided that this would be used as the optimum temperature value.

In another similar study reviewed by Paravantis et al. [30], Harlan et al. [12] investigated the mortality of males and females over and under 65 years of age from May to October of the years 2000–2008 in desert cities with an extremely hot climate in central Arizona (USA). To describe the relationship between temperature and mortality, cubic spline regressions were estimated. The authors discovered an increase in mortality above certain temperature thresholds for direct exposure to environmental heat. The daily maximum apparent temperature was found to be more strongly linked to mortality from direct exposure to high ambient heat.

Wolff et al. [36] estimated changes in all-cause mortality from changes in the mean daily temperatures using established heat mortality slopes. The authors used available heat mortality slopes for the Philippines and Vietnam (estimated by Lee et al. [19]), two countries that best match the area of study in Indonesia in the key drivers of heat-related mortality. These slopes, together with the estimated 2002–2018 temperature change, were used to calculate the percentage point increase in heat-related excess mortality between 2002 and 2018. Global Burden of Disease (GBD) mortality rates (at a province level) were then used to translate those percentage point increases into the actual numbers of excess heat-attributable deaths. The lack of daily time-step mortality data prevented a more nuanced modeling approach.

Azhar et al. [2] used 7-day moving averages, mortality rate ratios, and daily maximum temperature to analyze death counts during a 2010 heatwave in India. The period of the heatwave was found to have a 43.1% excess amount of mortality, with no lag time between temperatures and fatalities. In the hot months of April, May, and June, there were moderate to high correlations between mortality and temperature, particularly when the maximum temperature was over 43 °C and the mean temperature was over 36 °C, with significantly more female deaths.

The analysis of historical exposure–response functions of temperature and mortality, as well as consideration of future changes in climate, population, and acclimatization, were required for projecting heat mortality [14]. Karwat and Franzke [15] conducted a trend analysis of (projected) heat stress and mortality risk in major European cities using regional climate projections for the RCP2.6 and RCP8.5 scenarios. They looked at the years 1950–2018 and attempted to (1) evaluate daily maximum and minimum temperatures to investigate trends in quantiles, (2) analyze an empirically modeled wet bulb temperature in both observations and future projections, and (3) combine temperature and mortality observations to project heat deaths until 2050. Karwat and Franzke [15] used a regression model with long trends (unlike previous studies with short observational time series) to combine observations of maximum temperature and mortality to project excess heat deaths in RCP2.6 and RCP8.5 that were used for historical runs and future climate projections until 2050. Regional future climate projections were the result of combining global and regional climate models. Heat mortality increased by up to 7.9% in Spain, 1.7% in France, and 0.9% in Germany per decade. It is noted that, given that the regressions of Karwat and Franzke [15] were estimated using time series data, the low R square values were somewhat surprising.

Air mass/mortality relationships were developed for Philadelphia and multiple linear regressions were estimated based on the following meteorological and other variables: number of consecutive days the air mass was present; maximum temperature; and whether it was early or late in the warm season [32]. This was described by the authors as an inverse association, implying that early periods were linked to more mortality. Days with at least one estimated death were deemed oppressive. For a minor level of urban albedo increase of 0.1, the research predicted large regions with average daytime temperature depressions of 0.3–0.5 °C.

Honda et al. [13] considered mortality for people over the age of 65. A carry-on (lag) effect and mortality displacement (harvesting) were accounted for, in an attempt to improve the fit, precision, and robustness of projection for heat-related excess deaths. Their nonlinear distributed lag model was estimated in R (https://www.r-project.org/). Mortality projections for the 2030s and 2050s were produced by the World Health Organization (WHO). Their method involved a series of regression equations that quantified the current and historical relationship between mortality and independent variables such as gross domestic product (GDP) per capita; years of education; and time (assumed to be a proxy for health benefits arising from technological developments). The chronological trend was well controlled, and the disparity between areas was modest.

Kim et al. [16] investigated the relationship between social isolation levels and heat wave-related mortality risk in the elderly population of 119 urban administrative districts of the Republic of Korea. A two-stage analysis was conducted, including the heatwave mortality risk in the elderly population (aged over 65). A time series regression with a distributed lag model was followed by meta-regressions to pool the estimates across all the districts.

Gustin et al. [10] developed a recursive autoregressive time series model with exogenous inputs (ARX) to provide reliable short-term forecasts of the internal temperatures in three case study dwellings during the 2015 hot summer conditions. Those authors compared Autoregressive Models with Exogenous Inputs (ARX) and Autoregressive Moving Average models with Exogenous Inputs (ARMAX). Optimal predictor variables were selected based on the Akaike Information Criterion (AIC). Based on the literature, it was concluded that ARX models were more accurate than ARMAX models, although ARMAX models provided more reliable multi-step ahead forecasts.

Multiple linear regression has been used commonly, but the results obtained are frequently not robust because of the large variability of temperature extremes. Karwat and Franzke [15] used quantile regression, which allowed them to identify trends and detect changes in the maximum and minimum temperature distributions during the specified climate period. Quantile regression, as opposed to linear regression, can model changes in percentiles and identify extremes. Because climate time series are autocorrelated, identifying trends is difficult. Those authors employed a nonparametric test for monotonic trends, which is commonly used in the study of annual summer temperatures.

In a study of the correlation between temperature and cardiovascular or respiratory mortality of the elderly in the 20 largest metropolitan areas in the US reviewed by Paravantis et al. [30], Basu et al. [4] estimated conditional logistic and Poisson regression models with epidemiological time series data for one year (1992). Their study used the mean daily and dew point daily temperatures and took into account the confounding effects of particulate matter and ozone. Elevated cardiovascular and respiratory mortality was found to be associated with temperature exposure mostly in the summer.

The quasi-Poisson distribution was used to simulate relative mortality [13]. Zero lag days, followed by a lag of one, had the highest risk. Above the optimum temperature, heat-related mortality increased monotonically, whereas below the optimum temperature, the link between temperature and mortality was more complicated. May to September daily mortality counts (from 1980 to 2016) for Lisbon (Portugal) were modeled using Poisson and Negative Binomial Regression [27]. Cox proportional hazards models were used by Zhang et al. [38] to assess the effects of greenness, heatwaves, and their interaction on mortality, adjusted for covariates, with subgroup analyses conducted by residence, gender, and age. Kollanus et al. [18] analyzed the association between heatwave days and daily mortality with generalized estimating equations (GEE) for panel (longitudinal) data. Daily mortality counts were assumed to follow the Poisson distribution.

Various statistical procedures were used by Sailor et al. [32] to develop an air mass identification technique, including the Spatial Synoptic Classification (SSC). SSC is assigned each day to a certain mass type, allowing for the detection of air masses that have been linked to higher mortality in the past. Dry polar (DP), dry temperate (DM), dry tropical (DT), moist polar (MP), moist moderate (MM), and moist tropical (MT) were the six air mass types (MP). Excess deaths were caused by a particularly oppressive subset of MT (dubbed MT+).

Climate–mortality relationships have been studied in US cities, California, Washington State, and Australia, according to published research studies [14]. The synoptic approach quantified the effect of air and dew point temperatures, wind speed, cloud cover, barometric pressures, and other variables. The apparent temperature and humidity index (also known as Humidex, a measure of how hot we feel) combines temperature and humidity effects. Based on a review of the literature comparing maximum, mean, and minimum temperature values and apparent temperature, the authors concluded that these measures were strongly correlated with one another and have a similar predictive ability. According to those authors, the choice of model is probably more important than the choice of temperature measure. It is also important to select a baseline period for the temperature-mortality relationship.

Paravantis et al. [30] used independent sample t-tests and found statistically significant fatalities for people over 65 years of age below and above certain extreme temperature thresholds, both low (cold) and high (heat). Those authors also carried out a cluster analysis of 4018 daily fatalities of people over 65 years of age that occurred in Athens (Greece) from 2002 to 2012 and were due to cardiovascular and respiratory causes. Cluster variables included the daily death rate; maximum and minimum daily temperatures; heatwave days; maximum and minimum daily humidity; and ozone (O_3), sulfur dioxide (SO_2), and nitrogen dioxide (NO_2) concentrations. One of the five discovered clusters was the heatwave cluster, which had the highest daily mortality centroid value of any of the five clusters. The heatwave cluster included all heatwave days and had the highest minimum and maximum centroid temperatures (28.22 and 38.43 °C, with apparent values of 30.82 and 39.48 °C, respectively). Over 90% of the cases in this cluster occurred during the months of July and August.

Huang et al. [14] presented contradictory evidence regarding air pollution as a confounder and effect modifier of the temperature-mortality relationship. Air pollution appeared to have a minor confounding effect, and air pollution and temperature had independent effects on mortality. Urban air pollution is expected to rise as a result of climate change, increasing the combined exposure of urban populations to high temperatures and air pollutants in the future. Indirect causes, such as increased ozone concentrations (caused by rising temperatures) or the synergistic effect of heat and air pollution, may contribute to future heat-related deaths.

Papers written in English and published between January 2000 and April 2015 in PubMed, ProQuest, ScienceDirect, Scopus, and Web of Science were retrieved by

Xu et al. [37]. The following MeSH (US National Library of Medicine's Medical Subject Headings) keywords were used: "*climate change*", "*temperature*", "*heat wave*", "*heatwave*", "*death*", and "*mortality*". The initial search yielded 1608 papers, 60 of which were included in the final review (applying various exclusion criteria). The risks of total, cardiovascular, and respiratory mortality due to heatwaves were extracted from the papers. Meta-analysis revealed significant heterogeneity among studies due to differences in climate, study designs, statistical approaches, and demographics. Within the same region, mortality increased with heatwave intensity and duration. Heatwaves were found to have a significant negative effect on mortality, with the magnitude varying depending on the definition of a heatwave. Heatwave intensity affected mortality more than duration.

Zhu et al. [39] used a city-specific exposure–response function to project multiple temperature and population trends under different climate and urbanization scenarios, in order to assess non-optimum temperature-related mortality in China from 2000 to 2050. Non-optimum temperature is an aggregate of the burden attributable to low and high temperatures, corresponding to both heat and cold effects that occur. Data collected described the historical patterns of non-optimum temperature-related mortality from 2000 to 2010. Trends to 2050 were projected under multiple futures shared socioeconomic pathways (SSPs) and representative concentration pathway (RCP) scenarios across China.

Heat mortality projections from the literature for Los Angeles, New York City, Chicago, Washington State, cities in Canada, and Australia were provided [14]. Among the findings in the literature, increases in heat mortality for 44 US cities in 2050 were projected to range from 70 to 100% higher than the baseline summer mortality of 1964–1991. Climate change is expected to have the greatest impact on three Mediterranean cities (Barcelona, Roma, and Valencia) and two continental cities in Europe (Paris and Budapest). Annual heat-related deaths in Lisbon were projected to rise from 5.4 to 6 per 100,000 in 1980–1998 to 5.8 to 15.1 in the current decade (2020s) and 7.3 to 35.6 by the middle of the century, according to the literature (2050s).

Heat mortality projections were derived from daily maximum temperatures and excess deaths in Spain, France, and Germany during the summer months [15]. Excess mortality referred to the total number of deaths that occurred as a result of extreme conditions such as prolonged heatwaves. Germany had the highest excess mortality rates, followed by France and Spain. Almost all cities are expected to have more heat-related deaths in RCP8.5 than in RCP2.5. The findings supported previous research that concluded that heat mortality is higher in the Mediterranean than in northern and central Europe. Future projections on heat stress and mortality that take into account more European cities are scarce and mostly qualitative.

3.4 Algorithms and Scenarios

Further to (simulation) models and statistical analyses, other methods including algorithms and scenarios have been used in heat mortality research. Scenario-based projection is based on both quantitative and qualitative evidence.

A usual research goal is to predict heat stress under various greenhouse gas emission scenarios [15]. Those authors calculated the wet bulb temperature using an empirical formula and a gene expression algorithm (GEP). The algorithm generated a best-fit nonlinear formula for the wet bulb temperature, which included terms for the maximum or minimum temperature and relative humidity.

Park et al. [31] developed a heat-related health prediction model on the basis of a machine learning algorithm for early warning systems. The researchers used random forest (RF) regression, an ensemble machine learning method that combines several models that have been trained separately to create a powerful learner that can be used for classification and regression. The bootstrap approach, which involves random sampling with replacement, is the basis of RF, and overfitting may be avoided by generating many trees. In numerous disaster fields, RF techniques have been used to anticipate, forecast, and evaluate risk. Eighty percent of the total data examples were used for learning, while the remaining 20% was used for testing. The Boruta method (an algorithm designed to perform automatic identification of relevant features of a dataset) was used to confirm and rank variable relevance. The RF model was estimated using a Python package. There were 181 decision trees and 46 tree depths identified. The authors explained why they picked the mean absolute error (MAE) over the mean squared error (MSE) as a loss function, but their reasoning for the former was unclear. Also, a visualization of the model's observed against predicted values appeared to suggest that the data needed to be log-transformed before the regressions could be estimated, something the authors did not address. It was proposed that establishing learning algorithms for different Korean regions would help the model perform better.

Gustin et al. [10] mentioned that artificial neural networks (ANN) models may model more complex nonlinear relationships between the response variable and its predictors. On the other hand, ANN models have to be trained with large amounts of learning data; have no interpretability; and may give different results after repeated trials on the same data. Compared to ANNs, linear time series forecasting models are simpler to deploy and may be replicated.

As Huang et al. [14] mention, the IPCC has created a number of scenarios [28] that can be used to forecast future climate using general circulation models (GCMs). Different GCMs incorporate atmosphere or ocean dynamics, use different representations, and thus produce different climate projections (even when assuming the same pathway of future emissions). Different GCMs have been used in the literature to simulate the future of current climates, which adds to the uncertainty of emission scenarios. Downscaling methods are used because the spatial resolution of GCMs is too coarse to be directly used in the assessment of impacts at the local scale. GCM outputs can be used as initial and boundary conditions for finer-scale simulations by a regional climate model (RCM). The IPCC has suggested that no single GCM can be considered the best; instead, multiple GCMs should be used to account for modeling uncertainties.

Scenario-based projections have been used as a key approach for policy making and planning [14], particularly in the context of uncertain future conditions, as demonstrated by the IPCC's Special Report on Emissions Scenarios [28]. Such scenarios are not used to predict the future, but rather to better understand uncertainties and make more sound decisions. They should be viewed as potential futures that are dependent on demographic, technological, political, social, and economic development. Scenarios can be informed by quantitative and qualitative evidence, giving decision-makers insight into future trends, contexts, risks, and opportunities. They assist decision-makers in developing adaptation strategies and communicating future health risks associated with climate change to politicians and the general public.

Huang et al. [14] alleged that there were no guidelines for scenario-based projection research on heat mortality at the time of writing. Abadie and Polanco-Martínez [1] used the maximum daily temperatures from 21 climate circulation models under RCP8.5 and RCP4.5 representative concentration pathways. Zhu et al. [39] suggested that excess deaths are dependent on each specific scenario examined. Nevertheless, in their research the health burden in high-risk areas of China tended to intensify for the examined period. In another work, for both the baseline case and the altered climatic scenario, the annual number of fatalities in 2030 (as projected by WHO) was used [13]. The authors applied the projected 2030 temperature distribution produced with BCM2 (instead of the current climate) for the altered climate scenario. The excess number of fatalities linked to climate change was calculated as the difference in the number of excess deaths between the two cases.

4 Discussion

Karwat and Franzke [15] discovered that, despite the fact that capital cities have larger heat islands, they are likely to have fewer heat deaths (possibly due to more and greater public and medical facilities that are likely to mitigate climate risk to a larger extent). Regional mortality trends could also be attributed to regional conditions, such as socioeconomic conditions. To that end, projected regional mortality differences between France and Germany pointed to similar causes. Heat-related deaths were higher in southern Germany than in northern Germany. Eastern Germany had lower mortality rates, which could be attributed to smaller cities and lower population density.

The findings of Gustin et al. [10] suggested that detailed building information is not required to produce reasonable forecasts of indoor temperatures in dwellings without mechanical cooling and heating. This demonstrates the possibility of using time series forecasting with sufficient reliability at a low cost for buildings (particularly those housing vulnerable occupants or contents) and assisting in the preparation for an impending response.

During heatwaves, many people may spend more time at home or in air-conditioned spaces, implying that the role of indoor and outdoor temperatures is unclear [37]. As an example, more than 90% of residences in numerous mid-to-southern Japanese prefectures have air conditioning [13]. During heatwaves, the elderly, children, women, and people with socioeconomic disadvantages or chronic diseases should be protected [37].

As Kollanus et al. [18] pointed out, heatwaves are a significant burden on public health even in cold climates. Higher mortality risks were found for the elderly (at least 65 years of age) especially in social care facilities and among long-term inpatients in longer heatwaves. It was concluded that in Finland (a cold northern country), heatwaves increased mortality risk significantly among the elderly, with women more susceptible than men, and many chronic diseases being important risk factors. To reduce heatwave-related deaths, preparedness should be improved particularly in hospital and healthcare center wards, where the most vulnerable are long-term inpatients.

Kim et al. [16] suggested that social isolation is a critical factor of heat wave-related mortality for the elderly, especially males. Reduced social isolation and increased mutual aid levels reduced heatwave mortality risk. Single elderly households in detached houses were related to higher heatwave mortality risk. The relationship between temperature and mortality may also change over time [37]. For instance, as the population of region ages, the threshold temperature corresponding to the inflection point may decrease.

Zhang et al. [38] investigated the relationship between exposure to heatwaves combined with the protective action of the presence of vegetation (referred to as greenness) with adults over 65 years of age. A relationship between greenness, heatwaves, and mortality was observed, suggesting that vegetation is a protective factor for heat and heatwave-related mortality. Even if no further deforestation occurs, future temperature increases from climate change will almost certainly result in an even more serious public health concern [36]. These findings highlight the significant challenge posed by the combined effects of deforestation and climate change for populations in low-latitude, industrializing countries.

Embedding nature-based solutions (NBS) in cities and their quantifiable benefits were studied by Marvuglia et al. [23]. Among NBS, green roofs have an important role in temperature regulation in buildings and in lowering the damaging effects of heatwaves on human health.

It is important to remember that future heat mortality projections are subject to uncertainties that must be carefully examined for policy implications [14]. Changes in mortality risk may occur as a result of an aging population, acclimatization, socioeconomic development, and adaptation techniques. Failure to account for adaptation would almost certainly lead to an overestimation of future consequences. People's exposure patterns may change if they spend less time outside and thus are less exposed to heat. Air-conditioning and heat-health warning systems have also become more common, potentially reducing the health risks associated with heat waves.

Challenges also stem from the unknowns of future demographic shifts, which may alter the population's susceptibility to heat stress [14]. As the population ages, more older adults become vulnerable, increasing the proportion of the population at risk. Aside from having a reduced physiological ability to cope with heat, the elderly are more likely to live alone, have fewer social contacts, and be in poor health. Furthermore, heat appears to have a greater impact on mortality in elderly women. These factors complicate and confound the link between mortality and climate change. Future demographic trends should be examined to investigate the role of population susceptibility.

5 Conclusions

In conclusion, as Dubey et al. [9] argued, heat and heatwave hazard risks will significantly worsen in the future scenarios everywhere, under enhanced global warming. Heatwaves are expected to become more frequent and severe during the twenty-first century, as global warming accelerates [5, 24, 35]. Better information on the likelihood of heatwave occurrence improves the accuracy of projections about the health effects of climate change [14].

The ultimate goal of assessing the health impacts of heatwaves is to develop targeted public health interventions to prevent and reduce mortality [37]. Regional differences and local characteristics should be considered. City-specific heatwave definitions that combine meteorological and health perspectives will help develop early warning systems and reduce mortality.

Karwat and Franzke [15] stressed the importance of conducting systematic heat mortality risk assessments for adaptation planning, as well as quantifying the observed relationship between extreme temperatures and mortality in urban areas. Differences in heat mortality between RCP2.6 and RCP8.5 demonstrated that mitigation efforts have a positive effect on reducing heat-related deaths.

Acknowledgements The author wishes to express his gratitude to Ms. Nicoletta Kontoulis, doctoral candidate, and Mr. Vasileios Papakostas, research associate, both of the Department of International and European Studies of the University of Piraeus, for their assistance in collating published literature. Additionally, the author wishes to convey appreciation to Drs. T. Nadasdi and S. Sinclair for their online Spell Check Plus (https://spellcheckplus.com/) service, which was used to proofread this chapter.

References

1. Abadie LM, Polanco-Martínez JM (2022) Sensitivities of heat-wave mortality projections: moving towards stochastic model assumptions. Environ Res 204:111895. https://doi.org/10.1016/j.envres.2021.111895
2. Azhar GS, Mavalankar D, Nori-Sarma A, Rajiva A, Dutta P, Jaiswal A, Sheffield P, Knowlton K, Hess JJ, on behalf of the Ahmedabad HeatClimate Study Group (2014) Heat-related mortality in India: excess all-cause mortality associated with the 2010 Ahmedabad heatwave. PLOS ONE 9(3):e91831. https://doi.org/10.1371/journal.pone.0091831
3. Baccini M, Biggeri A, Accetta G, Kosatsky T, Katsouyanni K, Analitis A, Ross Anderson H, Bisanti L, D'Ippoliti D, Danova J, Forsberg B, Medina S, Paldy A, Rabczenko D, Schindler C, Michelozzi P (2008) Heat effects on mortality in 15 European cities. Epidemiology 19:711–719. https://doi.org/10.1097/EDE.0b013e318176bfcd
4. Basu R, Dominici F, Samet JM (2005) Temperature and mortality among the elderly in the United States—a comparison of epidemiologic methods. Epidemiology 16:58–66. https://doi.org/10.1097/01.ede.0000147117.88386.fe
5. Clark RT, Brown SJ, Murphy JM (2006) Modeling Northern Hemisphere summer heat extreme changes and their uncertainties using a physics ensemble of climate sensitivity experiments. J Clim 19:4418–4435. https://doi.org/10.1175/JCLI3877.1
6. Coates L, van Leeuwen J, Browning S, Gissing A, Bratchell J, Avci A (2022) Heatwave Fatalities in Australia, 2001–2018: an analysis of coronial records. Int J Disaster Risk Reduction 67:102671. https://doi.org/10.1016/j.ijdrr.2021.102671
7. D'Ippoliti D, Michelozzi P, Marino C, de'Donato F, Menne B, Katsouyanni K, Kirchmayer U, Analitis A, Medina-Ramón M, Paldy A, Atkinson R, Kovats S, Bisanti L, Schneider A, Lefranc A, Iñiguez C, Perucci CA (2010) The impact of heatwaves on mortality in 9 European cities: results from the EuroHEAT project. Environ Health 9(37):1–9. https://doi.org/10.1186/1476-069X-9-37
8. Díaz J, Garcia R, de Velazquez CF, Hernandez E, Lopez C, Otero A (2002) Effects of extremely hot days on people older than 65 years in Seville (Spain) from 1986 to 1997. Int J Biometeorol 46(3):145–149. https://doi.org/10.1007/s00484-002-0129-z
9. Dubey AK, Lal P, Kumar P, Kumar A, Dvornikov AY (2021) Present and future projections of heatwave hazard-risk over India: a regional earth system model assessment. Environ Res 201:111573. https://doi.org/10.1016/j.envres.2021.111573
10. Gustin M, McLeod RS, Lomas KJ (2018) Forecasting indoor temperatures during heatwaves using time series models. Build Environ 143:727–739. https://doi.org/10.1016/j.buildenv.2018.07.045
11. Hajat S, Armstrong B, Baccini M, Biggeri A, Bisanti L, Russo A, Paldy A, Menne B, Kosatsky T (2006) Impact of high temperatures on mortality: is there an added heatwave effect? Epidemiology 17:632–638. https://doi.org/10.1097/01.ede.0000239688.70829.63
12. Harlan SL, Chowell G, Yang S, Petitti DB, Morales Butler EJ, Ruddell BL, Ruddell DM (2014) Heat-related deaths in hot cities: estimates of human tolerance to high temperature thresholds. Int J Environ Res Public Health 11:3304–3326. https://doi.org/10.3390/ijerph110303304
13. Honda Y, Kondo M, McGregor G, Kim H, Guo Y-L, Hijoka Y, Yoshikawa M, Oka K, Takano S, Hales S, Sari Kovats R (2013) Heat-related mortality risk model for climate change impact projection. Environ Health Prev Med 19:56–63. https://doi.org/10.1007/s12199-013-0354-6
14. Huang C, Barnett AG, Wang X, Vaneckova P, FitzGerald G, Tong Sh (2011) Projecting future heat-related mortality under climate change scenarios: a systematic review. Environ Health Perspect 119:1681–1690. https://doi.org/10.1289/ehp.1103456
15. Karwat A, Franzke CLE (2021) Future projections of heat mortality risk for major European cities. Weather, Clim, Soc 913–931. https://doi.org/10.1175/WCAS-D-20-0142.1
16. Kim Y, Lee W, Kim H, Cho Y (2020) Social isolation and vulnerability to heatwave-related mortality in the urban elderly population: a time-series multi-community study in Korea. Environ Int 142:105868. https://doi.org/10.1016/j.envint.2020.105868

17. Knowlton K, Lynn B, Goldberg RA, Rosenzweig C, Hogrefe C, Rosenthal JK, Kinney PL (2007) Projecting heat-related mortality impacts under a changing climate in the New York City region. Am J Public Health 97(11):2028–2034. https://ajph.aphapublications.org/, https://doi.org/10.2105/AJPH.2006.102947
18. Kollanus V, Tiittanen P, Lanki T (2021) Mortality risk related to heatwaves in Finland—factors affecting vulnerability. Environ Res 201:111503. https://doi.org/10.1016/j.envres.2021.111503
19. Lee JY, Kim H, Gasparrini A, Armstrong B, Bell ML, Sera F, Lavigne E, Abrutzk R, Tong S, de Sousa Zanotti Stagliorio Coelho M, Nascimento Saldiva Paulo HP, Correa M, Valdes Ortega N, Kan H, Osorio Garcia S, Kyselý J, Urban A, Orru H, Indermitte E, Jaakkola JJK, Ryti NRI, Pascal M, Goodman PG, Zeka A, Michelozzi P, Scortichini M, Hashizume M, Honda Y, Hurtado M, Cruz J, Seposo X, Nunes B, Teixeira JP, Tobias A, Íñiguez C, Forsberg B, Åström C, Vicedo-Cabrera AM, Ragettli MS, Leon Guo Y-L, Chen B-Y, Zanobetti A, Schwartz J, Ngoc Dang T, Do Van D, Mayvaneh F, Overcenco A, Li S, Guo Y (2019) Predicted temperature-increase-induced global health burden and its regional variability. Environ Int 131:105027. https://doi.org/10.1016/j.envint.2019.105027
20. Leonard M, Westra S, Phatak A, Lambert M, van den Hurk B, McInnes K, Risbey J, Schuster S, Jakob D, Stafford-Smith M (2013) A compound event framework for understanding extreme impacts. Wiley Interdisc Rev: Clim Change 5(1):113–128. https://doi.org/10.1002/wcc.252
21. Lo S-H, Chen C-T, Russo S, Huang W-R, Shih M-F (2021) Tracking heatwave extremes from an event perspective. Weather Clim Extremes 34:100371. https://doi.org/10.1016/j.wace.2021.100371
22. Martens WJM (1998) Climate change, thermal stress and mortality changes. Soc Sci Med 46(3):331–344. https://doi.org/10.1016/S0277-9536(97)00162-7
23. Marvuglia A, Koppelaar R, Rugani B (2020) The effect of green roofs on the reduction of mortality due to heatwaves: results from the application of a spatial microsimulation model to four European cities. Ecol Model 438:109351. https://doi.org/10.1016/j.ecolmodel.2020.109351
24. Meehl GA, Tebaldi C (2004) More intense, more frequent, and longer lasting heatwaves in the 21st century. Science 305:994–997. https://doi.org/10.1126/science.1098704
25. Meier A (2006) Countermeasures to urban heat islands: a global view. Ernest Orlando Lawrence Berkeley National Laboratory: Berkeley, CA, USA
26. Mitchell D, Heaviside C, Vardoulakis S, Huntingford C, Masato G, Guillod BP, Frumhoff P, Bowery A, Wallom D, Allen M (2016) Attributing human mortality during extreme heatwaves to anthropogenic climate change. Environ Res Lett 11:074006. https://doi.org/10.1088/1748-9326/11/7/074006
27. Morais L, Lopes A, Nogueira P (2020) Which heatwave measure has higher predictive power to prevent health risks related to heat: EHF or Gato IV?—Evidence from modelling Lisbon mortality data from 1980 to 2016. Weather Clim Extremes 30:100287. https://doi.org/10.1016/j.wace.2020.100287
28. Nakicenovic N, Alcamo J, Grübler A, Riahi N, Roehrl RA, Rogner H-H, Victor N (2000) Special report on emissions scenarios: a special report of Working Group III of the IPCC. Cambridge University Press, Cambridge. http://pure.iiasa.ac.at/id/eprint/6101/. ISBN 0-521-80493-0
29. Nastos PT, Matzarakis A (2011) The effect of air temperature and human thermal indices on mortality in Athens, Greece. Theoret Appl Climatol 108:591–599. https://doi.org/10.1007/s00704-011-0555-0
30. Paravantis J, Santamouris M, Cartalis C, Efthymiou C, Kontoulis N (2016) Mortality impact of high ambient temperatures and urban air pollution associated to heatwaves and the urban heat island in Athens, Greece. Sustainability 9(4):606. https://doi.org/10.3390/su9040606
31. Park M, Jung D, Lee S, Park S (2020) Heatwave damage prediction using random forest model in Korea. Appl Sci 10(22):8237. https://doi.org/10.3390/app10228237
32. Sailor DJ, Kalkstein LS, Wong E (2002) The potential of urban heat island mitigation to alleviate heat-related mortality: methodological overview and preliminary modeling results for Philadelphia. In: 4th Symposium on the urban environment. https://coolrooftoolkit.org/wp-content/uploads/2012/05/UHI-and-mortality-sailor.pdf

33. Santamouris M, Kolokotsa D (eds) (2016) Urban climate mitigation techniques. Routledge, London, UK
34. Santamouris M (2016) Local climate change: assessing the impact and mitigating the heat. In: Padovani LM, Puglisi G (eds) Energia, ambiente e innovazione, vol 2. ENEA, Rome, Italy, pp 98–103
35. Schär C, Vidale P, Luthi D, Frei C, Haberi C, Liniger M, Appenzeller C (2004) The role of increasing temperature variability in European summer heatwaves. Nature 427:332–336. https://doi.org/10.1038/nature02300
36. Wolff NH, Vargas Zeppetello LR, Parsons LA, Aggraeni I, Battisti DS, Ebi KL, Game ET, Kroeger T, Masuda YJ, Spector JT (2021) The effect of deforestation and climate change on all-cause mortality and unsafe work conditions due to heat exposure in Berau, Indonesia: a modelling study. The Lancet Planet Health 5:e882–e892. https://doi.org/10.1016/S2542-5196(21)00279-5
37. Xu Z, FitzGerald G, Guo Y, Jalaludin B, Tong S (2016) Impact of heatwave on mortality under different heatwave definitions: a systematic review and meta-analysis. Environ Int 89–90:193–203. https://doi.org/10.1016/j.envint.2016.02.007. Models and forecasts on the future heat-related mortality under climate change
38. Zhang H, Liu L, Zeng Y, Liu M, Bi J, Ji JS (2021) Effect of heatwaves and greenness on mortality among Chinese older adults. Environ Pollut 290:118009. https://doi.org/10.1016/j.envpol.2021.118009
39. Zhu D, Zhou Q, Liu M, Bi J (2021) Non-optimum temperature-related mortality burden in China: addressing the dual influences of climate change and urban heat islands. Sci Total Environ 782:146760. https://doi.org/10.1016/j.scitotenv.2021.146760

Chapter 7
The Impact and Influence of Mitigation Technologies on Heat-Related Mortality in Overheated Cities

Mat Santamouris

Abstract Urban heat island increases the ambient temperature in cities and rises the magnitude of heat-related mortality. Several mitigation technologies including reflective materials for the urban fabric and additional green infrastructure are proposed and implemented to reduce the magnitude of urban overheating. The present work reviews the existing knowledge on the influence of two mitigation technologies, increase albedo and increase green infrastructure on the levels of the ambient temperature and the corresponding impact on heat-related mortality. It is reported that the increase of the urban albedo by 0.1 decreases in average the mean afternoon ambient temperature by 0.09 K. In parallel, higher urban albedo may reduce the magnitude of the heat-related mortality between 0.1 and 4 deaths per day, corresponding to an average decrease of deaths close to 19.8% per degree of temperature drop, or 1.8% per 0.1 increase of the albedo. Increase in the green infrastructure in cities is found to decrease the average maximum peak daily and night-time temperature drop by 1.8 and 2.3 °C, respectively, even for a maximum GI fraction. In parallel, it is found that a drop of the peak daily temperature by 0.1 °C, decreases then the percentage of heat-related mortality on average by 3.0%.

Keywords Urban overheating · Heat mitigation · Heat-related mortality

1 Introduction

Urban overheating is the most documented phenomenon of climate change. It affects the energy consumption of buildings, indoor and outdoor thermal comfort levels, the concentration of harmful pollutants and especially of the ground-level ozone, as well as the magnitude of heat-related mortality and morbidity.

Exposure of human beings to high ambient temperatures consists of a serious health hazard, [1–3], as the human thermoregulation system cannot offset the impact

M. Santamouris (✉)
Anita Lawrence Chair High Performance Architecture, University of New South Wales, Kensington, South Wales, Australia
e-mail: m.santamouris@unsw.edu.au

N. Aghamohammadi and M. Santamouris (eds.), *Urban Overheating: Heat Mitigation and the Impact on Health*, Advances in Sustainability Science and Technology,
https://doi.org/10.1007/978-981-19-4707-0_7

of temperatures above a certain threshold, resulting to increased hospital admissions and heat-related mortality. It is widely accepted that the current levels and the foreseen rise of the heat-related health problems caused by the regional and global climate change consists of one of the peak scientific topics as the predicted figures are extremely alarming, [4, 5]. It is characteristics that between 2000 and 2007, almost 59,114 persons all around the world, passed away during 52 extreme heat events [6].

Serious epidemiological studies carried out during the very recent years have shown that elderly consists of the most vulnerable population group together with people suffering from preexisting health problems like mental health, respiratory and cardiovascular diseases, [7, 8]. In parallel, vulnerable population includes those lacking access to support systems, those using medication affecting the thermoregulation system, socially isolated people and those living in hazardous places, [9–12]. Additionally, socioeconomic parameters are found to affect the levels of heat-related health problems. Given that low income and vulnerable populations use to live in deprived urban zones suffering from high magnitude of urban overheating and pollution, they experience an increased exposure to heat stress and heat-related problems, [13].

To counterbalance the impact of urban overheating, several mitigation technologies have been proposed, developed and implemented, [14]. Among them, the use of advanced and reflecting materials, as well as the increase of the green infrastructure in cities, are the more efficient systems that can decrease the peak ambient temperature up to 2.5 C and fight urban overheating, [15]. It is quite recently that the impact of urban mitigation technologies on heat-related mortality has been assessed in different cities. Most of the studies estimate the potential decrease of the mortality caused by urban overheating systems, using functions associating the levels of mortality against the corresponding ambient temperature, [16].

The present chapter aims to present the existing knowledge on the impact of two heat mitigation technologies, the use of increased urban infrastructure and the potential increase of the urban albedo, on heat-related mortality. Data from several relative studies are presented, while analytical expressions are proposed to estimate the potential decrease in the heat-related mortality caused by the implementation of urban heat mitigation technologies.

2 Heat Mitigation Technologies

To counterbalance the serious impact of urban overheating, several mitigation technologies, systems and strategies are proposed and used. Among them, the most implemented mitigation strategies are those: (a) aiming to rise the albedo of the cities in order to decrease the absorption of solar heat, (b) increase the green infrastructure of cities using additional trees and greenery on the building envelope to enhance evapotranspiration and solar control, (c) increase the evaporative losses in cities using water-based systems and (d) dissipate the excess urban heat into a low-temperature atmospheric heat sink like the ground and the sky.

Increase of the urban albedo using reflective materials is a well-studied mitigation strategy. Higher urban albedo decreases the absorption of heat from the urban structure and thus the urban surface temperature and reduces substantially the release of sensible heat from buildings and other urban structures that contribute to rise the ambient temperature. Several materials are developed and are commercially available to modify the urban albedo. White materials of high reflectance, infrared-reflective coloured materials, thermochromic and photonic structures are developed and tested in numerous urban mitigation projects. All materials present a high reflectance to the visible or infrared part of solar radiation, or in the whole spectrum combined with a relatively high emissivity that permits to emit the stored heat, [17]. Thermochromic materials can change their colour and reflectance as a function of the ambient temperature. Thus, can be very reflective during the warm period and absorbing during the winter, [18]. Photonic materials present a very high reflectance in the solar spectrum combined with a high emissivity in the atmospheric window, (8–13 μm). Photonic materials can achieve up to 20 C sub-ambient temperature under the sun and are considered as the most advanced materials for mitigation purposes, [19]. Fluorescent materials based on the use of nano quantum dots present an important potential for mitigation. Fluorescent materials re-emit part of the absorbed solar energy as fluorescent radiation keeping their mass at relatively low temperature. Development and testing of fluorescent materials for mitigation purposes, have shown that it is possible to decrease their surface temperature up to 13 C below the temperature of a conventional material of the same colour, [20]. A presentation of the various materials technologies for mitigation purposes is shown in Fig. 1.

Green infrastructure in cities including parks, forests, street trees and building integrated greenery can contribute substantially to decrease the ambient temperature of cities and mitigate urban overheating. Cooling is achieved mainly through evapotranspiration and solar control. Increase in the green infrastructure may decrease the daily ambient temperature up to 0.8 C and up to 1.7 C during the night, [21]. Attention should be taken to avoid excessive release of BVOC's from trees that can considerably urban pollution, while the use of water for irrigation has to be constantly monitored.

Other mitigation technologies that are widely used involve the use of evaporative systems as well as the dissipation of the excess urban heat in a low-temperature atmospheric heat sink, [22]. The cooling potential of the various mitigation technologies as calculated and monitored in various large-scale projects is assessed and reported in [23]. In general, the average maximum cooling potential of the existing mitigation technologies does not exceed 2.5 C during the peak, Fig. 2. However, the development and commercialisation of the advanced new materials for mitigation purposes, it is expected to increase the peak temperature cooling potential up to 4–5 °C.

Analysis of mortality records from the 2006 heatwave found that the daily peak ambient temperatures in outer, inner and central London were +0.3 °C, +0.4 °C and +0.5 °C higher, respectively, that those in a reference rural location. Heat-related mortality attributable to London's UHI was estimated to be 37.7, 46.6 and 47.2% higher in outer, inner and central London, respectively [18].

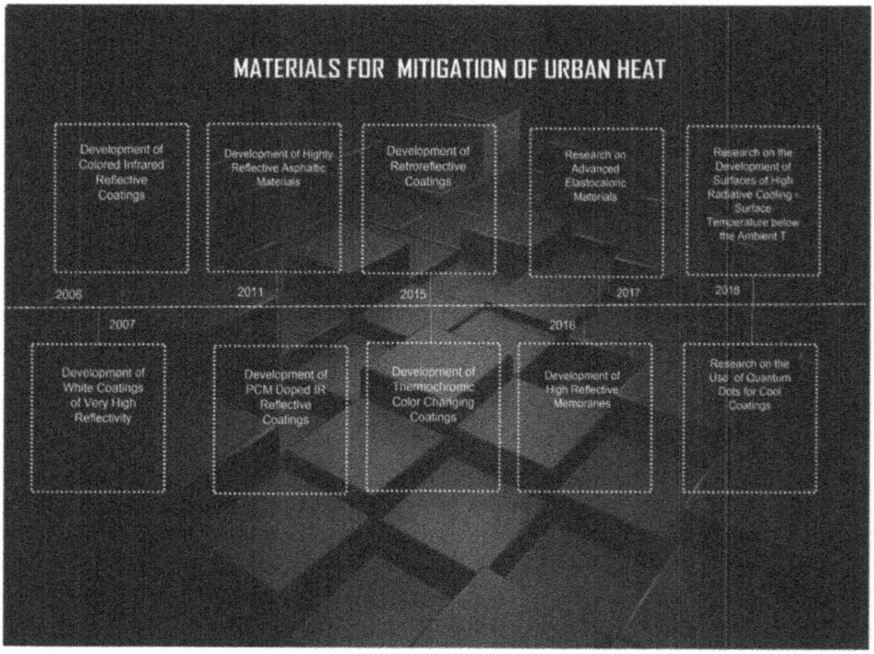

Fig. 1 Evolution of the recently developed material technologies for mitigation purposes in overheated cities (*Source* [PERSONAL DESIGN])

3 Heat-Related Mortality

It is well known that heat-related mortality is substantially increasing above a threshold ambient temperature. Existing studies showed that the threshold temperature differs substantially as a function of the local climatic conditions. Given that the human adaptation differs in different climates, threshold heat-related mortality temperatures are considerably higher in heating than in cooling-dominated climates, [16].

Research efforts focus to better document the impact of climate change and possible climatic mitigation on the magnitude of heat-related mortality using more precise data. Although there is a serious uncertainty around the assessment of the future health problems, the existing projections seem to be extremely alarming, [24]. Increasing urban population, local and global climate change levels, potential adaptation of humans and future levels of air pollution are among the main issues creating uncertainty.

Recent studies and analysis on the relation between the risk of cardiovascular mortality and heat exposure, have concluded that mortality is increasing by 1.3% for the heat exposure of the global population and 8.1% for the elderly population, [25]. In parallel, it is found that because of the urban overheating, the risk of mortality is much higher in cities than in rural areas while heat-related mortality

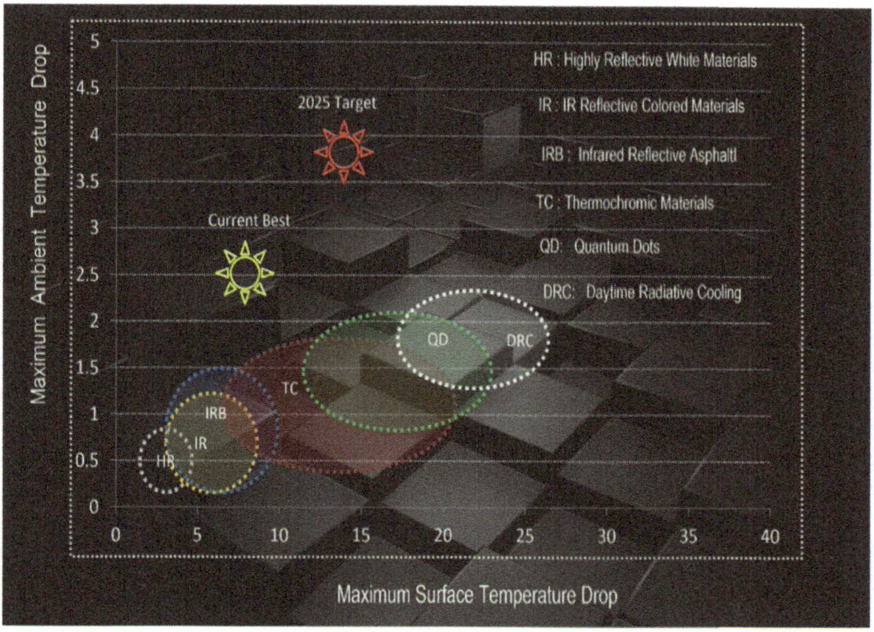

Fig. 2 Evolution of the mitigation potential of several material technologies when implemented in overheated cities (*Source* [PERSONAL DESIGN])

and morbidity increase the same or several days after the exposure to heat, [26], while demographic, biophysical and socioeconomic factors like vulnerability and deprivation affect strongly the levels of heat-related mortality, [27, 28].

Spatial variabilities of cities in terms of heat exposure and distribution of vulnerable populations are the sources of the significant intra—urban variability of heat mortality, [29]. The strength of urban overheating is completely heterogeneous in specific urban zones where ambient temperature is substantially higher, while demographic, socioeconomic and health problems may also be intensified resulting in a serious rise of the local deprivation levels and vulnerability. The impact of place on the magnitude of intra-urban mortality is studied by a high number of studies on heat-related local health outcomes, [30–32], Almost all studies agree that heat-related mortality at the neighbourhood level is influenced through four main pathways: The stresses in the physical and social environment, the availability of neighbourhood institutions and resources and the relative influence and impact of the local networks, [33].

4 The Impact of High Urban Albedo on Heat-Related Mortality

The impact of higher urban albedo on heat-related mortality has been analysed by several studies. A review of 14 relative studies investigating the ambient temperature drop and the corresponding decrease of the heat-related mortality is given in [34]. The same article has provided information on the potential influence and the synergies between the potential health benefits in cities under modified albedo and the socioeconomic urban characteristics and the main climatic and landscape data.

The study concentrates relative information provided for 13 cities and regions and in particular: Baltimore, [35], Detroit, [36], District of Columbia [37], Los Angeles [35, 36], New Orleans [36], New York [35] and Philadelphia in USA [36], Darwin [38] and Parramatta Sydney [39] in Australia, Montreal [40, 41] and Toronto [41] in Canada and West Midlands [42] in the UK. The studies for New Orleans, Philadelphia, Los Angeles. Detroit, Baltimore and District of Columbia reported the calculated impact of several heat waves on heat-related mortality. All the rest of the studies refer to the whole summer period. The full data of all reported studies is given in [34].

The range of the modified albedo varied between 0.1 to 0.7 with an average value close to 0.5. The impact of the modified albedo on the urban temperature has been calculated for most of the case studies using mesoscale climatic modelling. The initial reference mortality range was between 0.46 and 42.6 deaths per day and the final mortality was between 0.3 and 38.6 deaths/day and 100,000 people. The method used to estimate the impact of reduced ambient temperature on heat-related mortality is reported for each study in [34].

The methodology followed in [34], to investigate the magnitude of the temperature and heat-related mortality drop included the following steps:

(A) The average initial or reference heat-related mortality HRMin and final heat-related mortality, HRMfin, per day and 100,000 population were calculated for all days and heat events in each city.
(B) Given that in most of the studies, the drop in the heat-related mortality has been associated with the ambient temperature at 17:00 pm, the mean ambient temperature at 17:00 ($T17$), during all heat events and days, was used as the reference ambient temperature for each city
(C) The drop of the ambient temperature at 17:00, $\Delta T17$, was used to characterise the mitigation potential of reflective materials in cities.
(D) The possible relation and the potential influence of the main, landscape and socioeconomic parameters as well as of the climatic characteristics with the magnitude of HRMin was investigated using parametric analysis.
(E) The magnitude of the final heat-related mortality, HRMfin, was correlated with the initial mortality levels, HRMin, the landscape characteristics, the climatic data and the urban socioeconomic features and parametric relations have been developed.

It is reported that the temperature drop, ΔT17, can be well described by a linear multiparameter correlation of the following form:

$$\Delta T17 = a1 + \sum (\text{ajAlbParam}) + \sum (\text{akClimParam}) + \sum (\text{alLandParam}) \quad (1)$$

where, while ai are correlation coefficients and AlbParam, ClimParam and LandParam, are variables defining the albedo variability, the climate conditions and the landscape and layout characteristics of the cities, respectively.

Analysis has shown that the temperature drop, $\Delta T17$, is correlated at a statistically significant level with the increase of the albedo ΔAlb and percentage of greenery in a city, VC, while the association with the percentage of pavements in a city, PCS is almost statistically significant. In parallel, it is observed that although the correlation with the population density, PD, is not statistically significant, its use in the parametric equation is improving considerably its accuracy. Finally, the proposed parametric correlation takes the form:

$$\Delta T17 = a1 + a2\Delta\text{Albin} + a3\text{VC} + a4\text{PCS} + a5\text{PD} \quad (2)$$

The coefficients ai are:

coefficients: $a1 = -0.261$, $a2 = 0.935$, $a3 = 0.01$, $a4 = 0.013$ and $a5 = -0.000014$.

Equation (2) is valid for Albin in the range between 0.12 and 0.2, for Albfin in the range between 0.25 and 0.7 and for an albedo increase between 0.1 and 0.5.

A parametric correlation between HRMin and the considered climatic, geographic, landscape and socioeconomic data has the following form:

$$\text{HRMin} = f(\text{LAT}, T17, \text{PD}, \text{POV}, \text{VC}, \text{PCS}, \text{POP}) \quad (3)$$

where aj are coefficients, $T17$ the afternoon daily ambient temperature LAT is the latitude of the place, POV is the Poverty Rate (Percentage of People in Poverty) PD is the population density, VC is the percentage of the vegetation cover, PCS is the percentage of streets and POP is the total population.

It was observed that POP and POV, present the highest significance levels with HRMin and thus Eq. (3), has taken the below form

$$\text{HRMin} = a00 + a11\text{POV} + a22\text{POP} \quad (4)$$

If HRMin <0 then HRMin = 0.

where coefficients, (all data),: $a00 = -3.041$, $a11 = 0.236$, $a22 = 4.77\ \text{e}{-06}$.

The initial levels of heat-related mortality, HRMin was found to be in statistically significant association with the population POP and almost statistically significant association with poverty levels, POV.

It is found that the absolute drop in the heat-related mortality caused by the modified albedo varies between 0.06 and 4.0 deaths per day with an average and median value close to 0.9 and 0.4 deaths per day, respectively. In parallel, the percentage decrease of the initial heat-related mortality ranges between 1.1% and 22.6% with an average and median value close to 9.3% and 7.3%. Also, the potential decrease of the initial heat-related mortality, HRMin per degree of temperature drop ranges between 0.13 and 31.1 deaths per day and degree, with an average and median value close to 3.6 and 1.1 deaths per day and degree.

As it concerns the final heat-related mortality caused by the modification of the urban albedo is found to be a function of technical, climatic, landscape, socioeconomic and demographic parameters and should be treated as a multiparameter function.

It is found that the following relation can predict the final magnitude of the heat-related mortality:

$$\begin{aligned}\text{Log(DMort)} &= c0 + c1\text{HRMin} + c2\Delta T17 + c3\text{VC} \\ &\quad + c4\text{PCS} + c5\text{PR} + c6\text{POP} + c7\text{Lat}\end{aligned} \tag{5}$$

It is found that Eq. (5) can predict the decrease of mortality with considerable accuracy, R2 = 0.98. The corresponding coefficients are:

Coefficients: $c0 = -5.82$, $c1 = 0.157$, $c2 = -1.183$, $c3 = 0.0395$, $c4 = 0.098$, $c5 = 0.043$, $c6 = -4.13\text{E}{-}07$, $c8 = 0.031$.

When only the parameters presenting a higher statistical significance are considered then, the final heat-related mortality may be predicted by the following relation:

$$\text{Log(DMort)} = c00 + c11\text{HRMin} + c22\Delta T17 + c33\text{POV} \tag{6}$$

With $c00 = -3.844$, $c11 = 0.089$, $c22 = 1.59$, $c33 = 0.080$ and $R2$ equal to 0.94.

It is evident that the drop in the heat-related mortality is strongly associated with the initial heat-related mortality and the poverty ratio.

Using the previously developed correlations, it may be obtained that:

$$\text{Log(DMort)} = m0 + m1\Delta\text{Alb} + m2\text{HRMin} + m3\text{POV} \tag{7}$$

where $m0 = -3.88$. $m1 = 3.13$, $m2 = 0.084$, $m3 = 0.077$ and $R2$ close to 0.95.

When estimated values of HRMin, like the ones predicted by Eq. (3), are used as inputs in Eq. (7), DMort, may be predicted by the following relation:

$$\text{Log(DMort)} = m0 + m1\Delta\text{Albin} + m2\text{POV} + m3\text{POP} \tag{8}$$

where $m0 = -4.55$. $m1 = 4.16$, $m2 = 0.102$, $m3 = 4.39\text{E}{-}7$ and $R2$ close to 0.956.

5 The Impact of Additional Green Infrastructure on Heat-Related Mortality

The potential impact of increased green infrastructure in cities on heat-related mortality is assessed by a limited number of studies. A full review of five studies assessing the impact of urban greenery on heat-related mortality is presented in [43]. The study analyses and summarises results of five primary articles on the impact of additional tree infrastructure in cities, [22, 44–47]. The primary studies presented results for seven cities including New Orleans, Dallas, Philadelphia and Detroit, USA and Melbourne, Parramatta and Darwin, Australia, while 13 different scenarios are considered. Almost all studies were performed using mesoscale climatic modelling combined with empirical heat-related mortality statistics.

The whole analysis has concluded that the increase of the green infrastructure in cities reduces the magnitude of the heat-related mortality between 1.5% and 49%. It is also observed that there is a strong correlation between the potential drop of the urban temperature at 15:00 pm in a city and the percentage of HRM decrease (Fig. 3).

It is observed that on average drop of the maximum ambient temperature at 15:00 pm by 1 K, may reduce the magnitude of the heat-related by 30.45%. The following relation is proposed to evaluate the percentage decrease of the Heat-related mortality:

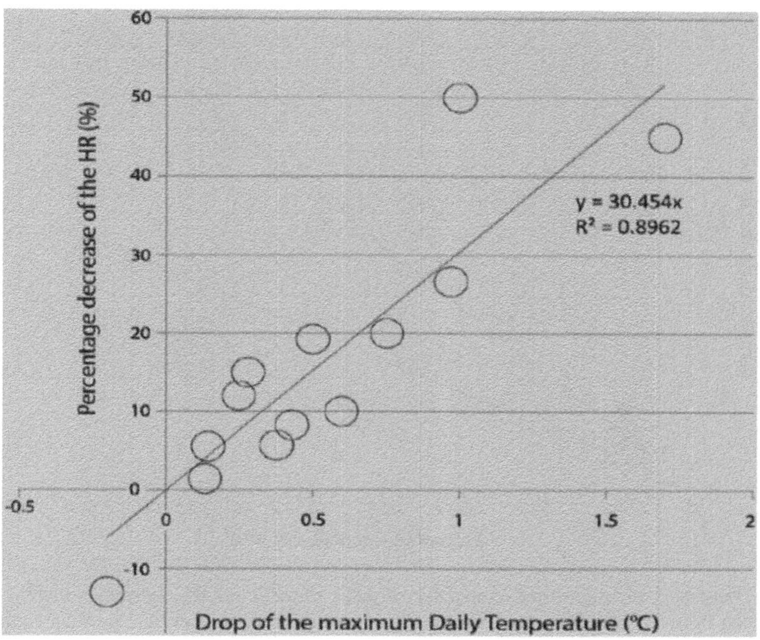

Fig. 3 Correlation between the percentage decrease of the HRM against the corresponding of the peak daily temperature at 15:00 p.m. (*Source* [43])

$$\text{Percentage Decrease of HRM} = 39.45\_\text{DTmax} \quad (9)$$

where DTmax is the decrease of the maximum daily temperature at 15:00 pm and *R*2 is equal to 0.896.

To calculate the DTmax, the following relation is proposed:

$$\Delta T15\max = 0.00013\Delta\text{GI}2 + 0.0079\Delta\text{GI} \quad (10)$$

where ΔGI the increase of the tree cover and an R2 equal to 0.98.

The study has investigated the impact of the additional urban green infrastructure and reduction of the ambient urban temperature and has proposed the relations given in Fig. 4.

The study has concluded to the following conclusions, [43]:

(a) The maximum potential drop of the daily ambient temperature caused by the increase of the green infrastructure may not exceed 1.8 °C even if the green infrastructure increases up to 100%.
(b) For a reasonable increase of the GI by 20%, the average expected daily peak temperature drop is close to 0.3 °C.

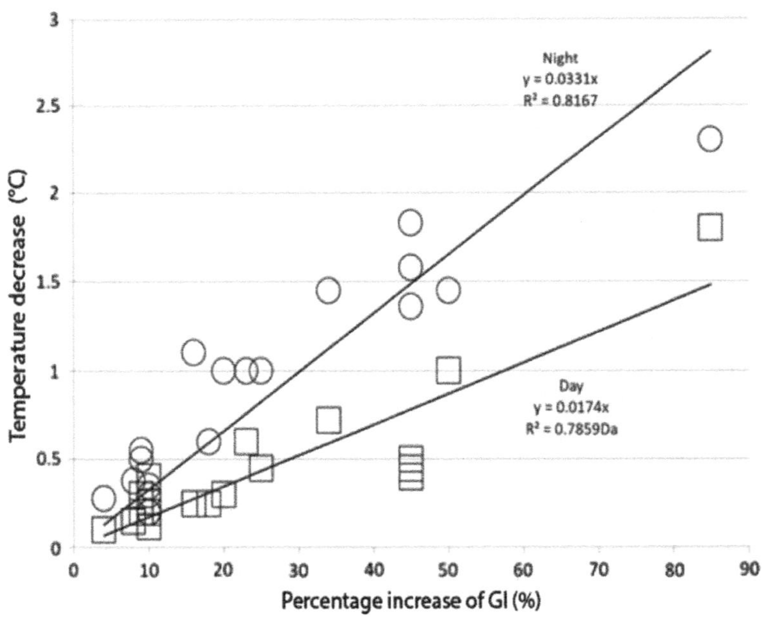

Fig. 4 Daily peak and night-time temperature drop as a function of the considered increase of the GI, (Green Infrastructure). Circles correspond to the night-time temperature drop and squares to the peak daytime decrease of the temperature in the studied cities. (*Source* [43])

(c) During the night, the maximum ambient temperature decreases corresponding to a GI rise of 80% may not exceed 2.3 °C.
(d) During the night the average temperature drop for a GI increase by 20% is close to 0.5 °C.
(e) Not a statistically significant relation is observed between the levels of heat-related mortality and socioeconomic and demographic parameters.

6 Conclusions

Local climate change has a serious impact on the ambient temperature while it affects considerably the magnitude of heat-related mortality. Urban mitigation technologies, involving the implementation of reflective materials at the city scale as well as the increase of the green infrastructure in cities, can seriously contribute to reduce the levels of heat-related mortality. Figures 5 and 6 summarise the main finding and results on the impact of reflective materials and additional urban green infrastructure on heat-related mortality, respectively.

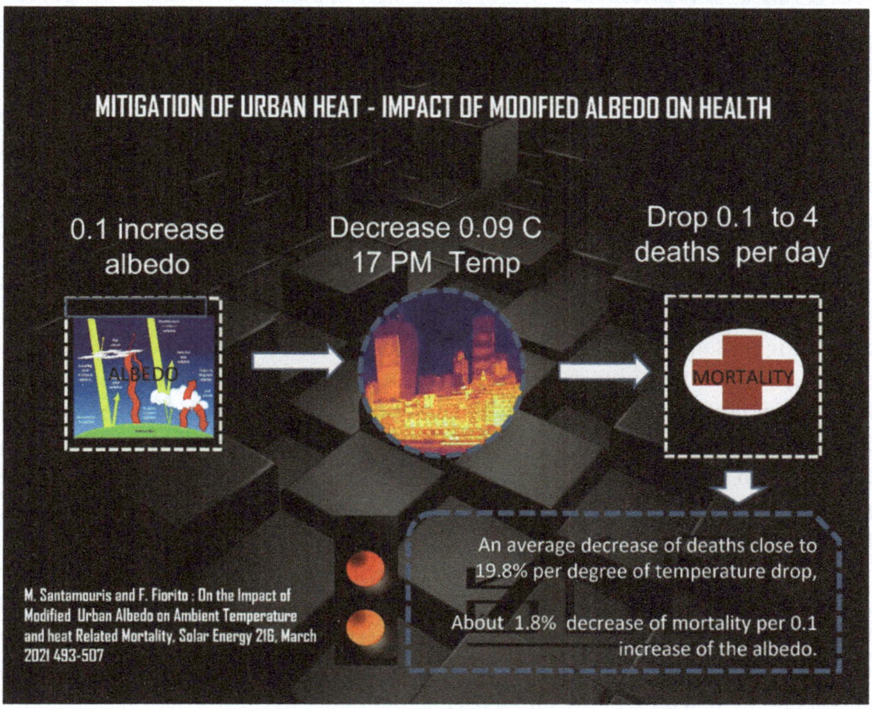

Fig. 5 Synopsis on the impact of the modified urban albedo on heat-related mortality. (*Source* [PERSONAL DESIGN])

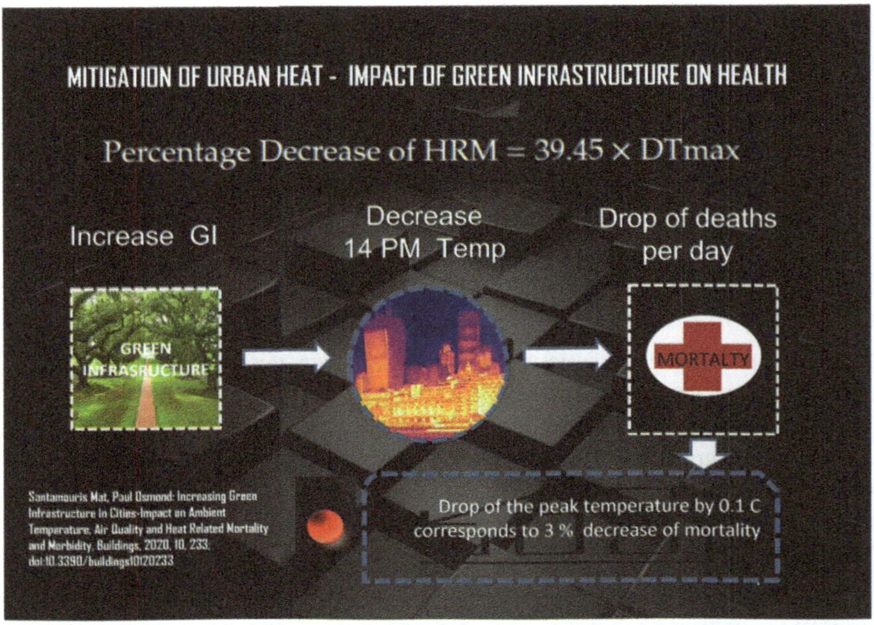

Fig. 6 Synopsis on the impact of additional urban green infrastructure on the magnitude of the heat-related mortality. (*Source* [PERSONAL DESIGN])

Research has shown that it is found that the potential temperature drop caused by the albedo increase in cities is a multiparameter function of the albedo increase and of landscape and socioeconomic characteristics of cities. It is important to note that as shown from the whole analysis, the mitigation potential of reflective materials is found to be in statistically significant relation with the percentage of urban greenery, which means that cities with high levels of green infrastructure benefit from a higher temperature drop when reflective materials are implemented at the city scale. This is because urban parts decrease the ambient temperature inside and around their boundaries, while it is well known that the higher the release of heat and the ambient temperature around them, the lower the penetration of the cool air from the parks into the city and the lower their mitigation capacity. Given those reflective materials decrease the ambient temperature there is an important synergy between the mitigation potential of reflective materials and urban green infrastructure.

Analysis of the data shows that the decrease of the HRM caused by the increase of the albedo in cities, it ranges between 9 and 10%, When normalised, it is obtained that a temperature drop by 1C, caused by the reflective materials, can in average reduce the initial HRM by about 2.08 deaths/day and degree of temperature decrease, corresponding to a drop of the initial average mortality by 19.8%, it is also obtained that in average an increase of the albedo by 0.1 decreases in average the initial mortality by 1.8. This is a very serious contribution of reflective materials on urban public health.

Increase in the vegetation fraction in cities offers serious benefits to the urban climate, pollution control and health. It is also associated with adverse phenomena that may affect air quality in cities. Successful deployment of additional greenery in cities requires a full knowledge of the phenomena and complete assessment of the potential benefits and drawbacks performed using advanced and detailed tools. In parallel, scientific knowledge should considerably improve, mainly through detailed and precise experimental studies.

References

1. Anderson B, Bell M (2011) Heat waves in the United States: mortality risk during heat waves and effect modification by heat wave characteristics in 43 U.S. Communities Environ Health Perspect 119(2):210–218
2. Kovats RS, Hajat S, Wilkinson P (2004) Contrasting patterns of mortality and hospital admissions during hot weather and heat waves in Greater London, UK. Occup Environ Med 61:893–898
3. Gasparrini A, Armstrong B, Kovats S, Wilkinson P (2012) The effect of high tem- peratures on cause-specific mortality in England and Wales. Occup Environ Med 69:56–61
4. Gasparrini A, Guo Y, Sera F, Vicedo-Cabrera AM, Huber V, Tong S, de Sousa Zanotti Stagliorio Coelho M, Saldiva PHN, Lavigne E, Correa PM, Ortega NV, Kan H, Osorio S, Kyselý J, Urban A, Jaakkola JJK, Ryti NRI, Pascal M, Goodman PG, Zeka A, Michelozzi P, Scortichini M, Hashizume M, Honda Y, Hurtado-Diaz M, Cruz JC, Seposo, Kim H, Tobias A, Iñiguez C, Forsberg B, A°ström DO, Ragettli MS, Guo YL, Wu CF, Zanobetti A, Schwartz J, Bell ML, Dang TN, Do Van D, Heaviside C, Vardoulakis S, Ha- jat S, Haines A, Armstrong B (2017) Projections of temperature-related excess mortality under climate change scenarios, Lancet 1
5. Patz JA, Campbell-Lendrum D, Holloway T, Foley JA (2005) Impact of regional climate change on human health. Nature 438:310–317
6. Gosling SN, Lowe JA, McGregor GR, Pelling M, Malamud BD (2009) Associations between elevated atmospheric temperature and human mortality: a critical review of the literature. Clim Change 92:299–341. https://doi.org/10.1007/s10584-008-9441-7
7. Page L, Hajat S, Kovats RS, Howard L (2012) Temperature-related deaths in people with psychosis, dementia and substance misuse, Br. J. Psychiatry
8. Chow WTL, Chuang W-C, Gober P (2012) Vulnerability to extreme heat in metropolitan phoenix: spatial, temporal, and demographic dimensions. Prof Geogr 64(2):286–302
9. Santamouris M, Alevizos S, Aslanoglou L, Mantzios D, Milonas P, Sarelli I, Karatasou CCS, Paravantis JA (2014) Freezing the poor—indoor environmen- talquality in low and very low income households during the winter period in Athens. Energy Build 70:61–70
10. Fouillet A, Rey G, Wagner V, Laaidi K, Empereur-Bissonnet P, Le Tertre A et al (2008) Has the impact of heat waves on mortality changed in France since the European heat wave of summer 2003? a study of the 2006 heat wave. Int J Epidemiol 37:309–317
11. Harlan SL, Brazel A, Jenerette GD, Jones S, Larsen L, Prashad L, Stefanov WL (2008) In the shade of affluence: the inequitable distribution of the urban heat island. Res Soc Probl Public Policy 15:173–202
12. Moghadamnia et al (2017) Ambient temperature and cardiovascular mortality: a systematic review and meta-analysis. Peer J 5:e3574. https://doi.org/10.7717/peerj.3574
13. Jesdale BM, Morello-Frosch R, Cushing L (2013) The racial/ethnic distribution of heat risk-related land covering relation to residential segregation. Environ Health Perspect 121(7):811–817. https://doi.org/10.1289/ehp.1205919

14. Akbari H, Cartalis C, Kolokotsa D, Muscio A, Pisello AL, Rossi F, Santamouris M, Synnefa A, Wong NH, Zinzi M (2016) Local climate change and urban heat Island mitigation techniques—the state of the Art. J Civ Eng Manag 22(1):1–16
15. Santamouris M, Ding L, Fiorito F, Oldfield P, Osmond Paul, Paolini R, Prasad D, Synnefa A (2017) Passive and active cooling for the outdoor built environment—analysis and assessment of the cooling potential of mitigation technologies using performance data from 220 large scale projects. Solar Energy 154:14–33
16. Chen D, Wang 7, Thatcher M, Barnett G, Kachenko A, Prince R (2014) Urban vegetation for reducing heat related mortality. Environ Pollut 192:275–284
17. Santamouris, Yun GY (2020) Recent development and research priorities on cool and super cool materials to mitigate urban heat island. Renew Energy 792–807
18. Karlessi T, Santamouris M, Apostolakis K, Synnefa A, Livada I (2009) Development and testing of thermochromic coatings for buildings and urban structures. Sol Energy 83(4):538–551
19. Santamouris M, Feng J (2018) Recent progress in daytime radiative cooling: is it the air conditioner of the future? Buildings 8(12):168. https://doi.org/10.3390/buildings8120168
20. Garshasbi S, Santamouris SHJVM (2020) Can quantum dots help to mitigate urban overheating? experimental and modelling study. Solar Energy 206:308–31622
21. Mat S, Osmond P (2020) increasing green infrastructure in cities-impact on ambient temperature. Air Qual Heat Related Mortal Morbidity Build 10:233. https://doi.org/10.3390/buildings10120233
22. Gao K, Santamouris M (2020) On the cooling potential of irrigation to mitigate urban heat island. Sci Total Environ 740:139754
23. Baccini M, Biggeri A, Accetta G, Kosatsky T, Katsouyanni K, Analitis A, Anderson HR, Bisanti L, D'Ippoliti D, Danova J et al (2008) Heat effects on mortality in 15 European cities. Epidemiol 19:711–719
24. Sanderson M, Arbuthnott K, Kovats S, Hajat S, Falloon P (2017) The use of climate information to estimate future mortality from high ambient temperature: a systematic literature review. PLoS ONE 12(7):e0180369
25. Moghadamnia et al (2017) Ambient temperature and cardiovascular mortality: a systematic review and meta-analysis. PeerJ 5:e3574. https://doi.org/10.7717/peerj.3574
26. Klein Rosenthal J, Kinney PL, Metzger KB (2014) Intra-urban vulnerability to heat-related mortalityin New York City, 1997–2006. Health Place 30:45–60. https://doi.org/10.1016/j.healthplace.2014.07.014
27. Reid CE, O'Neill MS, Gronlund CJ, Brines SJ, Diez-Roux AV, Brown DG, Schwartz JD (2009) Mapping community determinants of heat vulnerability. Environ Health Perspect 117:1730–1736
28. Smargiassi A, Goldberg MS, Plante C, Fournier M, Baudouin Y, Kosatsky T (2009) Variation of daily warm season mortality as a function of micro-urban heat islands. J Epidemiol Commun Health 63:659–664. https://doi.org/10.1136/jech.2008.078147
29. Laaidi K, Zeghnoun A, Dousset B, Bretin P, Vandentorren S, Giraudet E, Beaudeau P (2012) The Impact of Heat Islands on Mortality in Paris during the August 2003 Heat Wave. Environ Health Perspect 120(2)
30. Robert JR, Dino PR, Julio CS, Teresita MH, Linda MK (1999) The rate and risk of heat-related illness in hospital emergency departments during the 1995 Chicago heat disaster. J Med Syst 23(1)
31. Schinasi LH, Benmarhnia T, Roos AJD (2018) Modification of the association between high ambient temperature and health by urban microclimate indicators: a systematic review and metaanalysis. Environ Res 161:168–180
32. Goggins WB, Ren C, Ng E, Yang C, Chan EYY (2013) Effect modification of the association between meteorological variables and mortality by urban climatic conditions in the tropical city of Kaohsiung Taiwan. Geospatial Health 8(1):37–44
33. Ellen IG, Mijanovich T, Dillman KN (2001) Neighborhood effects on health: exploring the links and assessing the evidence. J Urban Aff 23(3–4):391–408. https://doi.org/10.1111/0735-2166.00096

34. Santamouris M, Fiorito F (2021) On the impact of modified urban albedo on ambient temperature and heat related mortality. Solar Energy 216:493–507
35. Vanos J, Kalkstein LS, Sailor D, Shickman K, Sheridan S Assessing the health impacts of urban heat island reduction strategies in the cities of Baltimore, Los Angeles, and New York
36. Kalkstein LS, Sheridan S (2003) The impact of heat Island reduction strategies on health-debilitating oppressive air massive in urban areas U.S. EPA Heat Island Reduction Initiative
37. Urban Heat Island Management Study. Dallas 2017; Texas Trees Foundation, 2017; Kalkstein LS, Sailor D, Shickman K, Sheridan S, Vanos J (2003) Assessing the health impacts of urban heat island reduction strategies in the District of Columbia. Global Cool Cities Alliance. Contract No.: DDOE ID#2013–10-OPS
38. Haddad S, Riccardo Paolini, Giulia Ulpiani, Afroditi Synnefa, Gertrud Hatvani- Kovacs, Samira Garshasbi, Jonathan Fox, Konstantina Vasilakopoulou, Lawrence Neild, and Mattheos Santamouris: Holistic approach towards urban sustainability: co-benefits of urban heat mitigation in a hot humid region of Australia, Scientific Reports, (2020) 10:14216
39. Santamouris M, Paolini R, Haddad S, Synnefa A, Garshasbi S, Hatvani-Kovacs G et al (2020) Heat mitigation technologies can improve sustainability in cities. An holistic experimental and numerical impact assessment of urban overheating and related heat mitigation strategies on energy consumption, indoor comfort, vulnerability and heat-related mortality and morbidity in cities. Energy Build 217. https://doi.org/10.1016/j.enbuild.2020.110002
40. Jandaghian Z, Akbari H (2018) The effects of increasing surface reflectivity on heat-related mortality in Greater Montreal Area Canada. Urban Clim 25:135–151. https://doi.org/10.1016/j.uclim.2018.06.002
41. Jandaghian Z, Akbari H (2020) Heat mitigation strategy to reduce heat death in toronto and montreal, Canada. Energy Build
42. Macintyre HL, Heaviside C (2019) Potential benefits of cool roofs in reducing heat-related mortality during heatwaves in a European city. Environ Int 127:430–441. https://doi.org/10.1016/j.envint.2019.02.065
43. Kalkstein LS, Sheridan S (2003) The impact of heat island reduction strategies on health-debilitating oppressive air massive in urban areas; U.S. EPA Heat Island Reduction Initiative: Berkeley, CA, USA
44. Chen D, Wang 7, Thatcher M, Barnett G, Kachenko A, Prince R (2014) Urban vegetation for reducing heat related mortality. Environ Pollut 192:275–284. [CrossRef]
45. Haddad S, Barker A, Yang J, Ilamathy D, Kumar M, Garshasbi S, Paolini R, Santamouris M (2020) On the potential of building adaptation measures to counterbalance the impact of climatic change in the tropics. Energy Build 229:110494. [CrossRef]
46. Yenneti K, Ding L, Prasad D, Ulpiani G, Paolini R, Haddad S, Santamouris M (2020) Urban overheating and cooling potential in Australia: an evidence-based review. Climate 8:126. [CrossRef]
47. Mills DM, Laurence Kalkstein L (2012) Estimating reduced heat-attributable mortality for an urban revegetation project. J Heat Island Inst Int 7:18–24

Chapter 8
Heat-Related Mortality in London

Anna Mavrogianni, Jonathon Taylor, Ai Milojevic, Eleni Oikonomou, Clare Heaviside, Ioanna Tsoulou, Giorgos Petrou, Phil Symonds, Mike Davies, and Paul Wilkinson

Abstract This chapter provides an overview of factors affecting heat-related morbidity and mortality in the Greater London Area. Evidence from past heatwave periods indicates that London's Urban Heat Island effect contributes substantially to heat-related mortality rates. During a warm period in 2006, the proportion of heat deaths attributable to this effect was estimated to be 37.7% in outer London, 46.6% in inner London and 47.2% in central London, compared to a reference rural location. The relative effect of the heat island needs to be also examined in conjunction with the influence of individual building geometry, fabric and system characteristics. Epidemiological analyses of London mortality records have demonstrated that the spatial variation of heat-related mortality in London reflects background mortality rates related to population age. As such, higher levels of excess heat-related mortality occur in areas towards the outskirts of London with larger proportions of older residents. Housing features, however, cause a larger variation in temperature exposure and associated risks than the heat island intensity alone. This chapter briefly discusses a series of London urban heat health assessment methods and decision-making support tools.

Keywords London · Urban environment · Cities · Heat · Overheating · Urban temperature · Urban warming · Urban climate · Urban heat island · Climate change · Thermal comfort · Health · Wellbeing · Heat-related mortality

---

A. Mavrogianni (✉) · E. Oikonomou · C. Heaviside · I. Tsoulou · G. Petrou · P. Symonds · M. Davies
Institute for Environmental Design and Engineering, University College London, London, UK
e-mail: a.mavrogianni@ucl.ac.uk

J. Taylor
Faculty of Built Environment, Tampere University, Tampere, Finland

A. Milojevic · P. Wilkinson
London School of Hygiene and Tropical Medicine, London, UK

N. Aghamohammadi and M. Santamouris (eds.), *Urban Overheating: Heat Mitigation and the Impact on Health*, Advances in Sustainability Science and Technology,
https://doi.org/10.1007/978-981-19-4707-0_8

1 Introduction

As a growing global city, London faces a multitude of pressures on its building stock, infrastructure, services, health and wellbeing of residents. Climate change will exacerbate these existing challenges and adverse health effects of excess heat exposure will be intensified by London's Urban Heat Island (UHI), one of the best-documented heat islands worldwide. London has been at the forefront of multi-stakeholder climate change adaptation and mitigation policy; its integrated *London Environment Strategy* outlines current and future risks associated with urban overheating [1]. According to current climate change projections [2], London heatwaves are expected to become increasingly common by the middle of the century. This will not only significantly increase indoor overheating risk in domestic and non-domestic properties; it is also likely to pose a considerable strain on critical infrastructure, public transport and the electricity grid if buildings increasingly rely on air conditioning. Droughts and limited water access are also likely to become an issue. All these factors will intensify health inequalities.

Drawing on existing research, this chapter summarises the evidence on the drivers of heat-related mortality in London and their interdependencies. The implications for planning and built environment design are discussed. While the focus is on the Greater London Area, some of the observations made in the chapter may also apply to other major cities in currently heating-dominated, temperate climates that face challenges related to urban overheating.

2 The London Heat Island

London's Urban Heat Island (UHI) constitutes one of the best-documented instances of inadvertent urban climate modification [3–8]. It was first described in the early nineteenth century by Luke Howard in the book *The Climate of London, deduced from Meteorological Observations, made at different places in the neighbourhood of the Metropolis* [8]. As early as 1952, mobile traverse data of urban and rural temperature across London were used to quantify London's UHI in the book *The Climate of London* [7]. Statistical analysis of temperature records indicates that London's nighttime UHI intensified between the late 1950s and 1980s, possibly due to a rise in frequency of occurrence of summer anticyclonic conditions [3] and urban densification trends. According to a monitoring campaign of London's heat island between 1999 and 2000 [9, 10], the average UHI intensity, i.e., the temperature difference between London's urban centre and surrounding rural areas was 2 °C, with a minimum of 1 °C observed during the daytime and a peak of 3–4 °C recorded after sunset. This is in accordance with more recent data obtained from monitoring and modelling studies, although there has also been some evidence of UHI intensities approaching 11 °C during the unprecedented 2003 heatwave [11]. However, the relationship between local urban temperatures and distance from the centre of London is not linear due to its highly heterogeneous urban environment and human activities taking place across the city. Large temperature variations over short distances reflecting varying local microclimatic characteristics have been observed [12].

The statistical analysis of daily death counts in Greater London between 1976 and 1996 demonstrated that historically excess heat-related mortality may be observed above relatively low thresholds of ambient temperature in London [13]. More recent work estimated that, between 1996 and 2013, 0.5% of total deaths and 0.4% of Years of Life Lost in London are attributable to heat [14].

London's heat-related mortality threshold, i.e., the ambient temperature threshold above which excess mortality is observed, is 19.6 °C (daily mean temperature 93rd centile). Past and projected annual heat and cold mortality burdens in London have been presented by Hajat et al. [15]; based on an ensemble of nine climate model realisations, heat-related deaths in the London regions/year/100,000 population of all ages range from on average 6.1 in the 2020s (min: 2.8, max: 10.8) and 17.5 in the 2080s (min: 8.4, max: 27.9). The corresponding cold-related death figures are on average 71 in the 2020s (min: 65.8, max: 83.1) and 48.8 in the 2080s (min: 35.7, max: 63). Similarly to temporal trends of urban heat-related mortality in other cities around the world, timing of heat events is crucial with higher excess mortality observed in London if hot spells occur earlier in the year as opposed to later on; this is attributed to the potentially higher levels of preparedness of the population and public health infrastructure. London heat-related mortality is also strongly correlated with the magnitude and duration of heat events. In terms of hospital admissions, heat is estimated to cause more than 100,000 patient-days in hospitals in London annually during a typical year under the current climate [16]. London's UHI has, however, a protective effect in winter, resulting in less snowfall and ice and a reduction in cold-related injuries, illnesses and deaths.

During more recent heatwave events, London's UHI was found to markedly increase heat-related mortality (Fig. 1): During the August 2003 heatwave, 616 excess heat-related deaths occurred in London, the majority of which were among people older than 75 years of age [17]. Mortality in London increased by 42% compared to the baseline; the corresponding % increase in England and Wales was 12%, clearly demonstrating the higher heat-related mortality risk facing London residents.

An analysis of mortality records from the 2006 heatwave found that the daily peak ambient temperatures in outer, inner and central London were +0.3 °C, +0.4 °C and +0.5 °C higher, respectively, that those in a reference rural location. Heat-related mortality attributable to London's UHI was estimated to be 37.7%, 46.6% and 47.2% higher in outer, inner and central London, respectively [18]).

Figures 2 and 3 illustrate modelling outputs of two different local urban climate models:

- London's UHI during a warm summer represented by mean midnight temperature during May–August 2006 (LondUM simulations) [19]
- London's UHI during an average summer represented by mean midnight temperature during May–September 2011 (UrbClim simulations) [20]

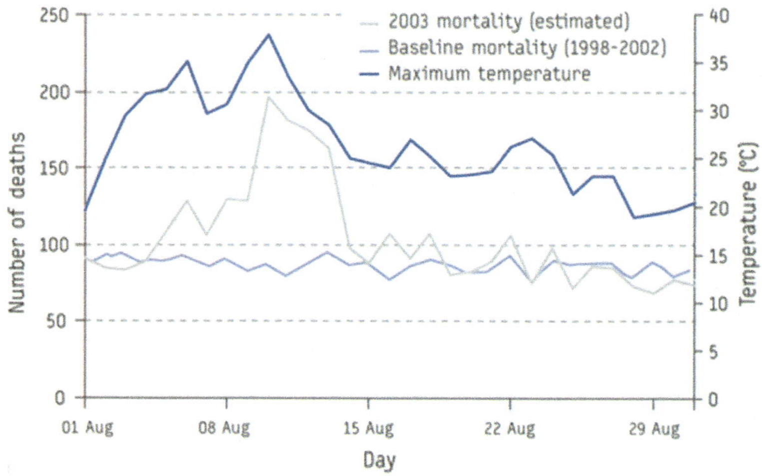

Fig. 1 Daily mortality, 75 + years, London, August 2003 (*Source* [17])

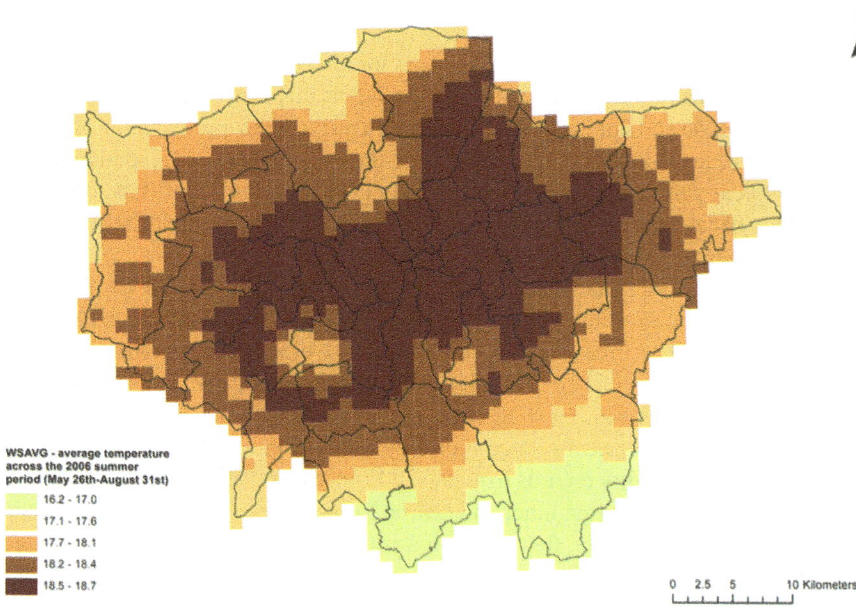

Fig. 2 Average air temperature in London, May 26th—August 21st 2006 (*Source* [19])

The heterogeneity of local urban temperatures illustrated in these maps possibly reflects the spatial variation of both building and urban fabric characteristics and likely affects the resulting heat-related morbidity and mortality risk. Research that attempts to quantify and map the heterogeneity of these factors is presented in the next two sections.

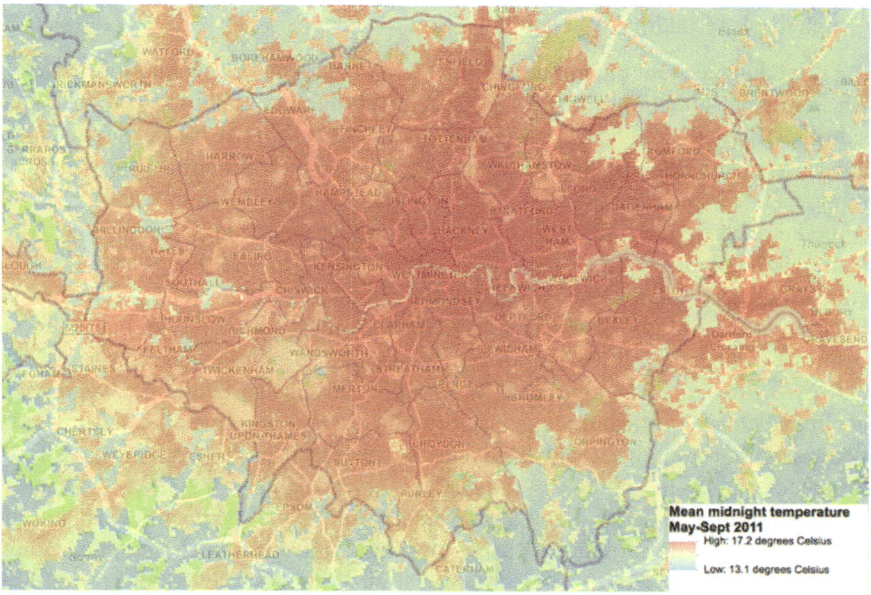

Fig. 3 Average midnight air temperature in London, May–September 2011 (*Source* [20])

3 Factors Influencing Heat Exposure in London

People's heat exposure levels depend not only on outdoor thermal conditions but also on the thermal characteristics of the indoor spaces they occupy. Building envelopes are important modifiers of heat exposure in temperate climates, such as that of London. Although London's UHI has a significant influence on the magnitude of heat-related mortality risk, as discussed in the previous section, the relative contribution on heat exposure of the building fabric characteristics of the dwelling an individual inhabits is appreciably higher than that of the dwelling's location [21]. In other words, when different dwellings are exposed to the same outdoor thermal conditions, the between-dwelling indoor temperature differences are much larger than the differences that are observed when dwellings of the same type are exposed to external temperatures of maximum vs. minimum UHI intensity, under typical UHI conditions.

Although more research is required to link building design features with health impacts, there is broad agreement across research studies that overheating risk is higher for homes with:

- very high or very low thermal insulation levels (in particular uninsulated or internally insulated top-floor flats),
- lack of solar heat gain control (e.g., no shading systems),
- large areas of glazing,
- poor ventilation characteristics (e.g., no cross-ventilation),

- high levels of internal heat gains.

Taking into account the importance of nighttime urban overheating for mortality risk, particular attention needs to be paid to the thermal conditions prevailing in bedrooms: The effect of heat exposure on summer mortality risk has been found to be much larger in homes where overheating is experienced in bedrooms during the nights of hot spells [18]. Analysis of 2001–06 summer data found that for each degree of temperature above the London heat-related mortality threshold, the estimated slope of the temperature-mortality curve was 3.3% (95% C.I. 2.5, 4.0). On the hottest days, this slope was only modified a little by daytime living room temperatures (interaction 0.1 (−0.3, 0.6)% per degree overheating), respectively), but bedroom nighttime overheating showed significant modification (interaction 1.2 (0.4. 2.0) %).

It is of fundamental importance that energy retrofit of London housing driven by climate change mitigation targets is combined with climate resilience strategies, such as cooling solutions, to avoid adverse health effects in the summer. It has been estimated that, by the middle of the century, approximately 261–269 cold-related deaths per million population could be avoided if ambitious energy efficiency retrofit took place across the London housing stock [22]. If these interventions are not, however, combined with overheating mitigation strategies, they would potentially result in an additional 12–13 heat-related deaths per million. The installation of external shading in homes, such as window shutters, could help avoid 38–73% of heat-related mortality during typical summers by the 2050s [22].

In addition to the importance of local climate and building fabric characteristics, individual susceptibility to heat varies widely, as discussed in Chap. 1. Population age is a key determinant of heat vulnerability, with people above 65 being at highest risk, alongside other factors, such as underlying chronic conditions, e.g., cardiovascular or respiratory disease. The next section presents research efforts to map those intersecting risk factors across the Greater London Area.

4 Mapping London Heat-Related Mortality Risk

The *Triple Heat Jeopardy* framework, produced by Taylor et al. [23] and illustrated in Figs. 4, 5, 6 and 7 [24], is the first comprehensive attempt to map, at a high spatial resolution, the London heat risk factors outlined above, in particular:

- local urban temperature variation as a result of London's UHI (based on LondUM modelling data for the period 26th May–19th July 2006, including a four-day heatwave period),
- indoor temperature predictions produced by a London housing stock model based on building performance simulation and housing stock survey data,
- population age as a proxy for heat vulnerability.

The main finding emerging from the triple heat jeopardy mapping work is that spatial variation patterns of heat-related mortality primarily reflect underlying

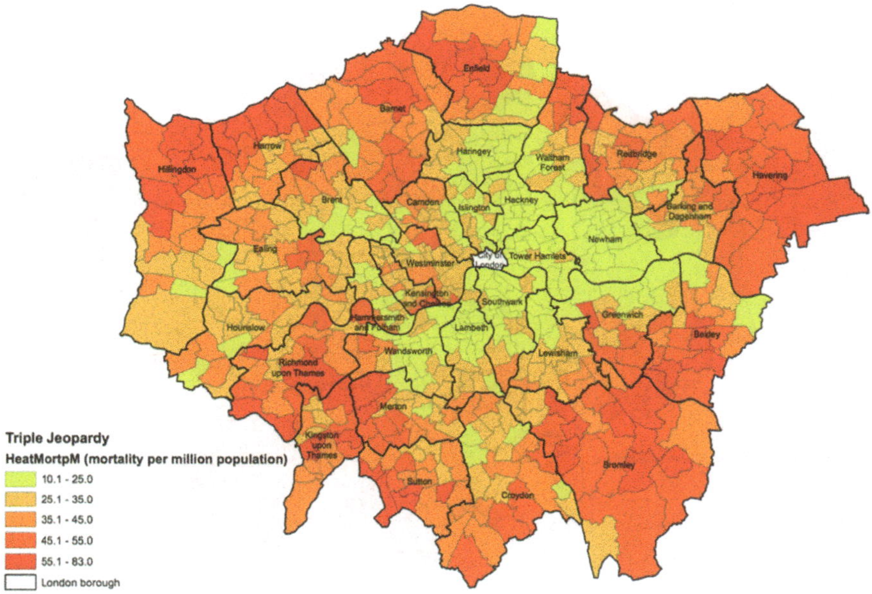

Fig. 4 Total estimated mortality due to heat (outdoor and indoor) per million population for the period 26th May—19th July 2006, inclusive of age effects (*Source* [24])

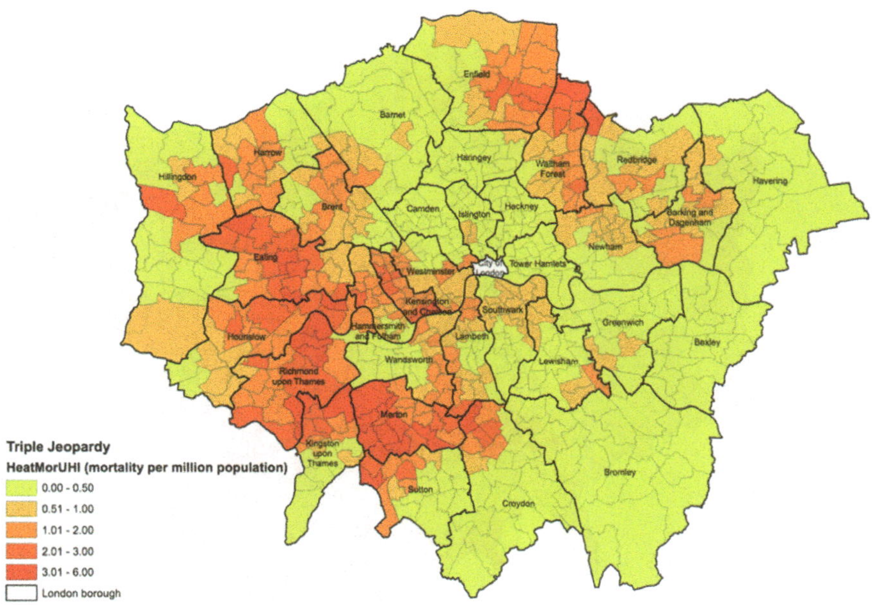

Fig. 5 Estimated mortality per million population due to increased outdoor temperature exposure caused by the UHI for the period 26th May—19th July 2006 (excluding the effect of overheating housing), inclusive of age effects (*Source* [24])

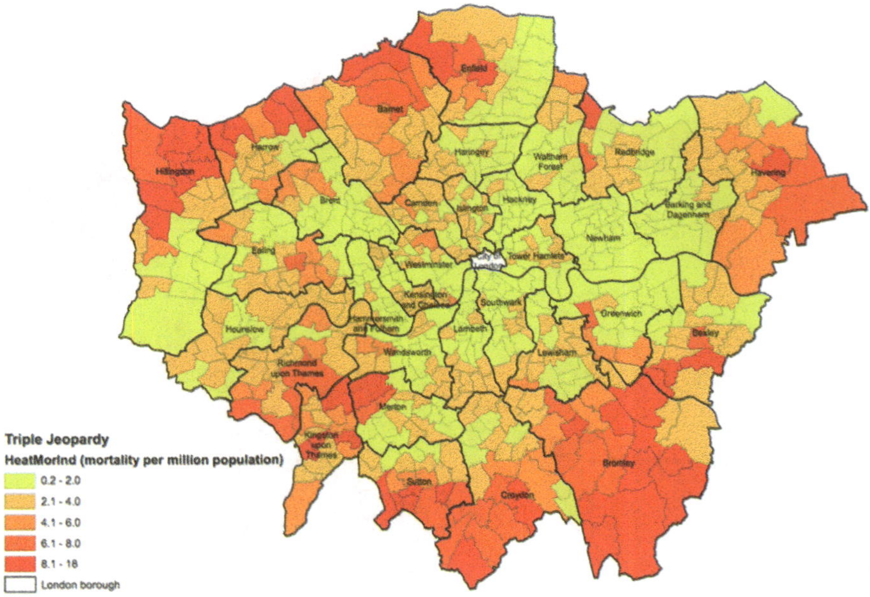

Fig. 6 Estimated mortality per million population due to increased temperature exposure caused by heat-vulnerable dwellings (excluding the effect of the UHI) for the period 26th May—19th July 2006, inclusive of age effects (*Source* [24])

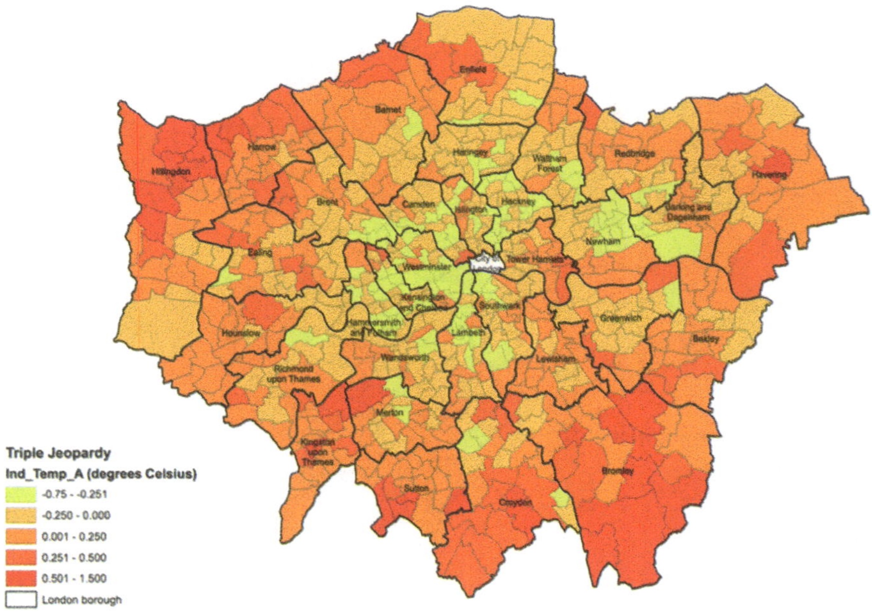

Fig. 7 Indoor temperature anomaly defined as the difference between the estimated indoor temperatures for dwellings and the average indoor temperature estimate for the whole of London for the period 26th May—19th July 2006, averaged by ward (*Source* [24])

mortality rates across London, which are in turn driven by population age. As a result, geographic clusters of high heat-related mortality appear to be concentrated in the outskirts of London that have older populations, but also in parts of East and Central London. In more central locations, overheating prone dwellings, such as top-floor flats in high-rise blocks, may not be highlighted in the map as the overall risk may be averaged out by lower-risk dwellings in the same block, e.g., on the ground floor.

A similar mapping approach was adopted for the London Heat Vulnerability Index (HVI) [25]; while this index did not include building fabric characteristics, it used an extended list of sociodemographic factors. Further work could also incorporate heat risk confounders, such as air pollution.

These tools have been used in the context of collaborative work with London Local Authorities and have the potential to be further developed into useful policy decision-making tools for heat risk prevention across Greater London.

5 Conclusions

While cold-related morbidity and mortality will remain the primary seasonal environmental health concern for London in the next decades, urban heat resilience is also an urgent research and policy priority. Estimating population risk at a fine spatiotemporal resolution needs to factor in the interdependencies between numerous confounding factors, including the UHI, ambient air pollution, building stock characteristics and individual susceptibility to heat. Interventions at the building and urban scale should integrate climate change mitigation and adaptation targets. A strong, evidence-based policy framework is required to ensure that London can successfully prepare for, respond to and recover from the adverse effects of urban overheating.

References

1. MoL (2018) London environment strategy. London, UK. Retrieved from https://www.london.gov.uk/what-we-do/environment/london-environment-strategy
2. MetOffice (2021) UK climate projections 2018 (UKCP18)
3. Wilby RL, Jones PD, Lister DH (2011) Decadal variations in the nocturnal heat island of London. Weather 66(3):59–64
4. Wilby RL (2003) Past and projected trends in London's urban heat island. Weather 58(7):251–260
5. Graves H, Watkins R, Westbury P, Littlefair P (2001) Cooling buildings in London: overcoming the heat island. London, UK
6. Lee DO (1992) Urban warming? An analysis of recent trends in London's urban heat island. Weather 47:50–56
7. Chandler TJ (1965) The climate of London. London, UK, Hutchinson
8. Howard L (1833) The climate of London, deduced from meteorological observations, made at different places in the neighbourhood of the metropolis. Harvey and Darton, London, UK

9. Watkins R, Palmer J, Kolokotroni M, Littlefair P (2002) The London heat island: results from summertime monitoring. Build Serv Eng Res Technol 23(2):97–106. https://doi.org/10.1191/0143624402bt031oa
10. Watkins R, Palmer J, Kolokotroni M, Littlefair P (2002) The balance of the annual heating and cooling demand within the London urban heat island. Build Serv Eng Res Technol 23(4):207–213. https://doi.org/10.1191/0143624402bt043oa
11. MoL (2011) Managing risks and increasing resilience, The Mayor's climate change adaptation strategy. London, UK
12. Kolokotroni M, Giridharan R (2008) Urban heat island intensity in London: an investigation of the impact of physical characteristics on changes in outdoor air temperature during summer. Sol Energy 82(11):986–998. https://doi.org/10.1016/j.solener.2008.05.004
13. Hajat S (2002) Impact of hot temperatures on death in London: a time series approach. J Epidemiol Community Health 56(5):367–372. https://doi.org/10.1136/jech.56.5.367
14. Arbuthnott K, Hajat S, Heaviside C, Vardoulakis S (2020) Years of life lost and mortality due to heat and cold in the three largest English cities. Environ Int 144(July). https://doi.org/10.1016/j.envint.2020.105966
15. Hajat S, Vardoulakis S, Heaviside C, Eggen B (2014) Climate change effects on human health: projections of temperature-related mortality for the UK during the 2020s, 2050s and 2080s. J Epidemiol Community Health 68(7):641–648. https://doi.org/10.1136/jech-2013-202449
16. LCCP (2012) A summary of climate change risks for London. London, UK. Retrieved from http://climatelondon.org/wp-content/uploads/2017/11/CCRA-London.pdf
17. Johnson H, Kovats S, Mcgregor G, Stedman J, Gibbs M, Walton H (2005) The impact of the 2003 heat wave on daily mortality in England and Wales and the use of rapid weekly mortality estimates. Eurosurveillance 10(7)
18. Milojevic A, Armstrong B, Taylor J, Mavrogianni A, Davies M, Wilkinson P (2014) Modification of outdoor heat-related mortality relationship by housing vulnerability to overheating in England and Wales. In International society for environmental epidemiology (ISEE) 26th conference, 24–28 August 2014. Seattle, Washington, USA
19. GLA (2021) London's urban heat island—during a warm summer. Greater London Authority (GLA). Retrieved from https://data.london.gov.uk/dataset/london-s-urban-heat-island
20. GLA (2021) London's urban heat island—average summer. Greater London Authority (GLA). Retrieved from https://data.london.gov.uk/dataset/london-s-urban-heat-island---average-summer
21. Oikonomou E, Davies M, Mavrogianni A, Biddulph P, Wilkinson P, Kolokotroni M (2012) Modelling the relative importance of the urban heat island and the thermal quality of dwellings for overheating in London. Build Environ 57(August 2003):223–238. https://doi.org/10.1016/j.buildenv.2012.04.002
22. Taylor J, Symonds P, Heaviside C, Chalabi Z, Davies M, Wilkinson P (2021) Projecting the impacts of housing on temperature-related mortality in London during typical future years. Energy Build 249. https://doi.org/10.1016/j.enbuild.2021.111233
23. Taylor J, Wilkinson P, Davies M, Armstrong B, Chalabi Z, Mavrogianni A, Bohnenstengel SI et al (2015) Mapping the effects of Urban Heat Island, housing, and age on excess heat-related mortality in London. Urban Climate 14:517–528
24. GLA (2021) Mortality risk from high temperatures in London (Triple Jeopardy Mapping). Greater London authority (GLA). Retrieved from https://data.london.gov.uk/dataset/mortality-risk-from-high-temperatures-in-london--triple-jeopardy-mapping-
25. Wolf T, McGregor G (2013) The development of a heat wave vulnerability index for London, United Kingdom. Weather Clim Extremes 1(August 2003):59–68. https://doi.org/10.1016/j.wace.2013.07.004

Chapter 9
The Impact of Greenery on Heat-Related Mortality in Sydney, Australia

Mahsan Sadeghi

Abstract Urban heat stress results from increased air temperature, surface temperature, and radiative fluxes including reflected shortwave fluxes and infrared emissions from the urban surfaces surrounding the pedestrian's body. In the urban context, due to the increased manmade surfaces, reduced vegetated and water surfaces, the adverse effects of this phenomenon on the body's heat balance are exacerbated. This study quantifies the benefits of greening infrastructure on urban heat stress and its associated mortality impacts by applying a Heat Health Impact (HHI) method to Sydney, Australia. Firstly, meteorological data from all the available weather stations throughout metropolitan Sydney were translated into the Universal Thermal Climate Index (UTCI) for each station—hereafter referred to as the observation study. Secondly, greening infrastructure was implemented throughout metropolitan Sydney in an urban climate model, and again, UTCI was calculated, this time by using air temperature predicted by simulating post-greening scenario—called intervention study. Finally, an extant Health Impact Assessment (HIA) model was applied in order to estimate the change in heat-attributable mortality post greenery intervention. This analysis suggests that urban greenery could reduce between 5 and 8 h of exposure to heat stress during a heatwave day in metropolitan Sydney. Greening could reduce UTCI between 1.3 and 3 °C during daytime in a heatwave, and between 0.3 and 2.7 °C during nighttime. Application of health impact assessment model to UTCI reductions of these magnitudes could reduce the attributable number of deaths by as many as 11.7 per day for Sydney's current population of 5.7 million.

Keywords Urban heat island · Greenery · UTCI · Heat health impact method · Health impact assessment · Heat-related mortality

M. Sadeghi (✉)
Centre for Air Pollution, Energy, and Health Research (CAR), NHMRC Centre of Research Excellence (CRE), Sydney 2037, Australia
e-mail: msad5726@uni.sydney.edu.au

School of Architecture, Design & Planning, The University of Sydney, Sydney 2006, Australia

N. Aghamohammadi and M. Santamouris (eds.), *Urban Overheating: Heat Mitigation and the Impact on Health*, Advances in Sustainability Science and Technology,
https://doi.org/10.1007/978-981-19-4707-0_9

1 Introduction

Australia has experienced extreme climate events and hazards such as bushfires, floods, and heatwaves in recent decades. It has been thoroughly demonstrated and largely accepted that, compared to current climatological norms, Australia faces a drier and hotter climate in 2050, largely due to anthropogenic forcing on the global climate system [1]. Heatwaves in Australia are already becoming more frequent and severe, killing more people than any other type of natural disaster [2, 3]. Apart from global climate impacts, local and mesoscale environmental conditions also play significant roles in modifying urban climate circumstances. Australia is ranks among the most urbanized countries in the South Pacific, and the processes of urbanization have exacerbated urban-scale climatic impacts, particularly urban heat islands (UHI) in which air and surface temperatures in built-up areas are significantly warmer compared to their rural surroundings [4].

Prompted by these urban environmental impacts, extensive research has been focused on the development of urban climatic mitigation technologies and strategies (i.e., [5–7]) with the specific aim being to provide governments, planning decision-makers, communities, and stakeholders with a scientific basis to manage urban heat island effects within the large cities. In addition to its environmental effects, urban heat also poses risk to public health, particularly that of vulnerable subpopulations [8]. Empirical correlations between exposure to extreme heat and risk of mortality and morbidity, such as cardiovascular disease, are reported extensively in the literature; this mostly happens when human body cannot manage to balance heat gains with adequate heat loss, which eventually results in dehydration, cardiovascular problems, and ultimately, death (i.e., [9–13]).

Direct impacts of urban heat on residents' health result in various indirect physical and mental health impacts, placing an even greater economic burden of heat-related disease on society. These secondary and tertiary impacts are far more complex and yet to be fully quantified [14]. Therefore, it is essential to correctly identify heat-related health impacts, understand the risk of exposure to extreme heat, and prioritize adaptation responses in order to mitigate extreme heat exposure. This chapter aims to assess the impact of enhanced greenery on urban heat stress and to predict the risk of heat-related mortality for a heat wave case study conducted in metropolitan Sydney, Australia. This chapter contains as the following sections: (a) climatological analysis of Sydney basin (observation study), (b) prediction of the impact of greening infrastructure on urban heat stress (intervention study), and (c) risk estimate of heat-related mortality attributable to heat stress reduction caused by an urban greening strategy across the Sydney basin. The observation scenario refers to the reference study, where intervention has not been applied. The intervention scenario refers to the greenery —implementation of greening infrastructure.

2 Climatological Study in Metropolitan Sydney—Observation Study

Located on the east coast of the Australian continent, where warm ocean currents exert moderating effects, the climate of Sydney is described as humid subtropical (Cfa) according to the Köppen-Geiger climate classification, with warm-to-hot summers and mild winters [15]. Proximity to the ocean moderates the seasonal contrasts, while bringing benefits of easterly sea breezes. However, Sydney's sprawling western suburbs, distanced from the maritime moderation, experience more temperature extremes, particularly during episodic incursions of hot and dry westerly air masses from the vast Australian continent's arid interior [16–18].

2.1 Meteorological Analysis

Hourly meteorological data were observed from 10 available meteorological stations across the Sydney metropolitan area for the entire year 2017 (total of 8,760 h). Meteorological data was measured by the Australian Bureau of Meteorology [19] and is available on the New South Wales (NSW) Department of Planning, Industry, and Environment (DPIE) website [20]. The location of meteorological stations is demonstrated in Fig. 1. The stations were named Richmond, St Marys, Prospect, Lindfield, Rozelle, Chullora, Liverpool, Earlwood, Randwick, Camden. The observed meteorological data obtained for the purpose of this study were temperature, humidity, wind speed, and solar radiation.

As a part of limitation of this study, the availability of solar radiation data was limited to only three observation stations, namely prospect (located in west inland Sydney), Chullora (in the southwest), and Rozelle (in the eastern coastal area). Solar radiation data from these three weather stations were assigned to the other seven stations. The method applied for data replacement was based on: (a) the closest distance between stations, and (b) their coastal or inland classification [21]. Meteorological data used in this study and the data collection method are demonstrated in Table 1.

2.2 Universal Thermal Comfort Index (UTCI) Simulations

Universal Thermal Climate Index (UTCI) is a state-of-the-art multi-node model which quantifies urban climate conditions into human body thermal physiology terms [22, 23]. UTCI integrates physiological effects of four environmental parameters; air temperature (Ta), mean radiant temperature (mrt), relative humidity (RH), and wind speed (Ws) [24]. For personal parameters, UTCI assumes a walking person with 2.3 metabolic rates (met), with an adaptive clothing algorithm (clo) [25, 26].

Fig. 1 Distribution of meteorological observation stations across metropolitan Sydney

Table 1 Meteorological parameters, data collection method, and meteorological stations observed data

Meteorological parameters	Unit	Data collection methods	Meteorological stations
Air temperature (Ta)	°C	Observation	Richmond, St Marys, Prospect, Lindfield, Rozelle, Chullora, Liverpool, Earlwood, Randwick, Camden
Relative humidity (RH)	%	Observation	Richmond, St Marys, prospect, Lindfield, Rozelle, Chullora, Liverpool, Earlwood, Randwick, Camden
Wind speed (Ws)	m/s	Observation	Richmond, St Marys, prospect, Lindfield, Rozelle, Chullora, Liverpool, Earlwood, Randwick, Camden
Solar radiation—global horizontal radiation (GHR)	Wh/m^2	Observation	Prospect, Chullora, Rozelle
Mean radiant temperature (mrt)	°C	Simulation	Not applicable (NA)

The complex section in calculation of human heat balance is related to mean radiant temperature [27, 28]. To be able to calculate UTCI, firstly we simulated mean radiant temperature (mrt) using RayMan© Version 1.2 [29] software by using observed global solar radiation (W/m2). To predict mrt we imported air temperature, relative humidity, day of the year, time, and albedo of the locations into RayMan© Version 1.2 [30]. With hourly data of four meteorological parameters for entire year 2107 (8760 h), hourly UTCI was predicted using RayMan© Version 1.2 [29] for all 10 stations throughout metropolitan Sydney. The results of UTCI calculations are demonstrated in Table 2.

Table 2 UTCI analysis for observation scenario

Stations	Total annual UTCI (°C)	Max UTCI of the year	Day/time of the year with max UTCI	Ave UTCI of day with max UTCI
Richmond	133,636.1	46.4	11 Feb. 15:00	31.1
Camden	128,205.4	45.5	11 Feb. 16:00	30.5
Chullora	133,627.6	45.1	10 Feb. 15:00	29
Earlwood	140,134.8	45.7	10 Feb. 15:00	29.2
Lindfield	142,822.8	44.9	10 Feb. 15:00	29.1
Liverpool	129,689.5	45.2	11 Feb. 15:00	30.4
Prospect	131,433.2	44.7	10 Feb. 15:00	29.5

The maximum UTCI for each station was calculated. In all the stations the maximum UTCI occurred mostly on 10 or 11 of February during year 2017. The time of the day and the average UTCI of that day is demonstrated in Table 2. According to UTCI equivalent temperature scale [31], heat stress occurs when UTCI exceeds 26 °C. In this study, a UTCI of 26 °C was considered as the heat stress threshold. Exposure to heat stress was calculated when UTCI was computed above 26 °C. The sum of all the exposure hours to heat stress was estimated for all the observation stations for the entire year 2017.

2.3 Spatial Distribution of UTCI—Observation Study

To demonstrate the spatial distribution of heat stress across metropolitan Sydney a Geographical Information System (GIS) has been applied to demonstrate the spatial distribution of heat stress across metropolitan Sydney using ARC GIS, ESRI 10.7.1 (2019). Figure 2 demonstrates the number of hours of exposure to heat stress in a heatwave day throughout metropolitan Sydney.

Since the purpose of this study was to estimate the extreme heat events, the analysis has been conducted during the Sydney's heatwave. According to [32] the 9th of February was the first day of heatwave in 2017. This chapter reports the results of analysis for 9th of February 2017 as a representative day of heatwave in Sydney. As demonstrated eastern suburbs across metropolitan Sydney were exposed to less heat stress in comparison with western inland districts. This could be explained by the geographical differential between eastern coastal area—exposure to sea breezes—and western inland terrain—exposure to dry westerly winds [17].

3 Impact of Urban Greening Infrastructure on Heat Stress—Intervention

Urban greening infrastructure so-called—greenery, has been studied largely in the literature of Urban Heat Island mitigation technologies (i.e., [33–35]). Urban trees can reduce peak ambient temperatures, and consequently impact on energy savings in cities (i.e., [36, 37]). Implementation of green infrastructure can improve urban climate conditions via evapotranspiration, shading, and change in urban air movements. The impact of greening technology on local urban heat depends on the characteristics and size of urban parks, green roofs, vegetation type, and rate of irrigation [38].

Implementation of green roofs also has been studied in greening infrastructure (i.e., [39, 40]). Application of green roofs in urban areas can reduce ambient air temperature between 0 and 3 °C, with average value of 0.6 °C [41].

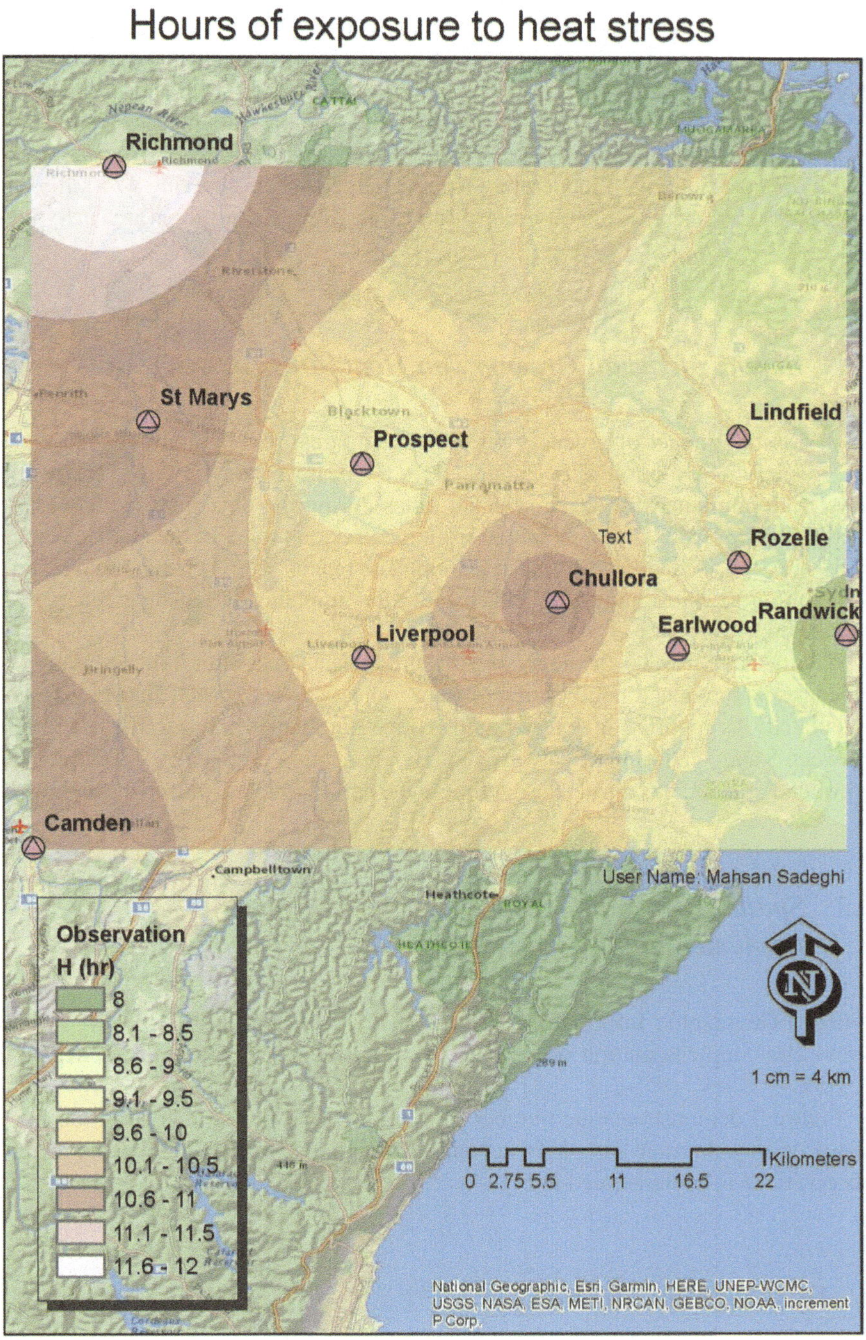

Fig. 2 Hours of exposure to heat stress during heatwave—observation

In this study, the urban greening simulations have been adopted from [37] and named as intervention study. The greenery simulations were conducted for a heatwave day in February 2017 (9 February 2017) using the Weather Research and Forecasting (WRF) model, version 3.9.1 [42]. The intervention scenario (greenery) was simulated as followings: implementation of two million well-irrigated trees (covering area of 200 km^2) combined with moderately irrigated green roofs (covering 474 km^2) and increase of 50% increase of evapotranspiration rate over existing green area in Sydney [37].

3.1 Impact of Urban Greenery on Heat Stress (UTCI)

To assess the impact of greenery on urban heat stress, temperature change under the greenery intervention were calculated for the heatwave day in 2017. The results have been computed for seven selected times during the day: 6:00, 12:00, 14:00, 16:00, 18:00, 21:00, 24:00. Obtained from the previous stage of the study. Average daily UTCI was calculated for each station. The results were benchmarked against observation scenario and the UTC change after implementation of greenery was calculated.

Table 3 demonstrates the hours of exposure to heat stress after implementation of greenery. The reduced number of exposure to heat stress for each station was also calculated. As reported greenery could reduce number of exposure to heat stress between 5 and 7 h during a heatwave day. By application of greenery UTCI reduced between 1.3 and 3.3 °C during the daytime and from 0.3 to 2.7 °C during the nighttime.

3.2 Spatial Distribution of UTCI After Greenery Implementation—Intervention Study

Using a Geographic Information System, spatial distribution of exposure to heat stress after implementation of greenery was plotted using ARC GIS, ESRI 2.7.1 (2019).

Figure 3 demonstrates the number of hours of exposure to heat stress during a heatwave day. Average daily UTCI change after implementation of greenery against observations is demonstrated in Fig. 4.

4 Heat-Related Mortality

To analyze the impact of reduced UTCI on population mortality, we defined our study region according to the Australian Bureau of Statistics [43], Sydney Greater Capital

Table 3 Exposure to urban heat stress during heatwave after intervention

Stations	Daily hours of exposure to heat stress (h)_Intervention	Daily hours of exposure change after intervention (h)	UTCI (°C) 12:00PM_Intervention	UTCI (°C) change 12:00PM	UTCI (°C) 12:00AM_Intervention	UTCI (°C) change 12:00AM
Camden	4	−7	28.2	−3	18.9	−1.5
Chullora	4	−7	30.4	−3.3	20.6	0.8
Earlwood	4	−5	32.9	−1.3	19.9	−0.3
Lindfield	3	−6	33.6	−1.7	2.5	0.3
Randwick	3	−5	29.6	−1.8	20.1	−0.3
Richmond	4	−8	31.5	−1.3	21.8	1.6
Rozelle	3	−6	31.2	−2.1	18.2	−2.7
Prospect	4	−5	27.7	−3	20.2	−0.9
St Mary	4	−7	31.2	−1.3	19.6	−0.3
Liverpool	4	−6	30.5	−1.4	20.3	−0.3

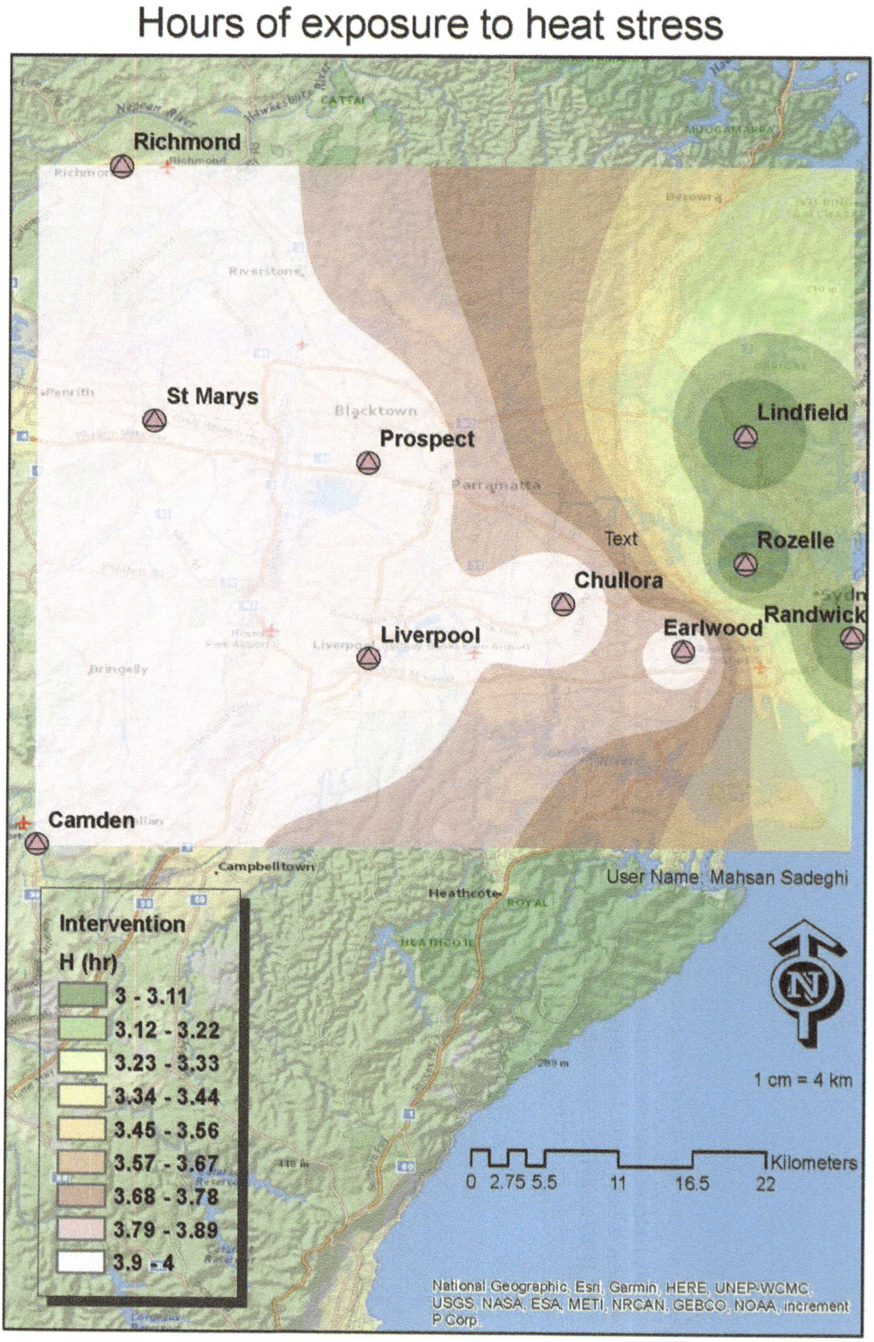

Fig. 3 Hours of exposure to heat stress during heatwave—intervention

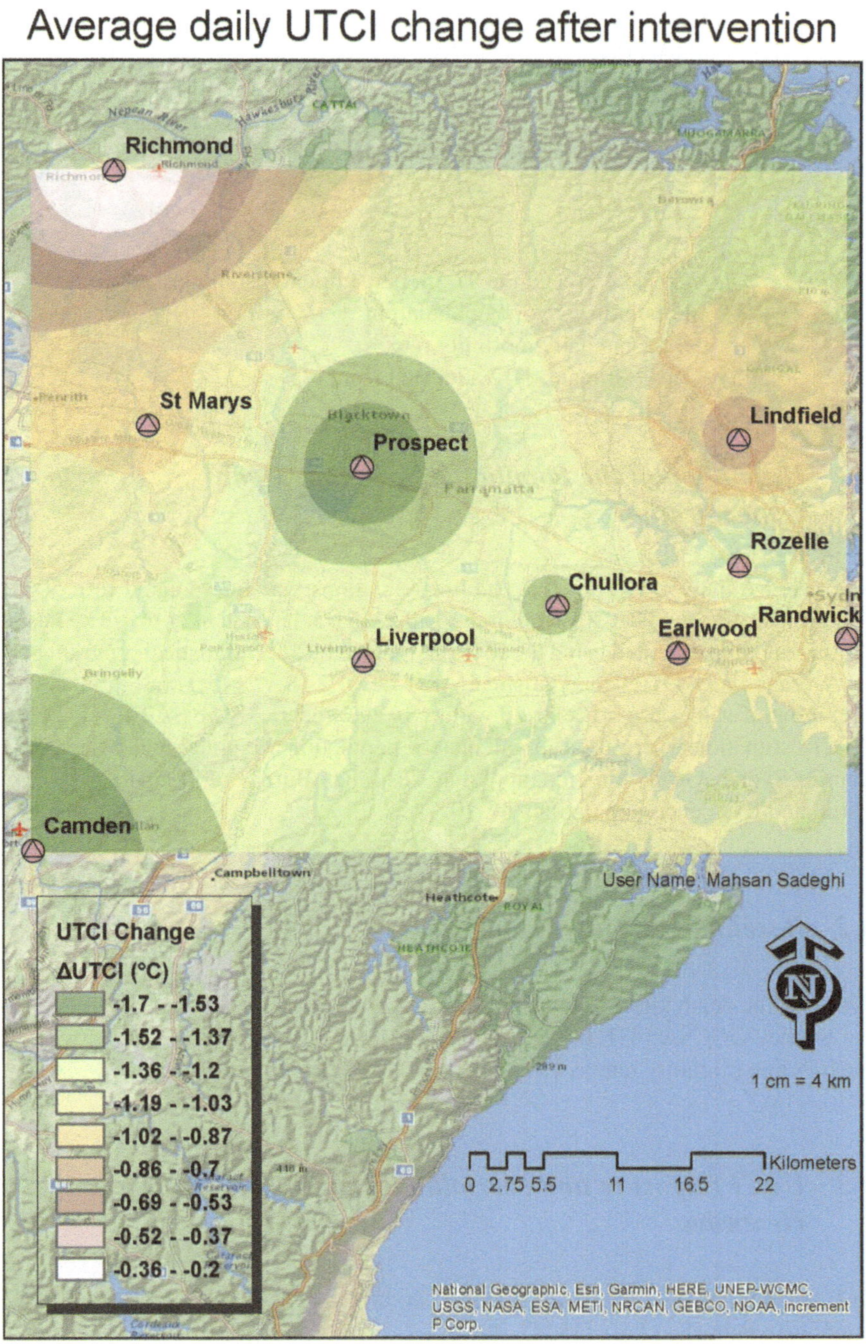

Fig. 4 Average daily UTCI change after intervention during heatwave

City Statistical Area (GCCSA) with population of 5.7 million. In this region, we used a well-established Health Impact Assessment method to estimate the number of premature deaths attributable to exposure to heat using Eq. 1 [44].

$$\text{Attributable Number (AN)} = \text{UTCI 97.5\%ile} \times P \times f \times (\text{RR} - 1)/\text{RR} \quad (1)$$

where,

UTCI 97.5%ile	If UTCI exceed the 97.5th percentile threshold is 1, if not, 0
P	Population, Sydney population = 5.7 million
f	Baseline daily mortality rate
RR	Risk estimate, 1.17 adopsted from [45].

4.1 Estimation of the Population Weighted-Mean Concentration

The average daily UTCI for the 10 climate stations was interpolated to a 500 × 500 grid for Sydney GCCSA (longitude 150.5–151.5, latitude −34 to −33.5) using kriging and implemented using the R © version 4.1.1 (2021) statistical program [46].

Average daily UTCI for Statistical Area Level 2 (SA2 [43]) was estimated by averaging of values at grid centroid within SA2 boundaries. The SA2-level average UTCI estimations were used to calculate a population weighted-mean concentration for Sydney GCCSA region, applying Australian Bureau of Statistics SA2-level Estimated Residence Populations for 2017.

4.2 Baseline Mortality Incidence

The baseline death rate was obtained from SA2-level ABS death data, aggregated to the Sydney GCCSA [43]. In the health Impact Assessment (HIA) method we used the baseline mortality rate as 1.6 deaths per 100,000 population per day.

4.3 UTCI Exposure and Mortality Risk Estimate During Heatwave

The association between daily changes in UTCI and daily mortality has been previously published (i.e., [45, 47–49].

We used the risk estimate from the study by [45] to predict the attributable number of deaths. While there is considerable uncertainty about the generalizability of this risk estimate to Sydney, we consider this risk estimate suitable for this proof-of-concept study. We used heat death risk estimate for days that UTCI exceeded the 97.5th percentile threshold as was reported by [45]. The lagged effects of heat on mortality were taken into account for the extreme UTCI day and the following 21 days [50]. UTCI attributable number of deaths for Sydney population was predicted for the observation study and after greenery intervention. It was estimated that the number of premature deaths attributable to UTCI could reduce by 11.7 in a heatwave day after implementation of urban greenery.

The most significant effect of greenery on UTCI would be attributed to reduction of surface temperature, and consequently, mean radiant temperature as a cooling effect of vegetation. The other direct impact of greenery on urban heat stress model would be through reducing the air temperature. These two important effects eventually would reduce urban heat stress imposed on human body, which would be accounting for the health benefits reported in this chapter.

Acknowledgements This case study was funded by Center of Air pollution, energy and health Research (CAR). The Author would like to acknowledge the research team who contributed to this study, Professor Bin Jalaludin, Associate Professor Geffrey Morgan, Professor Richard de Dear, Professor Mattheos Santamouris, Dr Timothy Chaston, and Dr Ivan Hanigan.

References

1. Steffen W (2009) Australia biodiversity and climate change. CSIRO Publishing
2. Mathers CD, Lopez AD, Murray CJ (2006) The burden of disease and mortality by condition: data, methods and results for 2001. Glob Burden Dis Risk Factors 45:88
3. Reid CE, O'neill MS, Gronlund CJ, Brines SJ, Brown DG, Diez-Roux AV, Schwartz J (2009) Mapping community determinants of heat vulnerability. Environ Health Perspect 117(11):1730–1736
4. Santamouris M (2020) Recent progress on urban overheating and heat island research. Integrated assessment of the energy, environmental, vulnerability and health impact. Synergies with the global climate change. Energy Build 207:109482
5. Founda D, Santamouris M (2017) Synergies between urban heat island and heat waves in Athens (Greece), during an extremely hot summer (2012). Sci Rep 7(1):1–11
6. Khan HS, Paolini R, Santamouris M, Caccetta P (2020) Exploring the synergies between urban overheating and heatwaves (HWs) in western Sydney. Energies 13(2):470
7. Wang Y, Berardi U, Akbari H (2016) Comparing the effects of urban heat island mitigation strategies for Toronto, Canada. Energy Build 114:2–19
8. Spickett J, Brown H, Katscherian D (2007) Health impacts of climate change: adaptation strategies for Western Australia. Department of Health, Government of Western Australia
9. Aghamohammadi N, Fong CS, Idrus MHM, Ramakreshnan L, Sulaiman NM (2021) Environmental heat-related health symptoms among community in a tropical city. Sci Total Environ 782:146611
10. Li M, Liu Z, Dong W, Shi P (2015) Mapping Heat Wave risk of the world. In: World atlas of natural disaster risk. Springer, Berlin, pp 169–188
11. Sadeghi M, Chaston T, Hanigan I, de Dear R, Santamouris M, Jalaludin B, Morgan GG (2022) The health benefits of greening strategies to cool urban environments—a heat health impact method. Build Environ 207:108546

12. Ebi KL, Capon A, Berry P, Broderick C, de Dear R, Havenith G, Honda Y, Kovats RS, Ma W, Malik A, Morris NB, Nybo L, Seneviratne SI, Vanos J, Jay O (2021) Hot weather and heat extremes: health risks. Lancet 398(10301):698–708
13. Jay O, Capon A, Berry P, Broderick C, de Dear R, Havenith G, Honda Y, Kovats RS, Ma W, Malik A, Morris NB, Nybo L, Seneviratne SI, Vanos J, Ebi KL (2021) Reducing the health effects of hot weather and heat extremes: from personal cooling strategies to green cities. Lancet 398(10301):709–724
14. Åström DO, Schifano P, Asta F, Lallo A, Michelozzi P, Rocklöv J, Forsberg B (2015) The effect of heat waves on mortality in susceptible groups: a cohort study of a mediterranean and a northern European City. Environ Health 14(1):1–8
15. Peel MC, Finlayson BL, McMahon TA (2007) Updated world map of the Köppen-Geiger climate classification. Hydrol Earth Syst Sci 11(5):1633–1644
16. Sadeghi M, de Dear R, Samali B, Wood G (2017) Optimization of wind tower cooling performance: a wind tunnel study of indoor air movement and thermal comfort. Procedia Eng 180:611–620
17. Sadeghi M, de Dear R, Wood G, Samali B (2018) Development of a bioclimatic wind rose tool for assessment of comfort wind resources in Sydney, Australia for 2013 and 2030. Int J Biometeorol 62(11):1963–1972
18. Sadeghi M, de Dear R, Samali B, Wood G (2016) Application of wind towers in the Australian residential context–a wind tunnel assessment of thermal comfort performance. The proceedings of 9th Windsor conference. NECUB. https://windsorconference.com/_archive/Proceedings_Windsor_Conference_2016.pdf. Accessed 10August 2022
19. Australian Bureau of Meteorology (2022). http://www.bom.gov.au/. Accessed 10 August 2022
20. NSW (2020) Department of planning, Industry, and environment (DPIE). https://www.planningportal.nsw.gov.au/opendata/dataset/environment-planning-instrument-local-environmental-plan-land-zoning. Accessed 20 Dec 2020
21. Sadeghi M, de Dear R, Morgan G, Santamouris M, Jalaludin B (2021) Development of a heat stress exposure metric–impact of intensity and duration of exposure to heat on physiological thermal regulation. Build Environ 200:107947
22. Jendritzky G, de Dear R, Havenith G (2012) UTCI—why another thermal index? Int J Biometeorol 56(3):421–428
23. Fiala D, Havenith G, Bröde P, Kampmann B, Jendritzky G (2012) UTCI-Fiala multi-node model of human heat transfer and temperature regulation. Int J Biometeorol 56(3):429–441
24. Matzarakis A, Muthers S, Rutz F (2014) Application and comparison of UTCI and PET in temperate climate conditions. Finisterra 49(98)
25. Havenith G, Fiala D, Błazejczyk K, Richards M, Bröde P, Holmér I, Rintamaki H, Benshabat Y, Jendritzky G (2012) The UTCI-clothing model. Int J Biometeorol 56(3):461–470
26. Sadeghi M (2017) Thermal comfort performance of wind towers in the Australian residential context. Doctoral dissertation. The University of Sydney. https://ses.library.usyd.edu.au/handle/2123/17121. Accessed 10 August 2022
27. Sadeghi M, Samali B, Wood G, de Dear R (2020) Comfort cooling by wind towers in the Australian residential context—experimental wind tunnel study of comfort. J Wind Eng Ind Aerodyn 196:104014
28. Sadeghi M, Wood G, Samali B, de Dear R (2020) Effects of urban context on the indoor thermal comfort performance of windcatchers in a residential setting. Energy Build 219:110010
29. Matzarakis A, Rutz F, Mayer H (2010) Modelling radiation fluxes in simple and complex environments: basics of the RayMan model. Int J Biometeorol 54(2):131–139
30. Matzarakis A, Rutz F (2010) Application of the RayMan model in urban environments. Meteorological Institute, University of Freiburg, Freiburg
31. Bröde P, Fiala D, Błażejczyk K, Holmér I, Jendritzky G, Kampmann B, Tinz B, Havenith G (2012) Deriving the operational procedure for the Universal Thermal Climate Index (UTCI). Int J Biometeorol 56(3):481–494
32. Santamouris M, Haddad S, Fiorito F, Osmond P, Ding L, Prasad D, Zhai X, Wang R (2017) Urban heat island and overheating characteristics in Sydney, Australia. An analysis of multiyear measurements. Sustainability 9(5):712

33. Anderegg WR, Anderegg LD, Sherman C, Karp DS (2012) Effects of widespread drought-induced aspen mortality on understory plants. Conserv Biol 26(6):1082–1090
34. He B-J, Ding L, Prasad D (2020) Relationships among local-scale urban morphology, urban ventilation, urban heat island and outdoor thermal comfort under sea breeze influence. Sustain Cities Soc 60:102289
35. Teskey R, Wertin T, Bauweraerts I, Ameye M, McGuire MA, Steppe K (2015) Responses of tree species to heat waves and extreme heat events. Plant Cell Environ 38(9):1699–1712
36. Akbari H, Cartalis C, Kolokotsa D, Muscio A, Pisello AL, Rossi F, Santamouris M, Synnefa A, Wong NH, Zinzi M (2016) Local climate change and urban heat island mitigation techniques–the state of the art. J Civ Eng Manag 22(1):1–16
37. Santamouris M, Paolini R, Haddad S, Synnefa A, Garshasbi S, Hatvani-Kovacs G, Gobakis K, Yenneti K, Vasilakopoulou K, Feng J (2020) Heat mitigation technologies can improve sustainability in cities. An holistic experimental and numerical impact assessment of urban overheating and related heat mitigation strategies on energy consumption, indoor comfort, vulnerability and heat-related mortality and morbidity in cities. Energy Build 217:110002
38. Taleghani M, Sailor DJ, Tenpierik M, van den Dobbelsteen A (2014) Thermal assessment of heat mitigation strategies: the case of Portland State University, Oregon, USA. Build Environ 73:138–150
39. Vahmani P, Ban-Weiss G (2016) Impact of remotely sensed albedo and vegetation fraction on simulation of urban climate in WRF-urban canopy model: a case study of the urban heat island in Los Angeles. J Geophys Res Atmos 121(4):1511–1531
40. Broadbent AM, Coutts AM, Tapper NJ, Demuzere M (2018) The cooling effect of irrigation on urban microclimate during heatwave conditions. Urban Clim 23:309–329
41. Santamouris M, Ding L, Fiorito F, Oldfield P, Osmond P, Paolini R, Prasad D, Synnefa A (2017) Passive and active cooling for the outdoor built environment—analysis and assessment of the cooling potential of mitigation technologies using performance data from 220 large scale projects. Sol Energy 154:14–33
42. Skamarock WC, Klemp JB, Dudhia J, Gill DO, Barker DM, Wang W, Powers JG (2008) A description of the advanced research WRF version 3. NCAR Technical note-475+ STR
43. ABS (2017) Australian bureau of statistics. https://www.abs.gov.au/
44. Broome R, Powell J, Cope M, Morgan G (2019) Source-specific impacts of PM2. 5 on mortality in the Greater Metropolitan Region of Sydney, Australia. Environ Epidemiol 3:275–276
45. Urban A, Di Napoli C, Cloke HL, Kyselý J, Pappenberger F, Sera F, Schneider R, Vicedo-Cabrera AM, Acquaotta F, Ragettli MS (2021) Evaluation of the ERA5 reanalysis-based Universal Thermal Climate Index on mortality data in Europe. Environ Res 198:111227
46. Pebesma E, Heuvelink G (2016) Spatio-temporal interpolation using gstat. RFID J 8(1):204–218
47. Matzarakis A, Nastos PT (2011) Human-biometeorological assessment of heat waves in Athens. Theoret Appl Climatol 105(1–2):99–106
48. Morabito M, Crisci A, Messeri A, Capecchi V, Modesti PA, Gensini GF, Orlandini S (2014) Environmental temperature and thermal indices: what is the most effective predictor of heat-related mortality in different geographical contexts? Sci World J 2014
49. Di Napoli C, Pappenberger F, Cloke HL (2019) Verification of heat stress thresholds for a health-based heat-wave definition. J Appl Meteorol Climatol 58(6):1177–1194
50. Gasparrini A, Guo Y, Sera F, Vicedo-Cabrera AM, Huber V, Tong S, Coelho MdSZS, Saldiva PHN, Lavigne E, Correa PM (2017) Projections of temperature-related excess mortality under climate change scenarios. Lancet Planet Health 1(9):e360–e367

Chapter 10
Impact of Increased Urban Albedo on Heat-Related Health: The Case Study of the Greater Toronto Area

Mohamed Dardir, Umberto Berardi, and Jeffrey Wilson

Abstract Applying urban heat island mitigation techniques requires a holistic understanding of their influences on the urban environment and community health. A novel approach is presented in this chapter by integrating microclimate simulations and correlation analyzes to define the impact of increasing the albedo of urban surfaces on urban health and thermal comfort. The application included enhancing the albedo of urban surfaces including roads, roofs, and building walls for a selected urban microclimate in the Greater Toronto Area. The study aimed to correlate the environmental and health responses by monitoring the historical meteorological and mortality and hospitalization data in the study region and to predict the outdoor thermal comfort due to changing albedo values. The results recommended wise implementation with careful considerations of albedo enhancements regarding urban thermal comfort. The hourly investigation proved the enhanced thermal comfort during specific periods of the day. The reported enhancements in outdoor thermal comfort led to an expected improvement in the instant health responses of the urban inhabitants.

Keywords Urban heat island · Heatwave · Albedo · Urban climate · Urban health · Mortality · Emergency visits · Microclimate simulations · Correlation analysis

1 Overview

The combination of climate change, urban heat island (UHI), and heatwave events leads to higher daytime temperatures, causes excessive heat stress for urban dwellers, and increases heat-related morbidity and mortality [1–4] mostly in vulnerable sections of the society such as elderly, homeless, and socially disadvantaged people

---

M. Dardir · U. Berardi (✉)
Department of Architectural Science, Faculty of Engineering and Architectural Science, Ryerson University, 350 Victoria Street, Toronto, ON M5B 2K3, Canada
e-mail: uberardi@ryerson.ca

J. Wilson
School of Environment, Enterprise, and Development, Faculty of Environment,, University of Waterloo, 200 University Avenue West, Waterloo, ON N2L 3G1, Canada

N. Aghamohammadi and M. Santamouris (eds.), *Urban Overheating: Heat Mitigation and the Impact on Health*, Advances in Sustainability Science and Technology,
https://doi.org/10.1007/978-981-19-4707-0_10

[5, 6]. Meanwhile, the consequences of the UHI and the frequency and duration of the heatwaves around the world are becoming more evident [4, 7, 8]. The heatwaves are coupled with high-ambient temperature and humidity. The severe impacts of heatwaves are associated with multi-day heat stress, warm nights, and increased relative humidity [4, 9]. The characteristics of heatwaves events are defined regionally according to the local climatic conditions. According to Health Canada [10] for Southern Ontario (Greater Toronto Area, *GTA*), a heatwave occurs daytime maximum temperature exceeds 31 °C, and the nighttime minimum temperature exceeds 20 °C for at least 2 consecutive days. Health Canada also reports a heatwave if the local heat stress index (temperature-humidity index, *Humidex*) in the above-mentioned region reaches 40 or more for at least 2 consecutive days. In the last 20 years, 60 heat warnings and 37 extended heat warnings were issued for the GTA region [11].

Severe health impacts were argued due to the expected increase in heatwaves intensity in future [1]. The Canadian Environment Health Atlas (CEHA) and the Toronto Public Health Department estimated that 120 people die in the GTA annually because of high temperatures. The predictions indicate that heat-related mortality will be doubled by 2050 [12]. Generally, significant correlations were reported between heat-related mortalities and hot ambient conditions, especially for elderly people [13–16], with a significant association for respiratory-related fatalities [8, 17]. Some investigations in North America reported strong causality for daily mortality due to exposure to excessive heat [18, 19]. Rainham and Smoyer-Tomic [15] confirmed a significant association between elderly mortality and extreme hot humid conditions in the City of Toronto. Within the hot weather in Montreal, Kolb et al. [20] reported a strong short-term relationship between hot weather (over 25 °C) and elderly cardiac-related daily mortalities. Berko et al. [18] reported that around 31% of the weather-related mortalities in various US regions were related to exposure to excessive heat.

The causality between mortality cases and hot weather was affirmed for cities where the heatwaves were less common [2]. Some studies reported the importance of considering the effects of temperature variations, not only the extremes, especially on the elderly mortalities. Smoyer-Tomic and Rainham [21] flagged an increase in elderly mortality cases in Toronto, Ontario related to a Humidex range of 30 to 35, which does not meet the local extreme heat stress conditions of the region. Paravantis et al. [8] observed an excessive increase in daily cardio-respiratory-related death of elderly people (> 65 years) associated with high-ambient temperature (even when heatwaves conditions were not met). Nastos and Matzarakis [22] studied the immediate causality between mortality and warm conditions on the same day. The authors reported an increase of 3% in mortality probability for each 10 °C increase in daily maximum air temperature. They also claimed the strong association between the 1-day lag effects on heat-related mortality.

Accordingly, applying heat mitigation strategies that protect human health and improve the urban environment is essential in the strategic planning for new and existing urban settlements [23, 24]. Given the large transformation and fast-growing population of the GTA, it is crucial to investigate the urban design influences on urban dwellers' heat responses. The enhancements in the urban microclimate include enhancing the urban surface physical properties including the surface albedo (SA).

Albedo is a measure of the surface reflectance of shortwave radiation. Increasing the SA reduces the absorbed solar radiation by the surface and magnifies the portion of the reflected solar radiation which controls the surface temperature and promotes the city cooling effect [24, 25]. Many studies proved the cooling potential of increasing the urban SA for UHI mitigation [4, 26–28]. However, a careful study of the reflected solar radiation within urban canyons is always recommended to avoid overheating and maximize the benefits. Increasing SA was discussed in the literature as a heat mitigation strategy promoting the city cooling effect.

The cooling potential of increasing urban SA as a heat mitigation strategy is highly influenced by the urban texture and density [24, 29] which could influence the urban thermal comfort. Jandaghian and Berardi [30] assessed the effect of increasing SA for building roofs and walls and urban grounds on urban microclimate in the urban region of the city of Toronto utilizing an urban mesoscale model coupled with a building energy module. They recorded a reduction in air temperature of up to 2 °C, a reduction in dew point temperature of up to 0.5 °C, and a slight increase in near-surface wind speed. They also recorded a reduction of almost 10% in cooling energy demand due to limited solar heat gains and reduced air temperature. Touchaei et al. [31] investigated the changes of urban SA including roofs, walls, and roads in the city of Montreal during a heatwave. They proved a reduction of ambient air temperature (with a maximum reduction of 0.7 °C) with an increase in SA of up to 0.65. They also reported a small reduction in pollutants' concentration due to changes in the height of the atmospheric layer. Morini et al. [26] conducted an application of increasing the urban SA of roofs, walls, and roads in an urban region in Rome. They reported a reduction of the ambient temperature of up to 4 °C promoting the mitigation of the UHI. Mohammed et al. [29] investigated the increase of urban SA including roofs, walls, and ground for an urban desert environment. They observed a maximum drop in the ambient temperature of 0.4 °C for each increase of 0.1 in urban SA. They reported that the cooling effect of increasing SA was greater at high-density urban areas with a higher total built-up area. The increased cooling potential was observed due to increased total convective losses and lower urban surface temperature in dense regions of the city. Whereas, Wang et al. [24] recommended that better outdoor thermal comfort can be obtained by integrating different heat mitigation techniques simultaneously.

Cool roofs with high-reflective materials were proved to contribute to the peak shaving of ambient temperature and decrease the intensity of the UHI [32–34]. The high-albedo materials are assumed an effective strategy for building roofs and walls as they efficiently reflect the incoming solar radiation. This reflected solar radiation was proved to enhance the thermal behavior indoors. Also, the overall convective thermal losses were proved by reducing the ambient air temperature. However, the outdoor comfort influenced by increasing the SA of building walls is still debatable. Some studies argued that the use of high-SA materials in urban canyon surfaces may affect outdoor thermal comfort due to the reflected solar radiation to the human body [35–39]. Lee and Mayer [38] studied the effect of increasing the albedo of building walls on pedestrians' thermal comfort during a heatwave. They showed that increasing the wall albedo leads to a systematic impairment in outdoor thermal comfort due to the reflected shortwave radiation. Schrijvers et al. [39] evidenced that the effect of changing urban SA on outdoor thermal comfort was limited compared to shading strategies. They claimed that although applying high-albedo materials could mitigate the atmospheric UHI effect, they contributed to increasing pedestrian heat stress. It is vital to carefully design the upgrading applications of SA considering urban form and shading features to comprehensively reveal the contradictions with the outdoor thermal comfort.

The expected health benefits from the UHI mitigation strategies are related to reducing ambient temperature, especially during extreme heat events. The heat-related mortality (HrM) was considered an indicator for community health responses against heatwaves and environmental changes. The impact of increasing urban SA on the HrM was limitedly discussed in the literature based on the ambient temperature. Jandaghian and Akbari [25] reported that a reduction of up to 2 °C in the ambient temperature was related to increasing the urban SA in the urban region of the city of Toronto. They claimed that the reported cooling effect enhanced the outdoor thermal comfort and reduced HrM by up to 7%. Santamouris and Fiorito [40] studied the relationship between ambient temperature and urban SA. They reported a reduction of 0.09 °C per 0.1 increase in SA. They found that there was an average decrease in HrM of 19.8% per one degree Celsius of temperature drop and 1.8% per 0.1 increase in SA. However, the effect of increasing urban SA, especially for building walls and roads, on outdoor thermal comfort is still controversial. It raises some doubts about the actual impact on human comfort and consequently, human health.

This chapter discusses the impact of increased urban albedo on heat-related mortality based on outdoor thermal comfort considering ambient temperature, wind speed, mean radiant temperature, etc. The study aims to assess the impact of increasing urban SA on an outdoor thermal comfort index and correlate the main comfort parameters (including ambient temperature, humidity, and solar radiation) with the health responses (mortality rates and emergency department visits). The study focuses on the GTA region as a typical urban morphology for the Canadian community.

2 Materials and Method

To conduct the link between health responses and thermal comfort, the proposed method combines a statistical approach with microclimate simulations to investigate the impact of increasing the SA on heat-related health responses. The microclimate simulations aimed to predict the changes in outdoor thermal comfort index based on different urban SA values. The study uses a specific spatial and temporal domain with the health records to test the validity of the method. The microclimate simulations were designed utilizing a developed simulation code to assess the UHI mitigation strategy on the urban microclimate. The outdoor thermal comfort index was represented by universal temperature climate index (UTCI). UTCI was selected because it includes ambient temperature, relative humidity, wind speed, global solar radiation, and surface temperature in the calculations of thermal comfort. This would better consider the effect of the changes in SA on a pedestrians' level. The study also correlated the behavior of the meteorological parameters with the health records in the region of study. The meteorological parameters include air temperature, relative humidity, and solar radiation. The health responses were measured by monitoring the total daily counts for mortalities (MOR) and emergency visits (EMR) in the region of study. The application was designed for daily records to assess the instant impacts and correlations (within the same day) among study variables. A statistical approach is introduced, including correlation analyzes for establishing the relationship between the study parameters. The results were discussed to conclude the impacts on human health response in the GTA.

The study domain, shown in Fig. 1, includes the Peel region focusing on Mississauga (*Mis*) and Brampton (*Brm*) municipalities. *Mississauga* is featured by higher urban densities, and *Brampton* is more featured by river valleys dispersal. The research dataset includes daily values for the investigated meteorological parameters. The weather data were obtained from Environment Canada [41] for 15 years from 2003 to 2017. This study focuses only on warm and hot seasons, which extend from May to September each year. The weather data were obtained from the international airport weather station (43.67N–79.63W), which represents the only available historical weather data in the region of study that cover the intended period of study.

The meteorological context of the region of the study is presented in Fig. 2 showing the monthly average values of the maximum and mean values during the period of investigation. The error bars in both figures indicate the maximum and minimum values of the monthly average values. It can be noticed that July represents the highest monthly average temperature with noticed high trends in August. While the relative humidity does not correlate with the behavior of the temperature.

The daily weather parameters included maximum, minimum, and mean values for ambient temperature and maximum and mean values for relative humidity. The solar

Fig. 1 Province of Ontario (left), Peel region regarding the Greenbelt (center), and municipalities of Brampton and Mississauga regarding Peel region (right), showing the locations of the weather station (W) and the air quality monitoring stations in Mississauga (A1) and Brampton (A2)

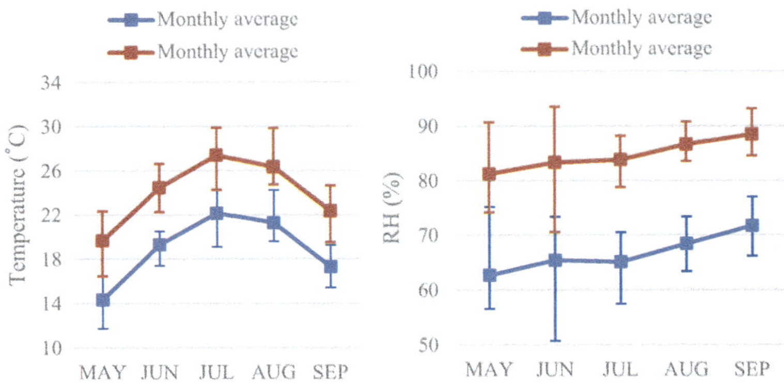

Fig. 2 Meteorological trends and monthly average values of ambient temperature (left) and RH (left) from 2003 to 2017

radiation data were obtained from the Prediction Of Worldwide Energy Resources (*POWER*) project by NASA [42]. The provided solar data by POWER are cumulative daily global radiation records. Accordingly, solar data were used directly in the dataset; no maximum or mean values were calculated.

The population health responses were retrieved from the Research Data Centers (RDCs) using the Statistics Canada microdata for health databases. The research used the Canadian Vital Statistics—Death Database (CVSD) for mortality counts including date of death and place of occurrence of death. It also used the Canadian Census Health and Environment Cohort (CanCHEC) linked to the National Ambulatory Care Reporting System (NACRS) to provide the daily counts of emergency department visits. The CanCHEC_NACRS database provided ambulatory care services and emergency departments registrations including the date of the medical service registration and postal codes of residence. The postal codes were only used to identify the municipality of the population. The conducted analyzes respected the data confidentiality and followed Statistics Canada guidelines and restrictions. The study focused on the daily counts of the health events for the period (2003–2017) from May to September (hot and warm seasons) each year for the municipalities of Mississauga and Brampton, Ontario. The total investigated mortality cases were around 28,000 cases, and the total emergency visits were around 581,500 cases within the study period in both municipalities. The daily counts for mortalities and emergency visits were included in the dataset as one observation daily; in total, 2295 days (observations) were included in the dataset. Such analysis is needed for policymakers to understand the health responses on a local scale; thus, specific policies and problem mitigation strategies could be promoted individually for each municipality.

A time series dataset was constructed with daily intervals for the designed period. Both maximum and mean values for weather, except for the solar radiation, were included in the dataset. The daily counts of mortalities and emergency visits were included in the dataset for both municipalities. The statistical approach included a descriptive analysis applying correlation analyzes among study variables. IBM SPSS Statistics software version 24 was used for the statistical analysis.

3 Microclimate Simulations

Most of the reviewed studies utilized numerical simulations to predict the effect of UHI mitigation strategies [43–45] using Weather Research and Forecasting Model (WRF) or Envi-met engines. Some limitations were reported regarding the commonly-used simulations platforms for urban thermal analysis. Envi-met was claimed to have some challenges regarding detecting the variation in model parameters, representing the heat transfer process for the proposed mitigation strategies, the analysis grid sensitivity, and underestimating the latent fluxes [46–48]. Also, using WRF is associated with some limitations on the microclimate scale regarding representing the surfaces properties, leaf area index of vegetation, evaporation cooling effect, and surface wind speed [49–51]. This ensures the need of developing a simulation tool that efficiently estimates the cooling effect of the green cover while considering the sequences of the warming effect on the urban microclimate.

An updated modeling approach was developed, based on the open-source Urban Weather Generator (UWG) [52, 53], and validated to assess the effects of UHI mitigation scenarios on the urban thermal quality, outdoor thermal indices, and building energy consumption considering also the economic benefits of the mitigation application. Detailed descriptions of model components, theoretical approaches, and model validation procedures are published in [54]. The microclimate simulations were designed utilizing the developed to evaluate the changes in urban SA within a proposed neighborhood within the GTA on the urban thermal behavior. The proposed time frame of the simulation (10 days) was an example of the urban thermal response applying the mitigation technique at a worst-case scenario (a heatwave period). A typical urban typology adjacent to a river valley was selected representing the Brampton residential neighborhood. Verification of the selection of the rural site, comparisons with nearby rural locations, and full characteristics of the rural site were provided by Dardir and Berardi [54]. The simulation was extended for 5 days to include the reported heatwave (from June 30 to July 5, 2018) during this period. Details of location, urban features, and assumptions are provided in Table 1.

The average albedo of the natural surfaces on earth is about 0.3, and it ranges from 0.1 to 0.85. The natural surfaces include low-albedo surfaces like rocks, woodland, and oceans, medium albedo surfaces like soil, sand, and ice, and high-albedo surfaces like snow [55]. For urbanized regions, construction materials vary with different values of SA. The behavior of newly applied materials differs from the used ones regarding the surface properties as the aging effect has to be considered for the long-term evaluation of SA enhancement. The average albedo of SA for building materials of a typical Canadian urban typology was assumed to be 0.2 [30]. The urban SA was mainly defined by the roofs, walls, and roads as the SA of these components can be controlled by replacing and upgrading the construction materials. The SA of the trees

Table 1 Specifications of the urban microclimate parameters

Location	Brampton, ON 
Coordinates	43°41′ N 79°49′ W
Distance from rural	35.5 km
Site area	528,000 m^2
Building footprint	24.6% (130,000 m^2)
Avr. building height	6 m
V-*to*-H ratio	0.25
Roads urban area	33.5% (133,500 m^2)
Water surface area	5.7% (22,500 m^2)
Tree canopy	20%
Vegetation cover	60%
Building types	100% residential

and vegetation was assumed to be 0.25 and was considered constant in the current study. The proposed enhancements of the SA include using high-albedo materials of 0.8 with a step increase of 0.2 to monitor the impact of increasing the SA on the urban environment. The application with the maximum enhancements assumed applying white paint to roofs and walls and applying concrete pavement tiles with highly reflective paint to roads to promote a pedestrian-friendly application. The used assumptions of enhancing the SA with the proposed construction materials are presented in Table 2.

The changes in SA affect the shortwave reflectivity of urban surfaces and thus their temperatures and the reflected solar radiation to other urban surfaces, green elements, and the human body outdoors. These factors were considered in the calculations of the urban microclimate in the updated version of UWG. The outdoor thermal comfort of the human body was measured using the UTCI which considers the urban surface temperature in index calculations, as illustrated in Fig. 3. Accordingly, the effects of increasing the SA of urban surfaces could be monitored. The UWG model was linked to RayMan 1.2 software [56] to assess the UTCI model.

Table 2 Proposed urban construction materials associated with assumed albedo [4, 56]

Albedo	Proposed construction materials		
	Roofs	Walls	Roads
0.2 (ref. case)	Red/brown tiles	Colored paint	Asphalt
0.4	Concrete tiles	Bricks/stone	Concrete
0.6	Light cement tiles	Gray paint	Light concrete
0.8	White paint	White paint	Highly reflective paint

The UTCI model follows a regression analysis approach based on measured physiological responses that consider meteorological parameters, clothing, and activity level. Precisely, the UTCI model considers ambient temperature, relative humidity, and wind velocity; the hourly values of these parameters were predicted using the UWG model. The UWG model considers dynamic surface temperature, shortwave and longwave radiation fluxes, and sensible heat fluxes from roofs, walls, and roads and aggregates the fluxes into the exchange of momentum and energy between the urban surface and atmosphere. In the updated version of the UWG model [54], the UWG promotes better prediction of evaporative cooling effect, more realistic shading behavior, further adaptability to urban surfaces variety, redefining the latent component of the air node, and defining new urban components. Specifically, all urban surfaces were defined separately for urban elements (roads, vegetation, water surfaces, rural area, walls, and roofs) with prominent properties of each component. This assumption well-estimates the effects of changes in SA for each element independently. Meanwhile, defining the features separately allows increasing the accuracy of the calculations of surface fluxes from urban elements to the canyon air node. Additionally, the solar bounces, reflections, and thermal fluxes were developed and adapted to the new model structure. Surface properties values for albedo and emissivity were defined for all urban and building surfaces as new model inputs. Also, the shading effect, regarding solar radiation, and blocking effect, regarding thermal radiation, of the tree canopy and roof vegetation were well-established. Also, the model considers the urban greenery cover in the calculations of urban and building surfaces and air nodes' gains and losses, ensuring the well-estimation of cooling and warming effects of the greenery coverage.

Other parameters were also considered in the calculations of UTCI; a clear sky condition was assumed, and an average urban surface temperature was calculated. The surface temperature is being used in the model to calculate the mean radiant temperature to which the human body is subjected. Accordingly, the average surface temperature was determined as the average of wall and road temperatures every time step. The enhancements in roof temperature, as well as wall and road temperatures, affect the heat fluxes to the air temperature and thus contribute to the convective cooling effect for UHI mitigation. For the calculations of UTCI, some assumptions were made. An average male adult (height of 1.75 m, weight of 75 kg, and age of

Fig. 3 Outdoor thermal comfort main components considered in UWG and UTCI calculations

35 years) was proposed for the study. A moderate activity level (standing; 80W) and a clothing factor of 0.9 clo (a measure unit of clothes thermal insulation; 0.9 clo represents typical business suit) were assumed for the outdoor condition.

The changes in SA influenced road, walls, and roofs temperatures. Figure 4 introduces the evolution of road temperature regarding increasing the SA from 0.2 to 0.8. As expected, the temperature changes were maximized during mornings and evenings, while during nighttime and early mornings, the changes were limited. It can be noticed that increasing the SA is an efficient strategy for peak temperature shaving for urban surfaces. The maximum reductions occurred around 2.00 pm every day when cooling the urban surfaces is highly needed. To quantify the average and maximum changes in urban thermal behavior, Table 3 presents the evolution of temperatures of urban surfaces and the corresponding changes in UTCI.

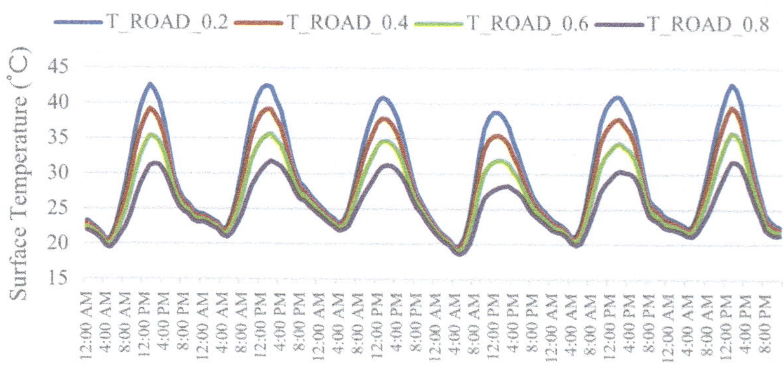

Fig. 4 Surface temperature of urban roads respecting the changes in SA

Table 3 UTCI and urban surfaces temperatures corresponding to increasing SA

Albedo	Roads			Walls			Roofs			UTCI		
	Average temp. (°C)	Maximum drop (°C)	Time of max. drop	Average temp. (°C)	Maximum drop (°C)	Time of max. drop	Average temp. (°C)	Maximum drop (°C)	Time of max. drop	Average index (°C)	Maximum drop (°C)	Time of max. drop
0.2	30.2	11.6	12.00 PM	21.0	1.8	12.00 PM	33.0	24.5	12.00 PM	23.17	1.0	8.00 AM
0.4	28.7			20.9			30.3			23.15		
0.6	27.1			20.7			27.5			23.13		
0.8	25.3			20.5			24.8			23.19		

The horizontal surfaces (roads and roofs) represented efficient thermal behavior and a better response to increasing SA. The reductions in average surface temperature during the whole application recorded 4.9 °C and 8.2 °C for roads and roofs, respectively. While the maximum reduction in surface temperature was noticed in roof temperature with a maximum drop of 24.5 °C. This also gives massive potential to improve buildings' thermal behavior and energy saving. For vertical surfaces (walls), the reductions were not as efficient as horizontal ones. With a maximum drop of only 1.8 °C in walls temperature, it can be concluded that increasing the SA for urban walls would not contribute efficiently to city cooling plans.

The enhancements in urban surface temperatures contribute to the reductions in ambient temperature and mean radiant temperature which affects the evolution of UTCI. Nevertheless, the average index showed slight reductions in UTCI related to increasing the urban SA, with a maximum reduction of 1 °C. An important notice is that the reductions in UTCI occurred with increasing the SA up to 0.6; increasing the SA to 0.8 achieved an increase in the UTCI affecting the outdoor thermal comfort. This counter effect of high-albedo materials on UTCI was also reported by [39]. This highly stresses the importance of the accurate design of applying albedo changes to urban surfaces for cooling purposes. An optimized application would be essential to maximize the cooling potential and urban thermal behavior.

The reported maximum reduction in UTCI occurred during the morning. While the maximum reductions in surface temperatures occurred around noon. To better understand the relationship with the comfort index, detailed illustrations were conducted during the mornings and evenings. Figure 5 shows days 2 and 5 of the application as examples for the whole period. It can be observed that UTCI responded efficiently during mornings and noon periods. During evenings, UTCI did not show enhancements, and in some cases, it showed a counter effect with an increase in UTCI with higher SA. However, in most cases, the UTCI reported a reduction during the peak period (2.00 pm–4.00 pm) which promotes the protection of urban dwellers. Table 4 illustrates in detail the hourly maximum temperature drop of UTCI by increasing SA from 0.2 to 0.8 during the mornings and early evenings. It can be noticed that the maximum temperature drop was distributed around noon. On the contrary, the increase in UTCI started to appear around 5.00 pm. This ensures that, regarding the thermal comfort index and not only the surface and ambient temperatures, increasing SA was efficient during mornings and afternoon periods. While the UHI mitigation strategy negatively affected the outdoor thermal comfort during nighttime. Accordingly, careful design and application procedures are always recommended for increasing SA regarding the thermal comfort of urban inhabitants.

4 Heat-Related Health Impacts

A statistical approach was followed by integrating data on ambient conditions with population health records to determine the impact on mortality (MOR) and emergency department visits (EMR). A historical dataset was created for 15 years for the

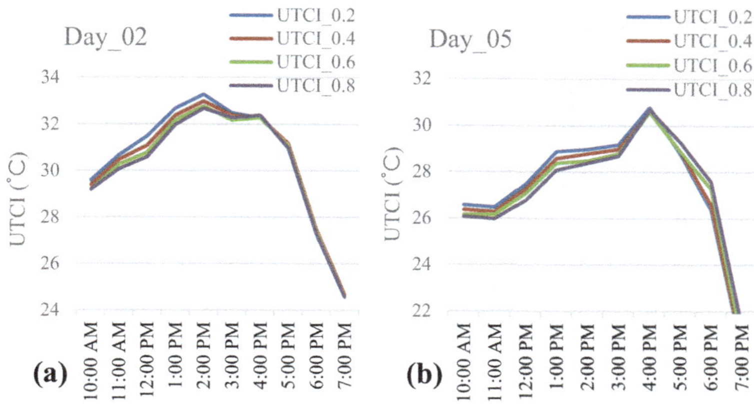

Fig. 5 UTCI behavior related to increasing SA during mornings and evenings on **a** day 2 and **b** day 5

Table 4 Hourly temperature drop of UTCI (°C)

	DAY_1	DAY_2	DAY_3	DAY_4	DAY_5	DAY_6
8:00 AM	1	0.2	−0.1	0	0	−0.1
9:00 AM	0.1	0.2	0.1	0.2	0.4	0.2
10:00 AM	0.5	0.4	0.3	0.5	0.5	0.5
11:00 AM	0.5	0.6	0.4	0.6	0.5	0.6
12:00 PM	0.9	0.9	0.6	0.7	0.7	0.7
1:00 PM	0.6	0.7	0.4	0.4	0.8	0.5
2:00 PM	0.4	0.6	0.2	0.3	0.6	0.6
3:00 PM	0.6	0.2	0.1	0.1	0.5	0.1
4:00 PM	0.3	−0.1	−0.1	−0.3	0.1	−0.1
5:00 PM	0.2	0.1	−0.6	−0.7	−0.5	−0.3

region of study to find the correlation between environmental changes and community health responses. Thus, further health responses can be anticipated when applying UHI mitigation scenarios that improve the environmental parameters. Correlation analyzes were conducted among daily maximum and mean values of ambient temperature, relative humidity, and solar radiation. Pearson's coefficients were utilized to represent the degree of correlations. The Pearson's coefficients vary from ± 1 for perfect correlation to zero for no correlation; values up to ± 0.3 are said to be a small correlation (low degree), values up to ± 0.5 are said to be a medium correlation (moderate degree), and values over ± 0.5 are said to be a strong correlation (high degree). The correlations of health records with ambient conditions are presented in Table 5. The health records were represented by the total cases of MOR and EMR in both municipalities. Correlations were built between total MOR and both maximum

Table 5 Correlations of community health responses with ambient conditions

		Max temp	Min temp	Mean temp	RH_Max	RH_Avr	Solar_RAD
MOR_Mis	*Pearson's coef*	0.011	−0.007	0.003	−0.036	−0.040	−0.011
	Sig.	*0.611*	*0.753*	*0.905*	*0.084*	*0.057*	*0.585*
MOR_Brm	*Pearson's coef*	−0.022	−0.036	−0.030	−0.096	−0.082	0.016
	Sig.	*0.282*	*0.088*	*0.150*	*0.000*	*0.000*	*0.454*
EMR_Mis	*Pearson's coef*	0.064	0.037	–	−0.065	–	0.029
	Sig.	*0.002*	*0.080*	–	*0.002*	–	*0.159*
EMR_Brm	*Pearson's coef*	0.071	0.048	–	−0.114	–	0.049
	Sig.	*0.001*	*0.022*	–	*0.000*	–	*0.020*

and mean daily values of environmental variables. Also, the analysis was conducted between total EMR and maximum daily values of environmental variables.

Regarding total MOR, there is a low degree of negative correlation with relative humidity with a statistically significant relationship, especially in Brampton. However, no significant correlation was proved with ambient temperature or solar radiation in both municipalities. Regarding total EMR, a statistically significant low-degree positive correlation was proved between total EMR and ambient temperature. Another statistically significant low-degree negative correlation was confirmed with relative humidity. A low degree of positive correlation was also confirmed with solar radiation with a statistically significant relationship in Brampton.

The daily health records showed no significant correlation with the ambient temperatures, yet a low degree of positive correlation was proved with total emergency visits. It is worth mentioning that the focus of the study is to look at the possible instant correlations (within the same day) between health records and environmental variables. The instant correlations were partially proved with the EMR but contradicted with the MOR cases that could happen some days after the heat event. Previously, Kolb et al. [20] and Rainham and Smoyer-Tomic [15] proved the correlation between the short-term effect of hot weather (2–3 days after the heat event) and both cause-based MOR and elderly caused-based MOR in Montreal, QC, and Toronto, ON, respectively. Accordingly, it is recommended to test the correlation for the short-term effect of environmental events on health responses. Additionally, Kolb et al. [20] did not find clear associations between health records and the short-term effect of humidity levels. In contrast, the current application reported a negative correlation between relative humidity and both total MOR and EMR. The correlations also did not confirm the relationship between solar radiation and total MOR; however, a positive correlation was noticed between solar radiation and total EMR. This indicates that with low levels of RH and high levels of solar radiation, some instant health responses tend to increase. This also implies that the impacts of the RH and solar radiation on health responses can be instant while the impact of ambient temperature could occur in a prolonged manner.

While relative humidity and solar radiation are important components in the calculations of UTCI; thus, a positive impact of enhancing the UTCI is expected on the instant health responses (most probably on EMR than MOR). Also, improving the UTCI promotes the short-term positive impact on health responses (especially MOR cases) based on the enhancements in ambient temperature. The reductions in UTCI associated with increasing the SA were observed during morning and afternoon periods. This promotes improving the instant health response, especially the EMR, during these periods associated with improving the urban surface properties (increasing the SA). However, increasing the SA could not be separately judged as an efficient strategy to reduce MOR and EMR rates. As the reflected solar radiation can cause an increase in the UTCI especially during evenings and nighttime when applying high-reflective materials (the case of applying SA of 0.8). Thus, careful implementation schemes and policies are recommended to apply albedo enhancements solutions to improve urban comfort and health.

5 Conclusion

The outdoor thermal comfort and health responses were analyzed in this chapter regarding the application of increasing urban surfaces albedo as a UHI mitigation strategy. Integrated microclimate simulations and statistical approaches were designed to investigate the impact of increasing the albedo on urban health and thermal behavior. An application was presented to monitor the daily records of the environmental and health parameters for the period (2003–2017) during hot seasons in the municipalities of Mississauga and Brampton, Ontario, CA. The microclimate simulations were developed to predict the changes in the outdoor thermal comfort index due to various values of urban surfaces albedo values. The enhancements in urban surfaces included roads, roofs, and building walls. The results confirmed the increased efficiency of road and roofs applications than walls application. Increasing the reflectivity of urban surfaces was claimed to contribute to reducing the heat fluxes to ambient air and reducing the mean radiant temperature of urban surfaces; thus, an increased cooling potential can be expected for UHI mitigation. The reductions in urban surfaces temperatures were noticed during mornings and afternoons, while during nighttime and early mornings, the changes were limited. It was concluded that increasing the albedo can be efficient for peak shaving for urban surfaces temperatures. Regarding the thermal comfort index, only a slight enhancement was monitored related to increasing the urban albedo, and a cutback in outdoor thermal comfort was observed when applying high-albedo materials. An hourly investigation proved that increasing urban albedo was efficient during the mornings and afternoon periods. While it affected the outdoor thermal comfort during nighttime. Accordingly, an optimized application with careful considerations was recommended for increasing the albedo to enhance the thermal comfort of urban inhabitants.

A statistical approach using Pearson's correlation was utilized to evaluate the correlation between environmental variables and community health responses (mortality and emergency department visits). The instant daily health records showed no significant correlations with the ambient temperatures and a low degree of negative correlation with relative humidity and air pollutants concentrations. However, an instant significant correlation was noticed between relative humidity, solar radiation, and emergency visits. This implies that the impacts of the relative humidity and solar radiation on health responses can be instant while the impact of ambient temperature could occur in a prolonged manner. Referring to the enhancements in outdoor thermal comfort, improving the instant health response during mornings and afternoons was predicted associated with improving the urban surface albedo.

References

1. Fischer EM, Schär C (2010) Consistent geographical patterns of changes in high-impact European heatwaves. Nat Geosci 3(6):398–403
2. Lee M, Nordio F, Zanobetti A, Kinney P, Vautard R, Schwartz J (2014) Acclimatization across space and time in the effects of temperature on mortality: a time-series analysis. Environ Health 13(1):89
3. Jandaghian Z, Akbari H. (2018) The effects of increasing surface reflectivity on heat-related mortality in Greater Montreal Area, Canada. Urban Climate 25
4. Santamouris M (2020) Recent progress on urban overheating and heat island research. Integrated assessment of the energy, environmental, vulnerability and health impact. Synergies with the global climate change. Energy Build 207:109482
5. Benmarhnia T, Deguen S, Kaufman JS, Smargiassi A (2015) Review article: vulnerability to heat-related mortality: a systematic review, meta-analysis, and meta-regression analysis. Epidemiology 26(6):781–793
6. Declet-Barreto J, Knowlton K, Jenerette GD, Buyantuev A (2016) Effects of urban vegetation on mitigating exposure of vulnerable populations to excessive heat in Cleveland, Ohio. Weather Clim Soc 8(4):507–524
7. Berardi U (2017) A cross-country comparison of the building energy consumptions and their trends. Resour Conserv Recycl 123:230–241
8. Paravantis J, Santamouris M, Cartalis C, Efthymiou C, Kontoulis N (2017) Mortality associated with high ambient temperatures, heatwaves, and the urban Heat Island in Athens, Greece. Sustainability 9(4):1–22
9. Santamouris M, Paolini R, Haddad S, Synnefa A, Garshasbi S, Hatvani-Kovacs G et al (2020) Heat mitigation technologies can improve sustainability in cities. An holistic experimental and numerical impact assessment of urban overheating and related heat mitigation strategies on energy consumption, indoor comfort, vulnerability and heat-related mortality and morbidity in cities. Energy Build 217:110002
10. Canada H. Health [Internet] 2017 [cited 2020 Jun 11]. Available from https://www.canada.ca/en/services/health.html
11. Heat Warning Statistics Archive [Internet]. City of Toronto. City of Toronto; 2018 [cited 2021 Jan 2]. Available from https://www.toronto.ca/community-people/health-wellness-care/health-programs-advice/hot-weather/heat-warning-statistics-archive/

12. Pengelly LD, Campbell ME, Cheng CS, Fu C, Gingrich SE, Macfarlane R (2007) Anatomy of heat waves and mortality in Toronto: lessons for public health protection. Canadian J Publ Health/Rev Can de Sante'e Publique 98(5):364–368
13. Azhar GS, Mavalankar D, Nori-Sarma A, Rajiva A, Dutta P, Jaiswal A et al (2014) Heat-related mortality in India: excess all-cause mortality associated with the 2010 Ahmedabad heat wave. PLoS ONE 9(3):e91831
14. Gronlund CJ, Zanobetti A, Schwartz JD, Wellenius GA, O'Neill MS (2014) Heat, heat waves, and hospital admissions among the elderly in the United States, 1992–2006. Environ Health Perspect 122(11):1187–1192
15. Rainham DGC, Smoyer-Tomic KE (2003) The role of air pollution in the relationship between a heat stress index and human mortality in Toronto. Environ Res 93(1):9–19
16. Tsekeri E, Kolokotsa D, Santamouris M (2020) On the association of ambient temperature and elderly mortality in a Mediterranean island-Crete. Sci Total Environ 738:139843
17. D'Ippoliti D, Michelozzi P, Marino C, de'Donato F, Menne B, Katsouyanni K et al (2010) The impact of heat waves on mortality in 9 European cities: results from the EuroHEAT project. Environ Health 9(1):37
18. Berko J, Ingram DD, Saha S, Parker JD (2014) Deaths attributed to heat, cold, and other weather events in the United States, 2006–2010. Natl Health Stat Rep 76:1–15
19. Harlan SL, Chowell G, Yang S, Petitti DB, Morales Butler EJ, Ruddell BL et al (2014) Heat-related deaths in hot cities: estimates of human tolerance to high temperature thresholds. Int J Environ Res Public Health 11(3):3304–3326
20. Kolb S, Radon K, Valois M-F, Héguy L, Goldberg MS (2007) The short-term influence of weather on daily mortality in congestive heart failure. Archives Environ Occup Health 62(4):169–176
21. Smoyer-Tomic KE, Rainham DG (2001) Beating the heat: development and evaluation of a Canadian hot weather health-response plan. Environ Health Perspect 109(12):1241–1248
22. Nastos PT, Matzarakis A (2012) The effect of air temperature and human thermal indices on mortality in Athens, Greece. Theor Appl Climatol 108(3):591–599
23. Berardi U, Wang Y (2016) The effect of a Denser City over the urban microclimate: the case of Toronto. Sustainability 8(8):822
24. Wang Y, Berardi U, Akbari H (2016) Comparing the effects of urban heat island mitigation strategies for Toronto, Canada. Energy and Buildings 114:2–19
25. Jandaghian Z, Akbari H (2021) Increasing urban albedo to reduce heat-related mortality in Toronto and Montreal, Canada. Energy Build 237:110697
26. Morini E, Touchaei AG, Rossi F, Cotana F, Akbari H (2018) Evaluation of albedo enhancement to mitigate impacts of urban heat island in Rome (Italy) using WRF meteorological model. Urban Clim 24:551–566
27. Sangiorgio V, Fiorito F, Santamouris M (2020) Development of a holistic urban heat island evaluation methodology. Sci Rep 10:17913
28. Santamouris M (2014) Cooling the cities—a review of reflective and green roof mitigation technologies to fight heat island and improve comfort in urban environments. Sol Energy 103:682–703
29. Mohammed A, Khan A, Santamouris M. (2021) On the mitigation potential and climatic impact of modified urban albedo on a subtropical desert city. Build Environ 206
30. Jandaghian Z, Berardi U (2020) Analysis of the cooling effects of higher albedo surfaces during heat waves coupling the weather research and forecasting model with building energy models. Energy Buildings 207:109627
31. Touchaei AG, Akbari H, Tessum CW (2016) Effect of increasing urban albedo on meteorology and air quality of Montreal (Canada)—episodic simulation of heat wave in 2005. Atmos Environ 132:188–206
32. Jacobs SJ, Gallant AJE, Tapper NJ, Li D (2018) Use of cool roofs and vegetation to mitigate urban heat and improve human thermal stress in Melbourne, Australia. J Appl Meteorol Climatol 57(8):1747–1764

33. Li X-X, Norford LK (2016) Evaluation of cool roof and vegetations in mitigating urban heat island in a tropical city, Singapore. Urban Clim 16:59–74
34. Pisello AL, Castaldo VL, Pignatta G, Cotana F, Santamouris M (2016) Experimental in-lab and in-field analysis of waterproof membranes for cool roof application and urban heat island mitigation. Energy Build 114:180–190
35. Erell E, Pearlmutter D, Boneh D, Kutiel PB (2014) Effect of high-albedo materials on pedestrian heat stress in urban street canyons. Urban Clim 10:367–386
36. Lee H, Mayer H, Schindler D (2014) Importance of 3-D radiant flux densities for outdoor human thermal comfort on clear-sky summer days in Freiburg, Southwest Germany. Meteorol Z 315–30
37. Lee H. (2015) Increasing heat waves require human-biometeorological analyses on the planning-related potential to mitigate human heat stress within urban districts. Freiburg University
38. Lee H, Mayer H (2018) Thermal comfort of pedestrians in an urban street canyon is affected by increasing albedo of building walls. Int J Biometeorol 62(7):1199–1209
39. Schrijvers PJC, Jonker HJJ, de Roode SR, Kenjereš S (2016) The effect of using a high-albedo material on the universal temperature climate index within a street canyon. Urban Clim 17:284–303
40. Santamouris M, Fiorito F (2021) On the impact of modified urban albedo on ambient temperature and heat related mortality. Sol Energy 216:493–507
41. Canada E and CC. Environment and natural resources [Internet]. 2018 [cited 2021 Jul 11]. Available from https://www.canada.ca/en/services/environment.html
42. POWER Data Access Viewer [Internet]. [cited 2021 Jan 2]. Available from https://power.larc.nasa.gov/data-access-viewer/
43. Berardi U, Jandaghian Z, Graham J (2020) Effects of greenery enhancements for the resilience to heat waves: a comparison of analysis performed through mesoscale (WRF) and microscale (Envi-met) modeling. Sci Total Environ 747
44. Jandaghian Z, Berardi U (2020) Comparing urban canopy models for microclimate simulations in weather research and forecasting models. Sustain Cities Soc 55:102025
45. Perini K, Chokhachian A, Dong S, Auer T (2017) Modeling and simulating urban outdoor comfort: coupling ENVI-Met and TRNSYS by grasshopper. Energy Build 152:373–384
46. Crank PJ, Sailor DJ, Ban-Weiss G, Taleghani M (2018) Evaluating the ENVI-met microscale model for suitability in analysis of targeted urban heat mitigation strategies. Urban Clim 26:188–197
47. Maggiotto G, Buccolieri R, Santo MA, Leo LS, Di Sabatino S (2014) Validation of temperature-perturbation and CFD-based modelling for the prediction of the thermal urban environment: the Lecce (IT) case study. Environ Model Softw 60:69–83
48. Liu Z, Zheng S, Zhao L (2018) Evaluation of the ENVI-Met vegetation model of four common tree species in a subtropical hot-humid area. Atmosphere 9(5):198
49. Glotfelty T, Ramírez-Mejía D, Bowden J, Ghilardi A, West JJ (2021) Limitations of WRF land surface models for simulating land use and land cover change in Sub-Saharan Africa and development of an improved model (CLM-AF v. 1.0). Geoscientific Model Devel 14:3215–3249
50. McMillen JD, Steenburgh WJ (2015) Capabilities and limitations of convection-permitting WRF simulations of lake-effect systems over the Great Salt Lake. Weather Forecast 30(6):1711–1731
51. Jiménez PA, Dudhia J (2012) improving the representation of resolved and unresolved topographic effects on surface wind in the WRF model. J Appl Meteorol Climatol 51(2):300–316
52. Bueno B, Norford L, Hidalgo J, Pigeon G (2013) The urban weather generator. J Build Perform Simul 6(4):269–281
53. Bueno B, Hidalgo J, Pigeon G, Norford L, Masson V (2013) Calculation of air temperatures above the urban canopy layer from measurements at a rural operational weather station. J Appl Meteorol Climatol 52(2):472–483

54. Dardir M, Berardi U (2021) Development of microclimate modeling for enhancing neighborhood thermal performance through urban greenery cover. Energy Build 252
55. Ramírez AZ, Muñoz CB (2012) Albedo effect and energy efficiency of cities. Sustainable development—energy, engineering and technologies—manufacturing and environment. IntechOpen, London. https://doi.org/10.5772/1404
56. Matzarakis A, Rutz F, Mayer H (2010) Modelling radiation fluxes in simple and complex environments: basics of the RayMan model. Int J Biometeorol 54(2):131–139

Chapter 11
Heat-Related Mortality in Cyprus

Andri Pyrgou and Mat Santamouris

Abstract This chapter includes a discussion of heat-related mortality in Cyprus. The circulatory and respiratory causes of death were investigated during the period 2007–2014 inclusive for the urban and rural areas of Nicosia, Cyprus, under urban heatwave and non-heatwave conditions. Also, the heat- and cold-related mortality risk was examined by using a generalized additive model (GAM) regression technique to quantify the effect of the stimulus of mortality but also the relative risk technique was used to evaluate variations in terms of residence area (urban or rural), gender and age group. The mortality risk decreased as the minimum temperature increased from the coldest days to a certain threshold temperature about 20–21 °C which varied for each age group and gender. Over this threshold temperature, based on the daily minimum temperature the mortality risk increased. The results also showed the increase in mortality risk particularly for men of ages 65–69 (RR = 2.38) and women of ages 65–74 (around RR = 2.54) in the urban area, showing that women were more vulnerable to high temperature extremities. The analysis of temperature variations and not just heat extremities showed an insignificant association of gender mortality, whereas the age-mortality association showed that the population over 80 was more vulnerable. It is still undetermined as to what degree a change in existing climatic conditions will increase the environmental stress to humans as the population is acclimatized to different climates with different threshold temperatures and minimum mortality temperatures.

Keywords Heatwave · Relative risk · Health · urban heat island · Ozone · PM10 · Absolute humidity · GAM regression · Minimum temperature · MMT · Gender · Age · Urban warming

A. Pyrgou (✉)
Department of Civil Aviation, Pindarou 27 str., 1429 Nicosia, Cyprus
e-mail: a_pyrgou@hotmail.com

M. Santamouris
School of Built Environment, University of New South Wales, Sydney 2052, Australia

N. Aghamohammadi and M. Santamouris (eds.), *Urban Overheating: Heat Mitigation and the Impact on Health*, Advances in Sustainability Science and Technology,
https://doi.org/10.1007/978-981-19-4707-0_11

1 Introduction

Respiratory and cardiovascular causes of mortality have been correlated with the prevalent hot and cold temperatures. The increase in temperature extremes and diminishment of outdoor thermal comfort alert governments and scientists [1, 2]. The population of every country is acclimatized to different climatic conditions; thus, exploration of regional climatic time series data reveals different temperature threshold [3] necessitating different adaptation measures for the avoidance of the climate-change impact and for the increase in each country's capacity to function at a forthcoming temperature [4–8].

Scientists in north-eastern Europe[9–12], USA[13, 14] and China[4] examined the mortality risk with respect to the temperature and by considering fixed variables such as gender and age resulted to contrasting findings of the threshold temperature and minimum mortality temperature (MMT). MMT is defined as the temperature with the lowest risk of mortality according to a probability risk assessment. Generally, MMT has been found to be lower for population living in colder climates and higher for population living in warmer climates [11, 14, 15] with the temperature–mortality relationship described as a J-, V- or U-shaped curve. The limited number of studies of MMT in the eastern Mediterranean showed relatively high MMT, between 29 and 32 °C [16]. Thus, we need to predetermine the fixed factors of gender, age groups and residence area.

The environmental stress due to sudden temperature changes even within a day may be a factor of increased cardiovascular and respiratory mortality. Diurnal temperature range (DTR) is the difference between daily minimum and maximum temperatures. Some studies correlate a high DTR with an increase in mortality risk [17] as people acclimatize to hotter or colder temperatures at varying rates and to a certain extent based on physiological parameters such as age, gender and other existing health conditions. Also, the effect of exposure to extreme cold or heat conditions is not limited to the specific day, but may be delayed in time [2, 14, 17]. To accurately estimate the impact of weather on mortality, it is important to observe the exposure–response curve, lag structure and temperature metric.

When exposed to extreme heat, several organ systems in the human body are inflamed. Heat-related illnesses include heat cramps, heat exhaustion or heat stroke. The human body may sufficiently regulate the heat to cope with the thermal stress, up to certain limits when physical and mental activities are impaired. Intense thirst, heavy sweating, weakness, paleness, discomfort, anxiety, dizziness, fatigue, fainting, nausea and headache are the most common symptoms [18–20] of heat stress. During conditions of higher relative humidity, present at extreme heatwaves, the human body's ability to cool down by sweating is reduced [18]. Moreover, the duration of extreme heat conditions increases the hyperthermia risks as a person's tolerance of high temperatures decreases over time [18]. Difficulty in sleeping was also positively correlated with high temperatures leading to fatigue and lack of concentration [21].

Older people are more vulnerable to heat-related complications because the body's regulator of temperature weakens with age [13, 19, 22–25]. The ability of the body to withstand heat as the indoor thermal conditions is also affected by the use of air conditioning or medications interfering with fluid balance and physical condition of a person.

Extreme weather conditions concern temperatures that are abnormal for a specific region. Heatwave is a climatological phenomenon associated with high temperatures and increased thermal stress imposed on human health. The impact of heatwaves on humans and natural systems includes the increase in energy consumption, decreased air quality, intensification of droughts and degradation of human health [26–31]. Definitions of heatwaves may depend upon a percentile threshold temperature of the existing dataset (daily minimum, mean or maximum) or a predetermined threshold temperature. Vulnerability to heatwaves is exacerbated through the urban heat island (UHI) phenomenon which describes the temperature difference between urban and rural areas, and attributes this difference to the contrasting response of urban impervious surfaces and the higher rural vegetation. That is, the existing temperature difference due to the urban heat island phenomenon increases during heatwave conditions [32, 33].

During heatwaves (high temperatures with duration of 3–4 consecutive days), the cardiovascular system is affected; drop of the blood pressure, faster and irregular heartbeat, heightening the risk for cardiac failure. According to Chen et al [22], heatwaves result in 24.6% increase in total mortality, 46.9% increase in cardiovascular mortality, 32% increase in respiratory mortality, 51.3% increase in stroke mortality, 63.4% increase in ischaemic heart disease mortality and 47.6% increase in chronic obstructive pulmonary disease mortality [22]. Women, elderly and people with lower education level are more vulnerable during heatwaves [13, 22–25]. Heaviside et al. [33] examined the heat-related mortality in Nicosia and found that for an increase of 1 °C over baseline temperature the estimated heat-related mortality increased by 24%, and for a 5 °C temperature increase the heat-related mortality increased by 133% [34].

Adaptation-related policies are required to adapt to the increasing temperatures and the increased frequency of extreme heat events. The development of early warning systems, improvement of health care services and change in infrastructure may assist in the mitigation of heat-related health effects [35, 36].

2 Study Area and Datasets

Cyprus is an island in the eastern basin of the Mediterranean Sea of area 9251 km^2. Nicosia is the capital of the island of Cyprus with population about 330 thousand, located in the eastern basin of the Mediterranean Sea with a hot summer Mediterranean climate and hot semi-arid climate (in the north-eastern part of island), according to the Köppen climate classification signs Csa (Mediterranean hot summer climates) and BSh (Hot semi-arid climates) [37], with warm to hot dry summers and wet winters. Urban areas were defined according to the geo-codes of the Statistical service of Cyprus as the centre of Nicosia city and the following suburban areas;

Agios Dometios (north-west of centre), Egkomi/Nicosia (west), Strovolos (south-west), Lakatamia (south), Latsia (south) and Geri (south-east). Rural areas include all of the villages in the rest of the district of Nicosia.

The following table (Table 1) summarizes the population for Nicosia urban and rural areas according to each age group [38]. According to the Table, 237,703 people lived in the urban area, specifically 113,833 males and 123,870 females, and 87,249 people lived in the rural area, specifically 43,474 males and 43,775 females.

Hourly weather data (temperature [°C] and relative humidity (%)), air quality data and daily mortality data for 2007 to 2014 inclusive were collected for two meteorological stations [39] in the urban (35.17° N, 33.36° E) and rural (35.05° N, 33.54° E) areas of Nicosia, Cyprus. Hourly air quality data (ozone (μg/m3) and PM10 (μg/m3)) were obtained from the Ministry of Labour, Welfare and Social Insurance. The daily mortality data (only for circulatory and respiratory causes) were provided by the Health Monitoring Unit of the Ministry of Health of Cyprus and were categorized by ischaemic heart diseases (I20–I25), cerebrovascular diseases (I60–I69), other heart diseases (I30–I51), other circulatory diseases (I00–I15, I26–I28, I70–I99), influenza (J00–J99), pneumonia (J12–J18), chronic lower respiratory diseases (J40–J47) and

Table 1 Male and female population in urban and rural areas of Nicosia ("Table 1" By Andri Pyrgou, & Mat Santamouris (Own work) [CC-BY-4.0, http://creativecommons.org/licenses/by/4.0/], via International Journal of Environmental Research and Public Health MDPI. Available at https://www.mdpi.com/1660-4601/15/8/1571)

Age group	Urban Area			Rural area		
	Male	Female	Total	Male	Female	Total
0–4	6204	5960	12,164	2557	2506	5063
5–9	5610	5417	11,027	2487	2339	4826
10–14	6046	5835	11,881	2808	2565	5373
15–19	7225	6827	14,052	3325	3038	6363
20–24	9645	9281	18,926	3914	3454	7368
25–29	11,314	11,409	22,723	3931	3729	7660
30–34	10,335	11,369	21,704	3419	3639	7058
35–39	8256	10,481	18,737	2879	3328	6207
40–44	7638	9817	17,455	2780	3054	5834
45–49	7263	8721	15,984	2644	2815	5459
50–54	7521	8480	16,001	2779	2716	5495
55–59	6417	7044	13,461	2576	2425	5001
60–64	6261	6642	12,903	2248	2203	4451
65–69	4709	5094	9803	1677	1710	3387
70–74	3840	4365	8205	1366	1561	2927
75–79	2705	3156	5861	953	1206	2159
80+	2842	3968	6810	1131	1486	2617
Total	113,833	123,870	237,703	43,474	43,775	87,249

Table 2 Percentage of circulatory and respiratory causes of death in urban and rural areas for 2007–2014 ("Table 3" By Andri Pyrgou, & Mat Santamouris (Own work) [CC-BY-4.0, http://creativecommons.org/licenses/by/4.0/], via International Journal of Environmental Research and Public Health MDPI. Available at https://www.mdpi.com/1660-4601/15/8/1571)

Area	Cause of death	Deaths (nr)	Percentage per urban/rural population
Urban	Ischaemic heart disease	1309	1.15
	Cerebrovascular disease	882	0.77
	Other heart diseases	1147	1.01
	Other circulatory diseases	657	0.58
	Influenza	9	0.00
	Pneumonia	249	0.22
	Chronic lower respiratory diseases	227	0.20
	Other respiratory diseases	469	0.41
Rural	Ischaemic heart disease	482	1.11
	Cerebrovascular disease	336	0.77
	Other heart diseases	514	1.18
	Other circulatory diseases	290	0.67
	Influenza	1	0.00
	Pneumonia	58	0.13
	Chronic lower respiratory diseases	119	0.27
	Other respiratory diseases	130	0.30

other respiratory causes (J00– J06, J20–J39, J60–J99), according to the ICD-10-CM (International Classification of Diseases, Tenth Revision, Clinical Modification).

Total mortality caused by circulatory and respiratory causes was 6880 people for the investigated period of 8 years in Nicosia, Cyprus. The initial analysis of the area of residency showed that about 70.5% (4849) of deceased lived in urban and suburban areas. Table 2 shows the percentage of number of deaths per cause (circulatory or respiratory) for the urban and rural sites, respectively. According to the Table, the leading circulatory cause of death was ischaemic heart disease at both areas and higher mortality rate due to respiratory causes were observed in the urban area compared to the rural area. Overall for the investigated time period (8 years), 4.34% and 4.43% of the urban and rural total population died from circulatory and respiratory causes, an average of 0.55% per year.

3 Mortality–Temperature Relationship

The issue of temperature-related mortality in Nicosia, Cyprus, was examined using the generalized additive model (GAM) regression technique [40], because the log

Table 3 Mean and maximum daily averaged values of investigated parameters under heatwave and non-heatwave conditions for months June–September ("Table 4" By Andri Pyrgou, & Mat Santamouris (Own work) [CC-BY-4.0, http://creativecommons.org/licenses/by/4.0/], via International Journal of Environmental Research and Public Health MDPI. Available at https://www.mdpi.com/1660-4601/15/8/1571)

Parameter	Non-heatwave conditions		Heatwave conditions	
	Maximum	Mean	Maximum	Mean
Temperature (°C)	33.74	28.15	35.08	31.67
Relative humidity (%)	87.33	51.64	75.33	42.42
Absolute humidity (μg/m^3)	2.86	1.51	2.94	1.77
Ozone levels (μg/m^3)	106.64	69.63	99.60	78.08
PM10 levels (μg/m^3)	156.52	40.26	80.70	40.96

Table 4 Relative risk (RR) of mortality for men and women of different age groups in urban and rural areas ("Table 4") By Andri Pyrgou, & Mat Santamouris (Own work) [CC-BY-4.0, http://creativecommons.org/licenses/by/4.0/], via International Journal of Environmental Research and Public Health MDPI. Available at https://www.mdpi.com/1660-4601/15/8/1571)

Age group	RR in urban areas		RR in rural areas	
	Men	Women	Men	Women
0–59	1.19	1.44	1.54	0.86
60–64	1.17	– (no deaths)	0.91	1.93
65–69	2.38	2.57	1.34	- (no deaths)
70–74	0.89	2.49	1.83	0.64
75–79	1.33	0.75	1.63	- (no deaths)
80+	0.96	1.48	0.55	1.19

mortality based on temperature was assumed to be smooth but not necessarily linear. Using the GAM, we obtained a high quality of prediction of the dependent variable (log mortality rate) from various distributions by estimating unspecific (non-parametric) functions of the predictor variables *xj* which were connected to the dependent variable (mortality rate) via a link function. The limitation of this analysis is that population characteristics such as health behaviours (smoking, drinking), comorbidities (hypertension, diabetes, cancer, etc.), medications and access to health care were not considered.

The generalized additive model (GAM) in combination with multiple linear regression allowed for the continuous values of minimum and maximum daily temperature and diurnal temperature range (DTR) to be modelled as smooth functions and by considering a lag period. The lag period was determined using the cross-correlation function (CCF) in RStudio [41], which shows the similarity of two series as a function of the displacement of one relative to the other.

Our findings indicated that the previous three days' temperatures were strongly predictive of mortality and that MMT should be calculated using daily minimum temperature values. The lag period of cross-correlation function (CCF analysis) that had an effect on mortality rate was 4 days for the months November–April and 0 days for the months May–October, agreeing with a similar study in Estonia [4, 9]. Our cross-correlation results agree with previous studies that the effects of heat on mortality rate shortly after temperatures start to increase, whereas the effects of cold may take longer to emerge, and, depending on the latitude and the local climate, these periods may vary [2, 9, 42].

According to other studies when heatwaves conditions were isolated and examined are usually more directly linked with increased mortality on the same day or a couple of days after a heatwave [2, 10]. Pyrgou et al. [2] for the investigated area showed an exacerbation of urban–rural temperature difference (urban heat island phenomenon) up to 2 °C difference during heatwaves, particularly at noon [32], that could lead to detriment effects on human health.

To further elaborate on the analysis of heatwaves, we defined them as the events when for four consecutive days the daily mean temperature exceeded 31.1 °C (90th percentile) in the urban area, resulting to 81 days in total for the investigated time period 2007–2014. During the defined heatwaves, higher cross-correlation with mortality rate with temperature was found even three days after exposure to heatwave conditions. High cross-correlation was also observed for absolute humidity with mean and maximum temperatures, reaching a maximum time lag of four days.

Preceding research in the area of investigation (Nicosia, Cyprus) showed higher air pollution, that is ozone concentration and PM10 concentration, during heatwave conditions [43]. Even though air pollution may influence mortality rates, the analysis for Nicosia showed that ozone and PM10 concentration levels were higher during heatwave conditions (see Table 3) but they were not correlated with mortality rate and with no time lag. The 0 days' time lag of ozone and temperature was expected because temperature acts as a catalyst for short-time chemical processes involving ozone formation. Despite the fact that air pollution was higher during heatwaves, the analysis revealed that air pollution was a low risk factor for mortality rate in the urban area of Nicosia.

The relative risk (RR) of mortality caused by circulatory and respiratory causes under heatwave and non-heatwave conditions was calculated using the followed expression:

$$RR = \frac{\frac{\text{peoplewhodiedunder HWconditions}}{\text{peopleexposed toHW}}}{\frac{\text{peoplewhodiedundernon}-\text{HWconditions}}{\text{peopleexposedtonon}-\text{HWconditions}}} \quad (1)$$

The investigated time period used for the relative risk estimation presumed as heatwaves the time frame 3 days after the recorded events to include the effect of the lag period and resulting in a total of 112 days. Non-heatwave days were considered the rest of the days from June–September.

During the presumed heatwave periods and the lag time of 3 days (total of 112 days), 432 deceased were reported due to circulatory and respiratory causes, whereas for the non-heatwave periods (June–September) (total of 864 days) 2595 deceased were reported due to circulatory and respiratory causes. The relative mortality rate (using Eq. 1) for all age groups and both genders under urban heatwave conditions and for a lag period of 3 days following heatwave conditions was calculated to be 1.28, showing a greater risk of mortality under heatwave conditions and the determined lag period. A relative risk of 0.88 was found for all age groups and both genders under urban heatwave conditions in the rural areas. This shows the higher risk of mortality for urban population which with the effect of urban heat island leads to higher urban temperatures which are more dangerous to human health.

The following figure (Fig. 1) shows the temperature–mortality relative risk function estimated for Nicosia using the GAM analysis for all the months of years 2007–2014. Six temperature parameters were examined: the maximum daily temperature (T_{max}), the weighted average of the maximum daily temperatures of the preceding three days ($T_{max[-3]}$), the minimum daily temperature (T_{min}), the weighted average of the minimum daily temperatures of the preceding three days ($T_{min[-3]}$), the diurnal temperature range (DTR) of the day and the weighted average of the DTR of the preceding three days ($DTR_{[-3]}$).

The relationship between maximum daily temperature and probability risk of mortality appeared linear, with a steeper slope when the average of the preceding three days' maximum temperature was considered. The analysis of the minimum

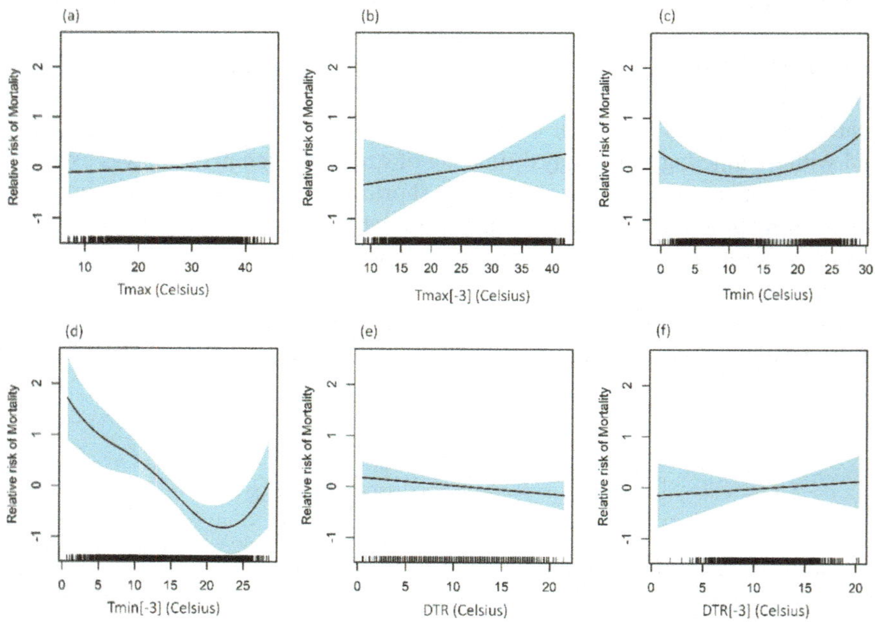

Fig. 1 Temperature–mortality relative risk functions for Nicosia, Cyprus, 2004–2014. T_{max}, $T_{max[-3]}$, T_{min}, $T_{min[-3]}$, DTR and $DTR_{[-3]}$ are shown in Figs. 1a, 1b, 1c, 1d, 1e and 2f, respectively. ("Fig. 2" By Andri Pyrgou, & Mat Santamouris (Own work) [CC-BY-4.0, http://creativecommons.org/licenses/by/4.0/], via Climate MDPI. Available at https://www.mdpi.com/2225-1154/8/3/40#)

daily temperature with respect to the mortality's probability risk (Fig. 1c and 1d) showed a U shape for same-day relationship and an inverse J shape when the weighted average of the minimum daily temperature of the preceding three days ($T_{\min[-3]}$) was used. That is, mortality risk decreased as minimum temperature increased from the coldest temperatures and rose as the temperature increases from a certain threshold temperature (about 22 °C). In warmer climates, such as Nicosia, the people are more acclimatized to high temperature conditions, and therefore, a steeper increase in the relative risk of mortality in colder conditions was observed.

The analysis of DTR (Fig. 1d and 1f) showed no significant results (mortality relative risk close to zero) insinuating that the human body could adapt to any DTR of Nicosia area within the same day. In contrast with what was previously thought, the presented research showed that DTR did not have a significant effect on mortality risk, suggesting more focussed research on environmental stress factors rather than within-day temperature variations.

4 Risks Per Gender, Age Group and Residence Area

Pyramid plots (Fig. 2) were utilized initially to show graphically the most sensitive age group and gender. Seasonal division was preferred to exclude deaths due to winter's lower temperatures (influenza and pneumonia) and to emphasize on summer mortality rates that could be caused by extreme heat conditions. According to the data (Fig. 2), recorded deaths from circulatory and respiratory causes were 1999, 1838, 1532 and 1511 for winter, spring, summer and autumn, respectively. Highest mortality for the entire year was observed for men of the age of 80–89 years old and women of the age of 70–79. The results showed increased mortality due to respiratory diseases in winter and spring, when several various and influenza prevail in colder temperatures. The high mortality rates of people over 80 years old during summer periods suggested an intolerance of human body to tolerate substantial heat [2].

Later, a similar procedure of the relative risk assessment for the mortality rate, explained in the preceding section, was performed for each age group and each gender in the urban and rural areas for heatwave and non-heatwave conditions. According to the results (Table 4), elderly and women have higher susceptibility to die under heatwave conditions. In urban area, men of age 65–69 under heatwave conditions had 138 times the risk of death compared to men of age 65–69 under non-heatwave conditions. The risk of death due to circulatory or respiratory causes among women of the same age (65–69) under heatwave conditions was 157 times as high as the risk of death among women under non-heatwave conditions. The relative risk of death at higher ages, over 70 for men and over 75 for women decreased. Summarizing, the most vulnerable age group of mortality in urban areas under heatwave conditions was 65–69 for men and 65–74 for women with a probability relative risk of 2.38 and 2.54, respectively. A similar analysis for rural areas resulted to insufficient conclusions due to the small rural population. An increased relative risk of mortality especially for the male population of ages 65–79 was observed under urban heatwave conditions.

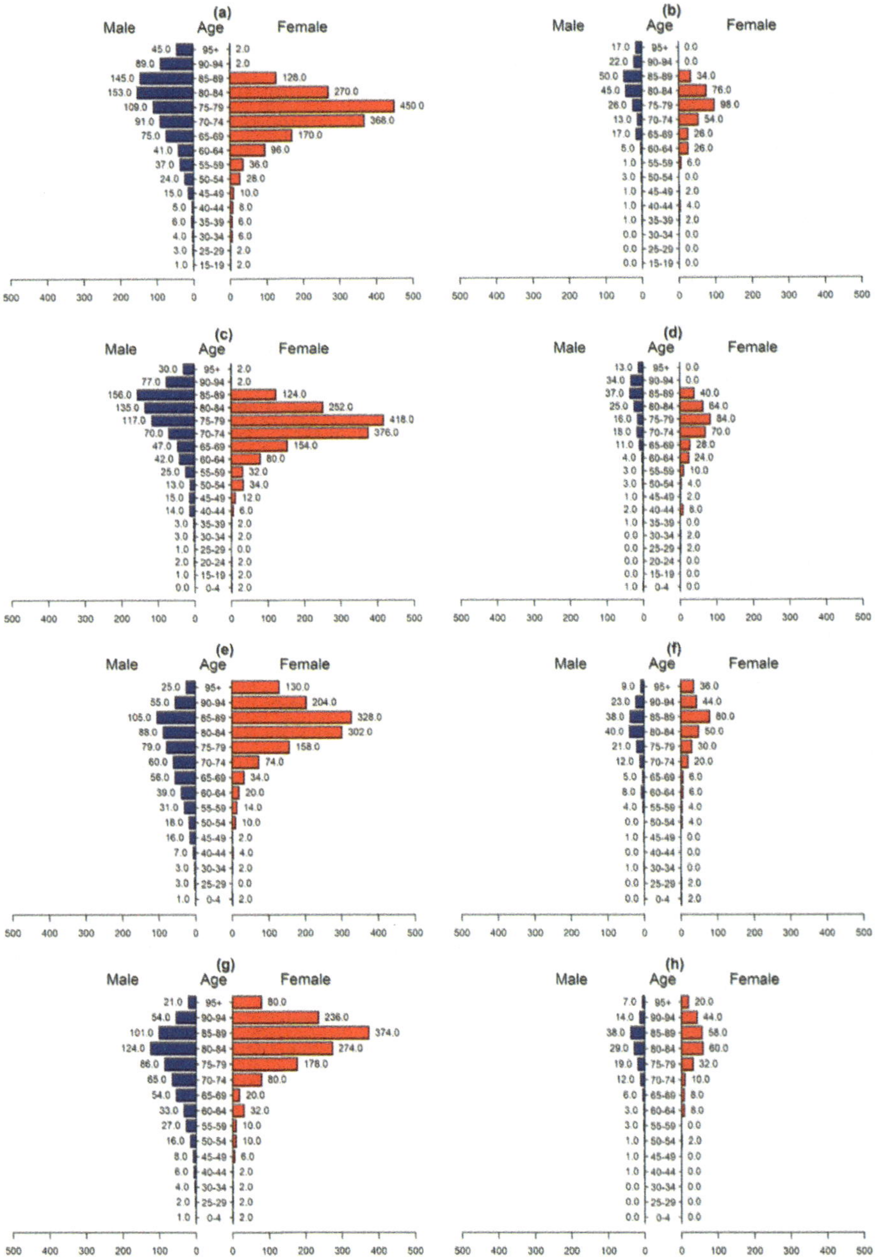

Fig. 2 Mortality by age group and gender (blue for male and red for female) for winter, (**a** circulatory causes, **b** respiratory causes), spring (**c** circulatory, **d** respiratory), summer (**e** circulatory, **f** respiratory) and autumn (**g** circulatory, **h** respiratory) for years 2007–2014. ("Fig. 1" By Andri Pyrgou, & Mat Santamouris (Own work) [CC-BY-4.0, http://creativecommons.org/licenses/by/4.0/], via International Journal of Environmental Research and Public Health MDPI. Available at https://www.mdpi.com/1660-4601/15/8/1571)

The comparison of the relative risk columns for male and for female population in urban and rural areas revealed a higher risk in urban areas which may be attributed to the exacerbated higher temperatures caused also by the synergy of heatwaves with urban heat island effect observed in the city of Nicosia [32].

To model the relationship between temperature and mortality, the generalized additive model (GAM) was implemented to assess the interaction between a fixed variable (gender, age group) and the observation time and to interpret quantitative results. Specifically, the analysis focussed on the Tmin of the current day and the weighted average of the preceding three days ($T_{\min[-3]}$) by subgrouping the risk by gender and age groups (Fig. 3a and b). According to Fig. 3a and b, cold temperatures imposed greater risk than hot temperatures (*y*-axis value > 0.5 for lower and higher temperatures), but other factors, such as respiratory epidemics, usually present in winter, made unclear the exact role of temperatures on increased mortality. A similar GAM analysis in Shanghai [17] found smaller relative risk for elderly people with values of 0.3. Therefore, in Nicosia where the population is more accustomed to higher temperatures, they are rather vulnerable to lower temperatures with relative risk over 0.5.

Figure 4 shows the relative risk of mortality for the DTR of the same day and the weighted average DTR of the preceding three days. According to Kan et al. [17], a large DTR change might be a source of additional environmental stress leading to greater mortality risk [17]. The GAM analysis's results showed greater risk at small

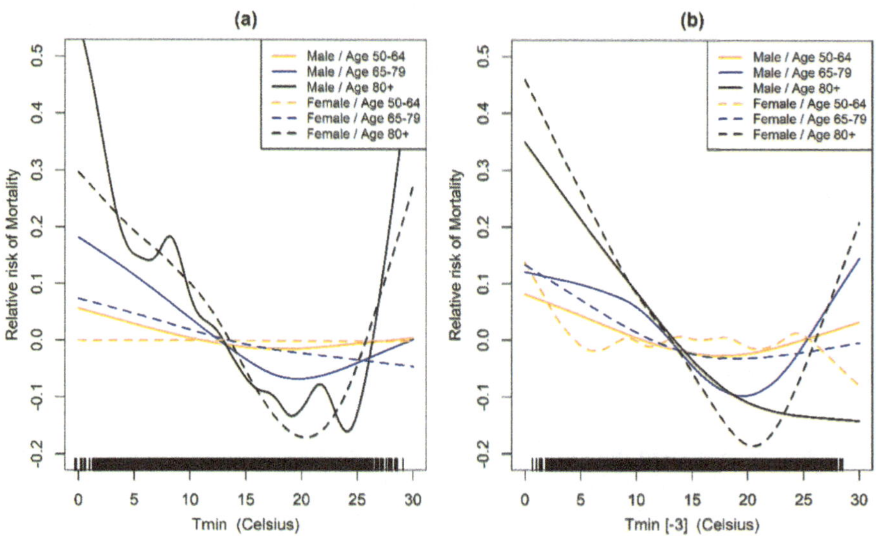

Fig. 3 Mortality relative risk functions for Nicosia, Cyprus, 2007–2014 with, **a** minimum temperature of the same day (Tmin), **b** weighted average minimum temperature of the preceding three days ($T_{\min[-3]}$). ("Fig. 3" By Andri Pyrgou, & Mat Santamouris (Own work) [CC-BY-4.0, http://creativecommons.org/licenses/by/4.0/], via Climate MDPI. Available at https://www.mdpi.com/2225-1154/8/3/40#)

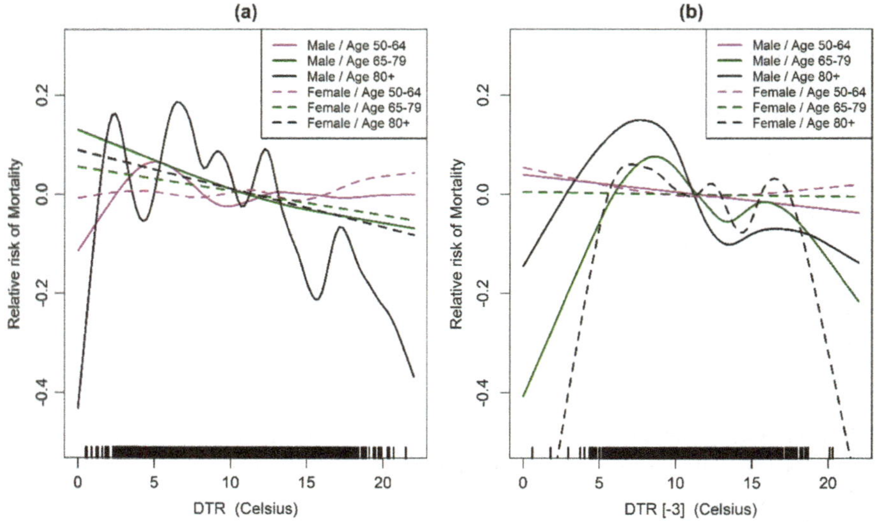

Fig. 4 Mortality relative risk functions for Nicosia, Cyprus, 2007–2014 with, **a** DTR of the same day, **b** Weighted average DTR of the preceding three days ($DTR_{[-3]}$). ("Fig. 4" By Andri Pyrgou, & Mat Santamouris (Own work) [CC-BY-4.0, http://creativecommons.org/licenses/by/4.0/], via Climate MDPI. Available at https://www.mdpi.com/2225-1154/8/3/40#)

DTR of the range of 6–8 °C and smaller risk at larger DTRs. Men and women of age 50–64 (magenta lines) were not affected by the variations in DTR throughout the day or the preceding three days. Men of age 65 and over had greater relative risk for DTR smaller than 5 °C, whereas for larger DTR there was negative relative risk showing that large DTR did not impose a risk factor for death in Nicosia. Overall, we found that DTR was independently associated with daily mortality in Nicosia and that fluctuations in DTR appeared to affect mostly people over 80 probably because they have reduced ability to regulate body temperatures, thus making them marginally more vulnerable.

5 Conclusion

This chapter examined the mortality risk in relation to high or low temperatures of different age groups and compared it between the two genders in the city of Nicosia, Cyprus, for the years 2007–2014 included. Mortality risk has only been evaluated for respiratory and circulatory causes. Not only the minimum and maximum daily temperatures were addressed, but also the diurnal temperature range (DTR) in order to examine the sensitivity for within the same day air temperature variations.

A combination of the cross-correlation analysis and the generalized additive model (GAM) regression technique was used to evaluate the mortality–temperature sensitivity and to observe differences between extremely hot and cold days. There was increased mortality during heatwaves and cold days, with a lag (delayed) effect up to 3–4 days. The existence of this lag implied that in warmer climates such as Nicosia, Cyprus, people are more acclimatized to high temperatures, and therefore, no precautions were taken to accommodate the extreme cold conditions or heatwaves.

Further analysis focussed during the warm months (June–September) in Nicosia revealing that the majority of male population died between ages 80 and 89 and of female population between ages 70 and 79 due to respiratory and circulatory causes. The lag effect of three days was included in the determination of the relative risk of mortality under heatwave conditions in urban and rural areas. Under heatwave conditions, people of ages 65–74 were more vulnerable to heat-related mortality and particularly women. Also, people residing in the urban areas had a greater risk of mortality probably due to the higher temperatures and the intensification of the temperatures due to the urban heat island phenomenon.

The identification of a threshold temperature per latitude and local climate will assist in the evaluation of the adaptation capacity of a specific population. This threshold temperature should be calculated using the daily minimum temperature, for which we observed the higher correlation with mortality. Nevertheless, even if humans become fully acclimatized to high temperatures, their health may still be negatively affected as a result of the poorer air quality associated with extremely high temperatures [2, 43]. It is fairly difficult to predict to what extent a population may adapt to temperature increase in future, but a good heat-health warning system may minimize these high mortality rates.

References

1. Hondula DM, Balling RC, Vanos JK, Georgescu M (2015) Rising temperatures, human health, and the role of adaptation. Curr Clim Chang Reports 1:144–154. https://doi.org/10.1007/s40641-015-0016-4
2. Pyrgou A, Santamouris M (2018) Increasing probability of heat-related mortality in a Mediterranean city due to urban warming. Int J Environ Res Public Health 15:1571. https://doi.org/10.3390/ijerph15081571
3. Coffel ED, Horton RM, De Sherbinin A (2018) Temperature and humidity based projections of a rapid rise in global heat stress exposure during the 21st century. Environ Res Lett 13. https://doi.org/10.1088/1748-9326/aaa00e
4. Chen R, Yin P, Wang L, Liu C, Niu Y, Wang W, Jiang Y, Liu Y, Liu J, Qi J, You J, Kan H, Zhou M (2018) Association between ambient temperature and mortality risk and burden: time series study in 272 main Chinese cities. BMJ 363. https://doi.org/10.1136/bmj.k4306
5. Rocklöv J, Forsberg B, Ebi K, Bellander T (2014) Susceptibility to mortality related to temperature and heat and cold wave duration in the population of Stockholm County. Sweden Glob Health Action 7:1–11. https://doi.org/10.3402/gha.v7.22737
6. Gronlund CJ, Berrocal VJ, White-Newsome JL, Conlon KC, O'Neill MS (2015) Vulnerability to extreme heat by socio-demographic characteristics and area green space among the elderly in Michigan, 1990–2007. Environ Res. https://doi.org/10.1016/j.envres.2014.08.042
7. Goggins WB, Ren C, Ng E, Yang C, Chan EYY (2013) Effect modification of the association between meteorological variables and mortality by urban climatic conditions in the tropical city of Kaohsiung. Taiwan Geospat Health. https://doi.org/10.4081/gh.2013.52

8. Milojevic A, Armstrong BG, Gasparrini A, Bohnenstengel SI, Barratt B, Wilkinson P (2016) Methods to estimate acclimatization to urban heat island effects on heat-and cold-related mortality. Environ Health Perspect 124:1016–1022. https://doi.org/10.1289/ehp.1510109
9. Orru H, Åström DO (2017) Increases in external cause mortality due to high and low temperatures: evidence from northeastern Europe. Int J Biometeorol 61:963–966. https://doi.org/10.1007/s00484-016-1270-4
10. Åström DO, Åström C, Rekker K, Indermitte E, Orru H (2016) High summer temperatures and mortality in Estonia. PLoS ONE 11:1–10. https://doi.org/10.1371/journal.pone.0155045
11. Oudin Åström D, Tornevi A, Ebi KL, Rocklöv J, Forsberg B (2016) Evolution of minimum mortality temperature in Stockholm, Sweden, 1901–2009. Environ Health Perspect 124:740–744. https://doi.org/10.1289/ehp.1509692
12. Huynen MMTE, Martens P, Schram D, Weijenberg MP, Kunst AE (2001) The impact of heat waves and cold spells on mortality rates in the Dutch population. Environ Health Perspect 109:463–470. https://doi.org/10.1289/ehp.01109463
13. Anderson BG, Bell ML (2009) Weather-related mortality: how heat, cold, and heat waves affect mortality in the United States. Epidemiology 20:205–213
14. Curriero FC, Heiner KS, Samet JM, Zeger SL, Strug L, Patz JA (2002) Temperature and mortality in 11 cities of the eastern United States. Am J Epidemiol. https://doi.org/10.1093/aje/155.1.80
15. Baccini M, Biggeri A, Accetta G, Kosatsky T, Katsouyanni K, Analitis A, Anderson HR, Bisanti L, D'Iippoliti D, Danova J, Forsberg B, Medina S, Paldy A, Rabczenko D, Schindler C, Michelozzi P (2008) Heat effects on mortality in 15 European cities. Epidemiology. https://doi.org/10.1097/EDE.0b013e318176bfcd
16. Yin Q, Wang J, Ren Z, Li J, Guo Y (2019) Mapping the increased minimum mortality temperatures in the context of global climate change. Nat Commun 10:4640. https://doi.org/10.1038/s41467-019-12663-y
17. Kan H, London SJ, Chen H, Song G, Chen G, Jiang L, Zhao N, Zhang Y, Chen B (2007) Diurnal temperature range and daily mortality in Shanghai. China Environ Res 103:424–431. https://doi.org/10.1016/j.envres.2006.11.009
18. Luber G, McGeehin M (2008) Climate change and extreme heat events. Am J Prev Med 35:429–435. https://doi.org/10.1016/j.amepre.2008.08.021
19. Bobb JF, Obermeyer Z, Wang Y, Dominici F (2014) Cause-specific risk of hospital admission related to extreme heat in older adults. JAMA J Am Med Assoc. https://doi.org/10.1001/jama.2014.15715
20. Hopp S, Dominici F, Bobb JF (2018) Medical diagnoses of heat wave-related hospital admissions in older adults. Prev Med (Baltim) 110:81–85. https://doi.org/10.1016/j.ypmed.2018.02.001
21. Sakka A, Santamouris M, Livada I, Nicol F, Wilson M (2012) On the thermal performance of low income housing during heat waves. Energy Build 49:69–77. https://doi.org/10.1016/j.enbuild.2012.01.023
22. Chen K, Bi J, Chen J, Chen X, Huang L, Zhou L (2015) Influence of heat wave definitions to the added effect of heat waves on daily mortality in Nanjing. China Sci Total Environ 506–507:18–25. https://doi.org/10.1016/j.scitotenv.2014.10.092
23. Schuman SH (1966) Patterns of urban heat-wave deaths and implications for prevention: data from New York and St. Louis during July, 1966. Environ Res 5:59–75. https://doi.org/10.1016/0013-9351(72)90020-5
24. Pascal M (2011) An analysis of the mortality risks associated with heat and heat waves in Ireland, to assist in planning for climate change. Dublin Institute of Technology
25. Gabriel KMA, Endlicher WR (2011) Urban and rural mortality rates during heat waves in Berlin and Brandenburg. Ger Environ Pollut 159:2044–2050. https://doi.org/10.1016/j.envpol.2011.01.016
26. Santamouris M, Assimakopoulos DN, Chrisomallidou N, Klitsikas N, Mangold D, Michael P, Tsangrassoulis A (2001) Energy and climate in the urban built environment. Routledge. ISBN 9781315073774

27. Herold N, Kala J, Alexander LV (2016) The influence of soil moisture deficits on Australian heatwaves. Environ Res Lett 11:064003. https://doi.org/10.1088/1748-9326/11/6/064003
28. Paravantis J, Santamouris M, Cartalis C, Efthymiou C, Kontoulis N (2017) Mortality associated with high ambient temperatures, heatwaves, and the urban heat island in Athens, Greece. Sustainability 9:606. https://doi.org/10.3390/su9040606
29. Pyrgou A, Castaldo VL, Pisello AL, Cotana F, Santamouris M (2017) On the effect of summer heatwaves and urban overheating on building thermal-energy performance in central Italy. Sustain Cities Soc 28:187–200. https://doi.org/10.1016/j.scs.2016.09.012
30. Smith S, Elliot AJ, Hajat S, Bone A, Bates C, Smith GE, Kovats S (2016) The impact of heatwaves on community morbidity and healthcare usage: a retrospective observational study using real-time syndromic surveillance. Int J Environ Res Pub Health 13. https://doi.org/10.3390/ijerph13010132
31. Xu Z, FitzGerald G, Guo Y, Jalaludin B, Tong S (2016) Impact of heatwave on mortality under different heatwave definitions: a systematic review and meta-analysis. Environ Int 89–90:193–203. https://doi.org/10.1016/j.envint.2016.02.007
32. Pyrgou A, Hadjinicolaou P, Santamouris M (2020) Urban-rural moisture contrast: regulator of the urban heat island and heatwaves' synergy over a mediterranean city. Environ Res 182:109102. https://doi.org/10.1016/j.envres.2019.109102
33. Li D, Bou-Zeid E (2013) Synergistic interactions between urban heat islands and heat waves: the impact in cities is larger than the sum of its parts. J Appl Meteorol Climatol 52:2051–2064. https://doi.org/10.1175/JAMC-D-13-02.1
34. Heaviside C, Tsangari H, Paschalidou A, Vardoulakis S, Kassomenos P, Georgiou KE, Yamasaki EN (2016) Heat-related mortality in Cyprus for current and future climate scenarios. Sci Total Environ 569–570:627–633. https://doi.org/10.1016/j.scitotenv.2016.06.138
35. Wang Y, Nordio F, Nairn J, Zanobetti A, Schwartz JD (2018) Accounting for adaptation and intensity in projecting heat wave-related mortality. Environ Res 161:464–471. https://doi.org/10.1016/j.envres.2017.11.049
36. Aström C, Orru H, Rocklöv J, Strandberg G, Ebi KL, Forsberg B (2013) Heat-related respiratory hospital admissions in Europe in a changing climate: a health impact assessment. BMJ Open. https://doi.org/10.1136/bmjopen-2012-001842
37. Peel MC, Finlayson BL, TAM (2007) Updated world map of the Koppen-Geiger climate classification. Hydrol Earth Syst Sci 11:1633–1644
38. Statistical Service of the Republic of Cyprus (2015) Population Census Final 2011. Retrieved from http://www.mof.gov.cy/mof/cystat/statistics.nsf/All/3D52D8F17C418CFAC2257EEA00346671/$file/POPULATION_CENSUS-2011_VolI-EL-261015.pdf?OpenElement
39. Cyprus, R. of Department of Meteorology, Cyprus. http://www.moa.gov.cy/moa/ms/ms.nsf/DMLannual_en/DMLannual_en?OpenDocument. Accessed 25 Apr 2019
40. Pyrgou A, Santamouris M (2020) Article probability risk of heat-and cold-related mortality to temperature, gender, and age using GAM regression analysis. Climate 8. https://doi.org/10.3390/cli8030040
41. R Core Team R: A language and environment for statistical computing. R Found. Stat. Comput. Vienna, Austria 2020
42. Brooke Anderson G, Bell ML (2011) Heat waves in the United States: mortality risk during heat waves and effect modification by heat wave characteristics in 43 U.S. communities. Environ. Health Perspect 119:210–218. https://doi.org/10.1289/ehp.1002313
43. Pyrgou A, Hadjinicolaou P, Santamouris M (2018) Enhanced near-surface ozone under heatwave conditions in a Mediterranean island. Sci Rep 8:9191. https://doi.org/10.1038/s41598-018-27590-z

Chapter 12
Urban Overheating and the Impact on Health in Melbourne

Elmira Jamei and Nigel Tapper

Abstract Climate change is already affecting health outcomes in Victoria. The health impacts of climate change present a systemic risk to Victorians that is largely hidden. Since Australia is a country with well-developed infrastructure, the direct exposure to climate hazards is often limited. Most impacts that do occur are indirect rather than direct. These impacts manifest through socio-economic factors that influence exposure and vulnerability to climate. This chapter reviews the impact of urban overheating on heat-related death and focuses on the Victorian context to discuss the factors that play a key role in vulnerability and exposure to heat. Furthermore, the key role of the government, especially at the municipal level, in promoting heat mitigation strategies is highlighted by listing some of the actions and initiatives undertaken by the Victorian government to cool the city.

Keywords Climate change · Urban heat island · Public health · Urban overheating

1 Introduction

According to a recent report by the United Nations, more than 55% of the world's population reside in urban areas; this number will increase to 60% by 2030 if the trend of rapid urbanisation continues [1]. People migrate from rural to urban areas mainly because cities account for almost 60% of the global gross domestic product [2]. This urban migration led to a significant increase in activities that generate anthropogenic heat and cause high-energy consumption in urban areas, which in turn contributes to 70% of global carbon emission [3]. Given the magnitude of anthropogenic activities and complex urban morphologies, the local climate in cities often features unique characteristics compared with that in rural areas [4]. One of the most well-known

E. Jamei (✉)
Institute for Sustainable Industries and Liveable Cities, Victoria University, Melbourne, VIC, Australia
e-mail: Elmira.Jamei@vu.edu.au

N. Tapper
School of Geography and Environmental Science, Urban Climate Research, Monash University, Melbourne, VIC, Australia

N. Aghamohammadi and M. Santamouris (eds.), *Urban Overheating: Heat Mitigation and the Impact on Health*, Advances in Sustainability Science and Technology,
https://doi.org/10.1007/978-981-19-4707-0_12

features of local urban climates, which have been observed in many cities across the globe, is the urban heat island (UHI) effect, a phenomenon wherein urban areas become warmer than their surrounding rural areas [5–9].

UHI adversely affects public health in various ways, such as by altering the rainfall pattern, increasing air pollution and flood risk, and reducing water quality; but the most significant and direct impact of UHI on public health is the occurrence of magnified extreme heat events (EHE) [10].

Australia is one of the most urbanised nations in the world; 86% of the country's population live in urban areas [11], and heat wave is a significant public health concern in Australian cities. The number of heatwave occurrences is rapidly increasing in Australian cities. For example, in 2009, the temperature threshold in the western suburbs of Melbourne exceeded 47.5 °C, and multiple heatwaves occurred in 2014 in Sydney, 2017 in Brisbane, and 2019 in Adelaide.

Heat waves account for more human fatality than any natural hazard in Australia; the percentage of heat-related deaths accounted for 55% since the 1900s, whereas flooding, which is the second largest cause of death, accounted for only 15% [12]. In contrast to most natural hazards, heatwaves occur in silence [13], as it is not possible to see, or hear them. Therefore, it does not lead to large-scale evacuations, but it affects each household individually and unequally and puts vulnerable groups at increased risks.

Victoria is Australia's second most populated state, and Melbourne is the country's second largest city. UHI studies in Melbourne indicated up to 7 °C temperature difference between Melbourne's CBD and surrounding rural districts [14]; moreover, UHI is often intensified during extreme heat waves [15]. Studies have also shown an increase in the intensity, frequency, and duration of heat waves in Melbourne (e.g. in 2009, the temperature threshold in Melbourne, particularly in the western suburbs, exceeded 47.5 °C) [16–18]. Heat waves affect communities differently. A study in Melbourne showed that hospital admissions of people who live in disadvantaged geographical areas increased during heat waves [19].

The urban consolidation in Melbourne, history of frequent heat waves exacerbated by UHI, and the adverse impacts of urban overheating on public health, namely heat-related mortality and morbidity, suggest the critical need to mitigate urban overheating and identify the most vulnerable urban areas to minimise the adverse impacts of urban overheating on health. This necessity has formed the main objectives of this chapter. First, this chapter discusses the heat vulnerability concept and the key factors that affect vulnerability to heat in Melbourne. The chapter then provides an overview of some of the key heat mitigation initiatives in Melbourne that contributed to a certain level of improvement in urban climate.

2 Urban Overheating and Heat-Related Deaths in Melbourne

The key causes of UHI formation are documented as the extensive use of impervious surfaces, such as construction materials, asphalts, pavements, complex urban geometries, lack of green infrastructures, and high level of anthropogenic heat emission in urban areas [20–24]. Impervious surfaces with high heat capacity feature elevated night-time temperatures, and those with low heat capacity exert increased daytime temperatures [24]. Urban geometries are mainly affected by street orientation and openness to the sky (Sky View Factor (SVF)) and play a critical role on radiative exchange between urban canopy and the atmosphere [25–28]. For instance, longwave radiation is often trapped within high-rise buildings and lead to increased night-time temperatures. However, during the day it causes a reduction in the surface temperature of adjacent buildings through shading [29–31]. Lack of green infrastructure is another key factor that intensifies the magnitude of UHI in urban areas. Greeneries reduce temperature through three mechanisms, namely evapotranspiration, shading, and altering wind behaviour [32, 33]. Another important cause of UHI is anthropogenic heat generated by human and industrial activities, transportation, and high level of energy consumption. Anthropogenic heat changes urban energy balance and increases the magnitude of UHI in cities [34–37].

A number of studies measured UHI intensity (UHII) in Melbourne and identified its spatial distribution and intensity across the metropolitan area [38, 39]. These studies identified UHII ranging from 2 to 4 °C, with daily maximum of 7 °C, depending on the time and geographical location. Torok found up to 7 °C temperature difference between Melbourne's CBD and surrounding rural districts [38]. This temperature difference is magnified during extreme heat wave events.

Prolonged period of excessive heat is known as heat wave. Heat waves occur when the average temperature reaches or exceeds the predetermined heat health temperature thresholds of a weather district. Heat waves are often magnified by UHI and have severe adverse impacts on human health. Some of the well-researched impacts of heat-related illnesses are disturbances of consciousness, cramps, fainting, and heat stroke. Heat discomfit also magnifies pre-existing chronic diseases, such as respiratory, cardiovascular, diabetes, neurological, and renal diseases [40].

Some studies have shown a link between extreme heat and mortality. For example in 2003, Europe experienced a heatwave that led to over 60,000 deaths over three months of summer [41]. Around 15,000 of these deaths occurred in France within only three weeks [42]. Other deaths were reported in Athens [43], England and Wales [44, 45], Chicago [46], Portugal [47], Netherlands [48], Spain [49], and California [50]. In January 2009, heat waves caused 32.4% excess mortality rate in Adelaide [51] and 374 excess deaths in Victoria [52]. A similar study conducted by Nicholls showed that the number of heat-related deaths increases by up to 21% when the daily minimum temperature exceeds 24 °C [53–56].

The relationship between temperature and mortality often follows as U-, V-, or J-shaped pattern, wherein the maximum risk is distributed at the upper end of the temperature scale [40]. Heat impact is not limited to the heat wave duration, as there is often a three-day lag effect [57, 58].

Post-event analysis of EHE and mortality/morbidity has shown clear thresholds or tipping points across all climate zones in Australia. Minimal adverse health response was observed below the threshold, but adverse health outcomes dramatically increase above the threshold. Temperature–mortality thresholds are identified for the majority of Australian capital cities [53, 59–64]. For example, Nicholls et al. [53] identified a temperature threshold for Melbourne, above which mortality increased for older persons (65+ years). In this instance, an average 24-hour (9 am to 9 am) temperature of 30 °C or higher (e.g. a 40 °C maximum day temperature followed by a 20 °C minimum night temperature) results in 15–17% increase in the number of deaths among the elderly.

3 Factors Affecting Heat Vulnerability

The extent that extreme heat events affect individuals and communities largely depend on their vulnerability [65]. In the context of climate change, vulnerability is defined as *"the degree to which a system is susceptible to and unable to cope with, adverse effects of climate change, including climate variability and extremes"* [66]. Vulnerability is also described as the combination of social, environmental, and economic factors and largely depends on the level of exposure, sensitivity, and adaptation capacity of individuals and communities [67].

The elderly were among the most vulnerable groups in all reported heat waves in Australia [63, 64, 68–71]. However, increased rates of mortality and morbidity were also reported among young people. For example, in the 2009 heatwave in Victoria, a 55% increase in mortality and morbidity rate was reported among people 64 years old and above, compared with those in the previous five years [72]. A similar finding was reported in the 2009 Adelaide heatwave, wherein the deaths reported were mainly among people aged 64 years or older [73].

According to a study conducted by Hansen [74], the majority of older female deaths during the heat waves in Melbourne and Adelaide were due to psychoactive substance use, acute myocardial infarction [75], and ischaemic heart disease [51]. In Melbourne, males were twice as much to be affected by extreme heat as women, as shown in the increased number of hospital admissions during the hot weather periods from 1999 to 2004 [75].

4 Outdoor Activity

Intense physical activity is a threat for heat stress and heat-related deaths [76] because the significant level of heat generated during physical activity, thereby increasing the risk for people who exercise outdoors or those reside within indoors that do not have a cooling equipment [77, 78]. Elderlies are the most affected cohorts by heat. In Melbourne, heat-related deaths increased by 55% among 5- to 64-year-olds and contribute to 64 excess deaths in the 2009 heatwave. This cohort is composed of different ages and health status. The risk is higher for obese cohorts or those who suffer from chronic illnesses.

5 UHI

UHI varies spatially and temporally based on background geographical characteristics and climatic conditions. Several studies reported more adverse impacts of extreme heat on heat-related deaths in urban areas than in surrounding rural areas [79, 80]. The reason is that the UHI does not allow for overnight relief of absorbed heat and contributes to extreme heat stress of citizens. This is a critical issue in Melbourne where the proportion of people who live in urban areas is predicted to increase by 67% in 2056 [81].

6 Land Use

It is well established in the literature that greeneries generate a cooling impact and thereby reduce the number of heat-related deaths, especially during heatwaves. Green spaces reduce air temperature through shading, evapotranspiration, and modifying the wind behaviour in urban areas [82, 83]. In Melbourne, proximity to green spaces has proved to be effective in reducing vulnerability to heat.

7 Urban Design

Urbanisation and urban sprawl increase vulnerability towards extreme heatwaves, as it prohibits citizens' accessibility to emergency services. A metropolitan-based combined index that includes residential density, land use division, and accessibility to street network, is strongly correlated with increased emergency service response time and delayed ambulance arrival [84].

8 Housing

The geographical positioning of buildings plays an important role in the intensity and duration of exposure to urban overheating. Furthermore, the poor insulation of old buildings leads to serious challenges in managing indoor thermal environments and energy use. Existing studies indicated that the risk of heat-related death and level of exposure are higher for city dwellers than those living in non-urban areas [80].

9 Access to Health Service

Several studies have shown that geographical location and travel distance play significant roles in access to emergency and health services during extreme heat wave events [85]. The majority of admissions were composed of people living within a 12-mile radius from medical centres. A spatial analysis of data revealed that the number of visits drop significantly as the distance from a medical centre increase. Therefore, accessibility to emergency services can be measured as the number of available e-Health service within a postcode area or the length of travel from an ambulance station to the postal address.

10 Age

Several studies showed that the elderly are at a high risk of heat-related deaths during extreme heat waves. Young children are the second vulnerable group in terms of heat-related illnesses because of their high surface-to-volume ratio, metabolic rate, less responsive cardiovascular system, and low level of sweat production [86]. Disabled children should stay in cooler environments and should be provided with liquids frequently. Ageing affects the body's physiological process of thermoregulation, thereby decreasing the rate of sweating, and older people also have lower blood circulation compared with younger cohorts [87]. These factors lead to a low level of dissipated heat through evaporation, conduction, and convection.

Another risk for the elderly is their social isolation, which puts them in increased risks during heat waves [88]. Elderly residing in aged care homes or nursing homes are at great risk and are identified as a vulnerable group [89] because of their dependence on the others for care and medications. During the 2003 heatwave in Europe, the number of heat-related mortalities was doubled among people aged over 75 years who were residing in aged care homes in France [79] and Netherlands [48].

11 Pre-existing Medical Conditions

Several studies indicated that pre-existing medical conditions increase the risk to heat-related illnesses and deaths during extreme heats. These pre-medical conditions vary from cardiovascular, respiratory, diabetes, neurological, to obesity and cancer [58, 90]. Additionally, the treatment of these illnesses requires certain types of medications that increase the vulnerability of individuals to heat by altering their metabolism and thermoregulatory systems, thereby delaying response to hot environments. Scholars also indicated that people with background mental conditions are highly vulnerable to extreme heat [73, 91, 92] due to the adverse impacts of medications on nervous systems, their social isolation, and lack of access to air-conditioning systems [93].

12 Socio-economic Status

The socio-economic status of individuals affects heat-related deaths by increasing the occurrence of heat-related illnesses, housing quality, access to air conditioning, medical services, and educational status [80, 94, 95]. A study in Australia showed that the most disadvantaged socio-economic groups are highly vulnerable to cardiovascular diseases and heat-related deaths [96]. Extreme heat events adversely affect disadvantaged groups in Chicago (1995) and Europe (2003). In Chicago, high percentage of elderly, crime, poverty, and lack of green infrastructure contributed to the increased number of heat-related deaths [97]. A study in Rome and Turin found that the highest level of heat-related deaths during extreme heat waves is associated with low socio-economic areas and educational status [98].

The discrepancy between Australian and US-European findings contribute to the homogeneity of wealth among Australians. A study in Melbourne showed that hospital admissions during heat waves increased among people who live in disadvantaged geographical areas. However, the age factor played a more important role than socio-economic status during certain periods.

13 Ethnicity

Unequal access to health information and low capacity for adaptation are related to the socio-economic background of people, as well as the language they speak. Studies have shown that ethnicity has a great impact on heat-related deaths [99, 100] because people who do not speak English have difficulty understanding instructions on how to use household appliances, such as air conditioning systems and electric fans. They also have difficulties understanding state guidelines, which increases their vulnerability towards heat. Therefore, there is a critical need for multilingual communication techniques to accommodate this cohort. A study in Adelaide examined the relationship between heat-related mortality deaths and people living in culturally and linguistically diverse communities. This study concluded that aged migrants in established neighbourhoods and refugees in new neighbourhoods have the highest risk of vulnerability [101].

14 Adaptability

According to the Fourth Assessment report from IPCC, adaptive capacity is defined as the ability or potential of a system to respond successfully to climate variability and change, and includes adjustments in behaviour and resources and technologies [66]. This capacity is affected by several factors, including political parties, socio-economic background of people, level of access to technology, information, and infrastructure [102]. The factors that affect this capacity include air conditioning and heat-health guidelines.

The factors discussed in literature that influence adaptive capacity include use of air conditioning and the development of heat-health plans.

15 Air Conditioning

A number of studies revealed that working environments equipped with air conditioning systems and the ability to visit the spaces with air conditioning systems are the most effective protective strategies against heat-related deaths [90, 103]. Other studies revealed that lack of access to air conditioning systems increases the risk of deaths during extreme hot weather [44, 104–106]. A study showed that 13% of heat-related strokes were among the elderly and those living on their own without access to air-conditioning systems [107]. A similar study found high rate of heat-related mortality among residents with African-American ethnicity [108].

16 Actions and Initiatives in Melbourne to Reduce Urban Overheating

It has been well established in the literature that even a small drop in temperature would contribute to saving many lives. Therefore, it is critical to understand the key technologies and approaches that are required to cool Melbourne. This section discusses some of the most effective urban overheating mitigation techniques in Melbourne, with a strong focus on water-sensitive urban design (WSUD) embedded methods. Urban overheating mitigation strategies have been proposed at a range of spatial scales using different methods, such as field measurements, remote sensing, and numerical simulations. At a micro-scale, WSUD strategies cool cities by reducing energy load via shading and decreasing surface temperature. Cooling at the local level occurs by decreasing the air temperature. Figure 1 provides a graphical illustration of these strategies from the micro-scale (household to street) to local scale (neighbourhood to city).

Studies have shown that the use of micro-scale mitigation techniques could cool down temperature by up to 2 °C during extreme heat wave events [109]. For example,

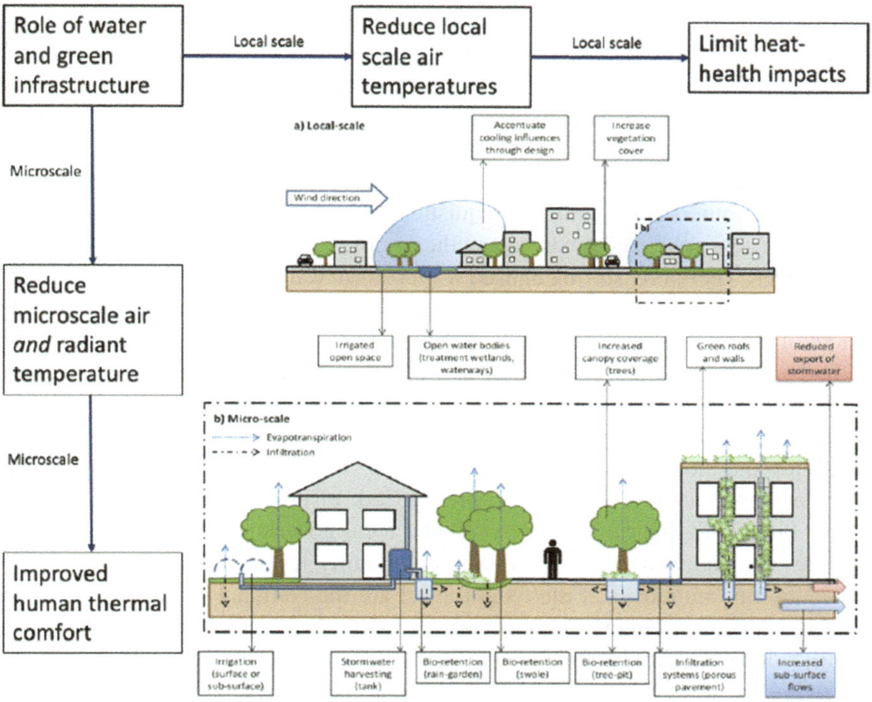

Fig. 1 Water sensitive urban design (WSUD) strategies at different scales [112]

isolated trees are common urban landscape features in cities, but their role in cooling cities is not commonly measured mainly due to difficulties in instrumentations. Coutts examined the role of an isolated tree on micro-scale cooling using mounted thermal sensors. Temperature distribution showed that the air temperature underneath tree canopy is 1 °C cooler than that in surrounding areas during the hottest time of the day. Irrigation becomes highly critical at the local scale because it provides a significant level of cooling during extreme heat waves. Studies that used the SURFEX climate model showed that large-scale irrigations could reduce temperature by an average of 2.5 °C, with varying degree of coolness depending on the perviousness of the substrate [110]. The study also showed that higher level of cooling is achievable if the water can infiltrate the soil. However, this method was not recommended as a long-term cooling method, especially during extreme heat waves, because of high cost and uncertainties attributed to water availability.

Studies at the city scale have shown that combination of approaches could further benefit urban climate [111]. For example, the citywide nocturnal irrigation of Melbourne was simulated and its impact on cooling the city during a heat wave event was quantified using the Weather Research Forecasting Model (WRF). In this simulation, the tree canopy was increased from 20 to 40% and roof albedo was increased from 0.2 to 0.7. The simulations were conducted for the existing scenario and future scenario of Melbourne, where future structural plans are implemented in the city. Tree canopies, cool roofs, and nocturnal irrigation provided cooling of up to 0.6 °C, 0.9 °C, and 0.3 °C, respectively. In overall, 1.5 °C cooling was achieved when all mitigation methods were applied together. This result confirmed the effectiveness of implementing multiple approaches, especially at the city scale.

For all the above-mentioned mitigation strategies, the presence and access to water during hot summers is the main pre-requisite. This requirement poses a serious issue in southern hemisphere contexts, where the summers are often characterised by hot-dry climates. Therefore, storing unused winter storm water resources is critical to the provision of an adequate level of water supply for hot summers. Notably, the cooling benefit of WSUD techniques in Melbourne can reach up to 2.5 °C, depending on the method used; this result is comparable to the cooling benefit reported in other studies conducted across the globe [113].

Several initiatives were proposed by the federal, state, and local government to develop the most sustainable heat mitigation strategies. Overarching guidelines were developed at the federal level, but the task of establishing detailed policies to cool cities was delegated to the state and local government, where most of the activities are underway at the local and community level. Two of the most well-known cooling initiatives at the federal level are the National Water Initiative and the National Urban Policy. These initiatives, respectively, highlighted the need for identification of environmental benefit outcomes in national and state water plans and improving the sustainability of Australian cities. The Cooperative Research Centre (CRC) programme established by the federal government in 1990 and the establishment of two key CRCs in the urban heat mitigation area in 2012 are some of these initiatives. The CRC for Low Carbon Living and Water Sensitive Cities, along with the National Centre for Climate Change Adaptation Research (established 2007), aimed to provide quality research and guidance to state and local governments.

The majority of urban cooling policies at the state and community levels aimed to enhance public health, heat emergency services and management, water shortage, and climate adaptation/mitigation. For example, policies that adopt green infrastructure as the key strategy to cool cities and mitigate UHI can be seen in a range of high-level assessments. However, studies are concerned that these policies are still immature and have not been appropriately integrated into urban planning policies [114]. Initiatives such as the urban cooling initiative of Melbourne Water, which was widely driven by local communities, changed this perception. This initiative aimed to transform 30 hectares of public land in Melbourne managed by Melbourne Water into cool and shady refuges [115].

Several initiatives are taking place at the municipal level to implement heat mitigation strategies. For example, in the city of Port Philip, a sustainable environment strategy called *Toward Zero was launched in 2007. This initiative contributed to several inter-linked policy developments, such as the City of Port Philip WSUD guidelines, Climate Adaptation Plan, the Greening Port Phillip, Open Space Water Management Plan, and Water Plan documents* [116]. The main aim of these initiatives is to promote the use of green spaces and storm water-supported WSUD to address UHI at the municipal scale. These successful initiatives were used as benchmark to develop a prioritisation framework for green infrastructures. This framework prioritises the implementation of green spaces based on the level of exposure to heat, heat vulnerability, and behavioural exposure.

17 Conclusion

The air temperature in cities is warmer than that in surrounding rural areas. This phenomenon is known as UHI and is considered a great concern for Australian government. Urban overheating is an additional burden to global climate change. Urban overheating affects public health differently depending on the vulnerability of individuals and communities to extreme heat. Factors that affect population vulnerability include features of the built environment, age of individuals, and socio-economic and health status. The impact of each factor on vulnerability has been thoroughly discussed in this chapter.

Issues concerning UHI, heat waves, and their adverse impacts on public health in Melbourne are the key engine towards the development of mitigation strategies at the federal, state, and local governments. These issues gained increasing concern after the heat wave incident in 2019 when the hottest year in Australian weather calendar was recorded. In this year, the mean maximum temperature was higher by 2.09 °C compared with those between 1961 and 1990 (Bureau of Meteorology 2020). This year was also identified as the driest year since any weather data was recorded. This heat wave incident increased the tendency of Australian cities to implement long-term adaptation and mitigation strategies against climate change and urban overheating. These long-term strategies aim to cool cities by enhancing best practices at different levels. Future studies should target further collaborations among stakeholders, community developers, and state governments, which can be achieved by highlighting the cost-environmental benefits of climate-sensitive urban design and would pave the way for climate-proof and liveable cities in the future.

References

1. D. o. E. a. S. A. United Nations, Population Division (2018) The world's cities in 2018—data booklet (ST/ESA/ SER.A/417). New York: United Nations
2. D. o. E. a. S. A. United Nations, Population Division (2019) The world's cities in 2018—data booklet (ST/ESA/ SER.A/417). New York: United Nations
3. Cities C (2020) A global opportunity for cities to lead. https://www.c40.org/why_cities. Accessed 5 Sept 2020
4. Oke TR, Mills G, Christen A, Voogt JA (2017) Urban climates. Cambridge University Press
5. Arnfield AJ (2003) Two decades of urban climate research: a review of turbulence, exchanges of energy and water, and the urban heat island. Int J Climatol 23(1):1–26. https://doi.org/10.1002/joc.859

6. Li X, Zhou W (2019) Spatial patterns and driving factors of surface urban heat island intensity: a comparative study for two agriculture-dominated regions in China and the USA. Sustain Cities Soc 48:101518. https://doi.org/10.1016/j.scs.2019.101518
7. Peng S et al (2012) Surface urban heat island across 419 global big cities. Environ Sci Technol 46(2):696–703. https://doi.org/10.1021/es2030438
8. Rizwan AM, Dennis LYC, Liu C (2008) A review on the generation, determination and mitigation of Urban Heat Island. J Environ Sci 20(1):120–128. https://doi.org/10.1016/S1001-0742(08)60019-4
9. Roberge F, Sushama L (2018) Urban heat island in current and future climates for the island of Montreal. Sustain Cities Soc 40:501–512. https://doi.org/10.1016/j.scs.2018.04.033
10. Heaviside C, Macintyre H, Vardoulakis S (2017) The urban heat island: implications for health in a changing environment. Curr Environ health Rep 4(3):296–305
11. W. Bank (2020). https://data.worldbank.org/indicator/SP.URB.TOTL.IN.ZS?view=chart
12. Coates L, Haynes K, O'brien J, McAneney J, De Oliveira FD (2014) Exploring 167 years of vulnerability: an examination of extreme heat events in Australia 1844–2010. Environ Sci Policy 42:33–44
13. Coates EL (1996) Conference of natural disaster reduction. Gold Coast, Australia
14. Loughnan ME, Tapper NJ, Lynch K, McInnes J, Phan T (2012) A spatial vulnerability analysis of urban populations during extreme heat events in Australian capital cities. Natl Clim Change Adapt Res Facility
15. Rogers CD, Gallant AJ, Tapper NJ (2019) Is the urban heat island exacerbated during heatwaves in southern Australian cities? Theoret Appl Climatol 137(1):441–457
16. Alexander LV, Arblaster JM (2009) Assessing trends in observed and modelled climate extremes over Australia in relation to future projections. Int J Climatol J R Meteorol Soc 29(3):417–435
17. CSIRO (2012) Australian Bureau of Meteorology. State of the Climate. http://www.csiro.au/en/Outcomes/Climate/Understanding/State-of-the-Climate
18. Field CB, Barros V, Stocker TF, Dahe Q (2012) Managing the risks of extreme events and disasters to advance climate change adaptation: special report of the intergovernmental panel on climate change. Cambridge University Press
19. Loughnan ME, Nicholls N, Tapper NJ (2010) When the heat is on: threshold temperatures for AMI admissions to hospital in Melbourne Australia. Appl Geogr 30(1):63–69
20. Akbari H, Kolokotsa D (2016) Three decades of urban heat islands and mitigation technologies research. Energy Buildings 133:834–842
21. Arnfield AJ (2003) Two decades of urban climate research: a review of turbulence, exchanges of energy and water, and the urban heat island. Int J Climatol J R Meteorol Soc 23(1):1–26
22. Grimmond SU (2007) Urbanisation and global environmental change: local effects of urban warming. Geogr J 173(1):83–88
23. Mohajerani A, Bakaric J, Jeffrey-Bailey T (2017) The urban heat island effect, its causes, and mitigation, with reference to the thermal properties of asphalt concrete. J Environ Manage 197:522–538
24. Phelan PE et al (2015) Urban heat island: mechanisms, implications, and possible remedies. Annu Rev Environ Resour 40:285–307
25. Abreu-Harbich LV, Labaki LC, Matzarakis A (2014) Thermal bioclimate in idealized urban street canyons in Campinas, Brazil. Theoret Appl Climatol 115(1):333–340
26. Ali-Toudert F, Mayer H (2006) Numerical study on the effects of aspect ratio and orientation of an urban street canyon on outdoor thermal comfort in hot and dry climate. Build Environ 41(2):94–108
27. Bourbia F, Boucheriba F (2010) Impact of street design on urban microclimate for semi arid climate (Constantine). Renew Energy 35(2):343–347
28. Lobaccaro G, Acero JA, Sanchez Martinez G, Padro A, Laburu T, Fernandez G (2019) Effects of orientations, aspect ratios, pavement materials and vegetation elements on thermal stress inside typical urban canyons. Int J Environ Res Pub Health 16(19):3574

29. Harman IN, Best MJ, Belcher SE (2004) Radiative exchange in an urban street canyon. Bound-Layer Meteorol 110(2):301–316
30. Kolokotroni M, Giridharan R (2008) Urban heat island intensity in London: an investigation of the impact of physical characteristics on changes in outdoor air temperature during summer. Sol Energy 82(11):986–998
31. Theeuwes N, Steeneveld G, Ronda R, Heusinkveld B, Van Hove L, Holtslag A (2014) Seasonal dependence of the urban heat island on the street canyon aspect ratio. Q J R Meteorol Soc 140(684):2197–2210
32. Akbari H, Pomerantz M, Taha H (2001) Cool surfaces and shade trees to reduce energy use and improve air quality in urban areas. Sol Energy 70(3):295–310
33. Bowler DE, Buyung-Ali LM, Knight TM, Pullin AS (2010) A systematic review of evidence for the added benefits to health of exposure to natural environments BMC Pub Health 10(1):1–10
34. Doan VQ, Kusaka H, Nguyen TM (2019) Roles of past, present, and future land use and anthropogenic heat release changes on urban heat island effects in Hanoi, Vietnam: Numerical experiments with a regional climate model. Sustain Cities Soc 47:101479
35. Du H et al (2016) Influences of land cover types, meteorological conditions, anthropogenic heat and urban area on surface urban heat island in the Yangtze River Delta Urban Agglomeration. Sci Total Environ 571:461–470
36. Hamilton IG, Davies M, Steadman P, Stone A, Ridley I, Evans S (2009) The significance of the anthropogenic heat emissions of London's buildings: a comparison against captured shortwave solar radiation. Build Environ 44(4):807–817
37. Sailor DJ (2011) A review of methods for estimating anthropogenic heat and moisture emissions in the urban environment. Int J Climatol 31(2):189–199
38. Torok SJ, Morris CJ, Skinner C, Plummer N (2001) Urban heat island features of southeast Australian towns. Aust Meteorol Mag 50(1):1–13
39. Walker C (2004) Assessing the temporal and spatial variability of Melbourne's urban heat island. Geogr Environ Sci:146
40. Kovats RS, Hajat S (2008) Heat stress and public health: a critical review. Annu Rev Public Health 29:41–55
41. Robine J-M et al (2008) Death toll exceeded 70,000 in Europe during the summer of 2003. CR Biol 331(2):171–178
42. Fouillet A et al (2008) Has the impact of heat waves on mortality changed in France since the European heat wave of summer 2003? A study of the 2006 heat wave. Int J Epidemiol 37(2):309–317
43. Katsouyanni K, Trichopoulos D, Zavitsanos X, Touloumi G (1988) The 1987 Athens heatwave. Lancet 2:573
44. Rooney C, McMichael AJ, Kovats RS, Coleman MP (1998) Excess mortality in England and Wales, and in Greater London, during the 1995 heatwave. J Epidemiol Community Health 52(8):482–486
45. Johnson H, Kovats S, McGregor G, Stedman J, Gibbs M, Walton H (2005) The impact of the 2003 heat wave on daily mortality in England and Wales and the use of rapid weekly mortality estimates. Eurosurveillance 10(7):15–16
46. Whitman S, Good G, Donoghue ER, Benbow N, Shou W, Mou S (1997) Mortality in Chicago attributed to the July 1995 heat wave. Am J Public Health 87(9):1515–1518
47. Nogueira P, Paixao E (2008) Models for mortality associated with heatwaves: update of the Portuguese heat health warning system. Int J Climatol J R Meteorol Soc 28(4):545–562
48. Garssen J, Harmsen C, De Beer J (2005) The effect of the summer 2003 heat wave on mortality in the Netherlands. Eurosurveillance 10(7):13–14
49. Simón F, Lopez-Abente G, Ballester E, Martínez F (2003) Mortality in Spain during the heat waves of summer 2003. Eurosurveillance 10(7):9–10
50. Hoshiko S, English P, Smith D, Trent R (2010) A simple method for estimating excess mortality due to heat waves, as applied to the 2006 California heat wave. Int J Public Health 55(2):133–137

51. Nitschke M, Tucker GR, Hansen AL, Williams S, Zhang Y, Bi P (2011) Impact of two recent extreme heat episodes on morbidity and mortality in Adelaide, South Australia: a case-series analysis. Environ Health 10(1):1–9
52. Sutton B, Mulvenna V, Voronoff D, Humphrys T (2020) Acting on climate change and health in Victoria. Med J Aust 212(8):345
53. Nicholls N, Skinner C, Loughnan M, Tapper N (2008) A simple heat alert system for Melbourne, Australia. Int J Biometeorol 52(5):375–384
54. Anderson GB, Bell ML (2011) Heat waves in the United States: mortality risk during heat waves and effect modification by heat wave characteristics in 43 US communities. Environ Health Perspect 119(2):210–218
55. Curriero FC, Heiner KS, Samet JM, Zeger SL, Strug L, Patz JA (2002) Temperature and mortality in 11 cities of the eastern United States. Am J Epidemiol 155(1):80–87
56. D'Ippoliti D et al (2010) The impact of heat waves on mortality in 9 European cities: results from the EuroHEAT project. Environ Health 9(1):1–9
57. Conti S et al (2005) Epidemiologic study of mortality during the Summer 2003 heat wave in Italy. Environ Res 98(3):390–399
58. Basu R, Samet JM (2002) Relation between elevated ambient temperature and mortality: a review of the epidemiologic evidence. Epidemiol Rev 24(2):190–202
59. Banwell C, Dixon J, Bambrick H, Edwards F, Kjellström T (2012) Socio-cultural reflections on heat in Australia with implications for health and climate change adaptation. Glob Health Action 5(1):19277
60. Williams S et al (2012) Heat and health in Adelaide, South Australia: assessment of heat thresholds and temperature relationships. Sci Total Environ 414:126–133
61. Williams S, Nitschke M, Weinstein P, Pisaniello DL, Parton KA, Bi P (2012) The impact of summer temperatures and heatwaves on mortality and morbidity in Perth, Australia 1994–2008. Environ Int 40:33–38
62. Tong S, Ren C, Becker N (2010) Excess deaths during the 2004 heatwave in Brisbane, Australia. Int J Biometeorol 54(4):393–400
63. Vaneckova P, Beggs PJ, De Dear RJ, McCracken KW (2008) Effect of temperature on mortality during the six warmer months in Sydney, Australia, between 1993 and 2004. Environ Res 108(3):361–369
64. Guest C et al (1999) Climate and mortality in Australia: retrospective study, 1979–1990, and predicted impacts in five major cities in 2030. Climate Res 13(1):1–15
65. Canada H (2011) Adapting to extreme heat events: guidelines for assessing health vulnerability. Health Canada, Ottawa, ON, 43. http://www.hc-sc.gc.ca/ewhsemt/pubs/climat/adapt/index-eng.php
66. IPCC CC (2007) Impacts, adaptation and vulnerability. Contribution of working group II to the fourth assessment report of the intergovernmental panel on climate change
67. D. o. E. a. H. Australian Greenhouse Office, Commonwealth of Australia (2005) Climate change, risk and vulnerability. http://www.sfrpc.com/Climate%20Change/24.pdf
68. McMichael AJ et al (2003) Human health and climate change in Oceania: a risk assessment 2002. Canberra, Commonwealth Department of Health and Ageing
69. Gosling SN, McGregor GR, Páldy A (2007) Climate change and heat-related mortality in six cities Part 1: model construction and validation. Int J Biometeorol 51(6):525–540
70. Vaneckova P, Hart MA, Beggs PJ, De Dear RJ (2008) Synoptic analysis of heat-related mortality in Sydney, Australia, 1993–2001. Int J Biometeorol 52(6):439–451
71. Wang XY et al (2012) The impact of heatwaves on mortality and emergency hospital admissions from non-external causes in Brisbane, Australia. Occup Environ Med 69(3):163–169
72. V. G. D. o. H. Service (2009). http://www.health.vic.gov.au/chiefhealthofficer/downloads/heat_impact_rpt.pdf
73. Nitschke M, Tucker GR, Bi P (2007) Morbidity and mortality during heatwaves in metropolitan Adelaide. Med J Aust 187(11–12):662–665
74. Hansen A, Bi P, Nitschke M, Ryan P, Pisaniello D, Tucker G (2008) The effect of heat waves on mental health in a temperate Australian city. Environ Health Perspect 116(10):1369–1375

75. Loughnan ME, Nicholls N, Tapper NJ (2010) The effects of summer temperature, age and socioeconomic circumstance on acute myocardial infarction admissions in Melbourne, Australia. Int J Health Geogr 9(1):1–11
76. Yip FY et al (2008) The impact of excess heat events in Maricopa County, Arizona: 2000–2005. Int J Biometeorol 52(8):765–772
77. Silverthorn DU (2010) Human physiology: an integrated approach. Books a la Carte Edition, Benjamin-Cummings
78. Hanna EG, Kjellstrom T, Bennett C, Dear K (2011) Climate change and rising heat: population health implications for working people in Australia. Asia Pac J Pub Health 23(2_suppl):14S-26S
79. Fouillet A et al (2006) Excess mortality related to the August 2003 heat wave in France. Int Arch Occup Environ Health 80(1):16–24
80. Smoyer KE, Rainham DG, Hewko JN (2000) Heat-stress-related mortality in five cities in Southern Ontario: 1980–1996. Int J Biometeorol 44(4):190–197
81. A. B. o. Statistics (2011). http://www.abs.gov.au/websitedbs/D3310114.nsf/home/Australian+Standard+Geographical+Classification+(ASGC)
82. C. o. Melbourne (2012) Urban forest strategy: making a great city greener 2012–2032
83. Jonsson P (2004) Vegetation as an urban climate control in the subtropical city of Gaborone, Botswana. Int J Climatol J R Meteorol Soc 24(10):1307–1322
84. Trowbridge MJ, Gurka MJ, O'connor RE (2009) Urban sprawl and delayed ambulance arrival in the US. Am J Prev Med 37(5):428–432
85. Henneman PL, Garb JL, Capraro GA, Li H, Smithline HA, Wait RB (2011) Geography and travel distance impact emergency department visits. J Emerg Med 40:333–339
86. Grubenhoff JA, du Ford K, Roosevelt GE (2007) Heat-related illness. Clin Pediatr Emerg Med 8(1):59–64
87. Kenney WL, Munce TA (2003) Invited review: aging and human temperature regulation. J Appl Physiol 95(6):2598–2603
88. Vandentorren S et al (2006) August 2003 heat wave in France: risk factors for death of elderly people living at home. European J Pub Health 16(6):583–591
89. Klenk J, Becker C, Rapp K (2010) Heat-related mortality in residents of nursing homes. Age Ageing 39(2):245–252
90. Naughton MP et al (2002) Heat-related mortality during a 1999 heat wave in Chicago. Am J Prev Med 22(4):221–227
91. Hansen AL, Bi P, Ryan P, Nitschke M, Pisaniello D, Tucker G (2008) The effect of heat waves on hospital admissions for renal disease in a temperate city of Australia. Int J Epidemiol 37(6):1359–1365
92. Khalaj B, Lloyd G, Sheppeard V, Dear K (2010) The health impacts of heat waves in five regions of New South Wales, Australia: a case-only analysis. Int Arch Occup Environ Health 83(7):833–842
93. Bark N (1998) Deaths of psychiatric patients during heat waves. Psychiatr Serv 49(8):1088–1090
94. C. f. D. Control and Prevention (2004) Impact of heat waves on mortality—Rome, Italy, June-August 2003. MMWR: Morb Mortal Wkly Rep 53(17):369–371
95. Johnson DP, Wilson JS (2009) The socio-spatial dynamics of extreme urban heat events: The case of heat-related deaths in Philadelphia. Appl Geogr 29(3):419–434
96. AIHW (2005) Living dangerously, Australians with multiple risk factors for cardiovascular disease. AIHW bulletin no. 24 http://www.aihw.gov.au/publicationdetail/?id=6442467692
97. Klinenberg E (2002) Heat wave: a social autopsy of disaster in Chicago
98. Michelozzi P et al (2005) The impact of the summer 2003 heat waves on mortality in four Italian cities. Eurosurveillance 10(7):11–12
99. Basu R, Ostro BD (2008) A multicounty analysis identifying the populations vulnerable to mortality associated with high ambient temperature in California. Am J Epidemiol 168(6):632–637

100. Harlan SL, Brazel AJ, Prashad L, Stefanov WL, Larsen L (2006) Neighborhood microclimates and vulnerability to heat stress. Soc Sci Med 63(11):2847–2863
101. Hansen A, Bi P, Saniotis A (2013) Extreme heat and climate change: adaptation in culturally and linguistically diverse (CALD) communities
102. D. o. E. a. H. Australian Greenhouse Office, Commonwealth of Australia (2005) Department of Environment and Heritage, Commonwealth of Australia . Climate change, risk and vulnerability. http://www.sfrpc.com/Climate%20Change/24.pdf
103. Bouchama A, Dehbi M, Mohamed G, Matthies F, Shoukri M, Menne B (2007) Prognostic factors in heat wave–related deaths: a meta-analysis. Arch Intern Med 167(20):2170–2176
104. Nunes B, Paixão E, Dias CM, Nogueira P, Falcão JM (2011) Air conditioning and intrahospital mortality during the 2003 heatwave in Portugal: evidence of a protective effect. Occup Environ Med 68(3):218–223
105. Semenza JC et al (1996) Heat-related deaths during the July 1995 heat wave in Chicago. N Engl J Med 335(2):84–90
106. McGeehin MA, Mirabelli M (2001) The potential impacts of climate variability and change on temperature-related morbidity and mortality in the United States. Environ Health Perspect 109(suppl 2):185–189
107. Margolis H, Gershunov A, Kim T, English P, Trent R (2008) 2006 California heat wave high death toll: insights gained from coroner's reports and meteorological characteristics of event. Epidemiology 19(6):S363–S364
108. O'Neill MS, Zanobetti A, Schwartz J (2003) Modifiers of the temperature and mortality association in seven US cities. Am J Epidemiol 157(12):1074–1082
109. Matzarakis A, Rutz F, Mayer H (2010) Modelling radiation fluxes in simple and complex environments: basics of the RayMan model. Int J Biometeorol 54(2):131–139
110. Broadbent AM, Coutts AM, Tapper NJ, Demuzere M, Beringer J (2018) The microscale cooling effects of water sensitive urban design and irrigation in a suburban environment. Theoret Appl Climatol 134(1):1–23
111. Jacobs SJ, Gallant AJ, Tapper NJ, Li D (2018) Use of cool roofs and vegetation to mitigate urban heat and improve human thermal stress in Melbourne, Australia. J Appl Meteorol Climatol 57(8):1747–1764
112. Coutts AM, Tapper NJ, Beringer J, Loughnan M, Demuzere M (2013) Watering our cities: the capacity for water sensitive urban design to support urban cooling and improve human thermal comfort in the Australian context. Prog Phys Geogr 37(1):2–28
113. Santamouris M et al (2017) Passive and active cooling for the outdoor built environment–Analysis and assessment of the cooling potential of mitigation technologies using performance data from 220 large scale projects. Sol Energy 154:14–33
114. Bush J, Aye L, Hes D (2015) Cooling cities with green space: a policy analysis framework
115. M. Water (2020) https://www.melbournewater.com.au/community-and-education/about-our water/liveability-and-environment/urban-cooling-program. Accessed 5 Jun 2020
116. C. o. P. Phillip (2020) http://www.portphillip.vic.gov.au/sustainability.htm. Accessed 5 Jun 2020

Chapter 13
The Potential for Urban Canopy Cover to Reduce Heat-Related Mortality in Adelaide

Bartesaghi-Koc Carlos, Soebarto Veronica, Hawken Scott, and Sharifi Ehsan

Abstract In Australia, heatwaves and extremely hot weather present the greatest climate-related threat resulting in hundreds of deaths every year and killing more people than any other natural hazard does. There are an increasing number of studies evaluating the influence and benefits of urban greenery on human health; however, few studies have thoroughly investigated the explicit relationships between urban vegetation and heat-related mortality (HRM), morbidity and vulnerability levels and no such studies have been completed for Adelaide so far. To address this need, this chapter provides a quantitative investigation of the potential influence of urban canopy cover on reducing HRM rates using the Greater Adelaide Metropolitan area as case study. The study employs spatial analytic techniques to evaluate the spatial variability and relationships between increased tree coverage, potential reduction of HRM rates and various socio-economic and demographic risk factors. Results indicates dramatic public health benefits possible from increasing tree canopy coverage, emphasising an urgent need for close alignment of public health policy with urban planning and greening policies for the whole region. Furthermore, we found that there is a solid case for tailoring urban greening and tree canopy enhancement at the district level based on the socio-economic and cultural characteristics of local populations. This study is ultimately intended to support decision-making by local governments for critical urban greenery interventions and initiatives that can increase liveability and thereby help reduce the health burden and risk of deaths, particularly among vulnerable populations across the Greater Adelaide Metropolitan in the future.

B.-K. Carlos (✉) · S. Veronica · H. Scott · S. Ehsan
School of Architecture and Built Environment, Faculty of Science, Engineering and Technology, The University of Adelaide, Adelaide 5005, Australia
e-mail: carlos.bartesaghikoc@adelaide.edu.au

S. Veronica
e-mail: veronica.soebarto@adelaide.edu.au

H. Scott
e-mail: scott.hawken@adelaide.edu.au

S. Ehsan
e-mail: ehsan.sharifi@adelaide.edu.au

N. Aghamohammadi and M. Santamouris (eds.), *Urban Overheating: Heat Mitigation and the Impact on Health*, Advances in Sustainability Science and Technology,
https://doi.org/10.1007/978-981-19-4707-0_13

Keywords Heat-related mortality · Heat-related morbidity · Public health · Tree canopy · Green infrastructure · Heat stress · Human thermal comfort · Spatial statistics

1 Introduction

Global climate change is one of the greatest concerns of the twenty-first century and presents significant risks to human health and habitats, economic prosperity, as well as natural and modified ecosystems. Climate change-related stressors occur across a variety of scales including the local, district, regional, national and international levels [1]. In particular, cities and their urban ecosystems are under threat from climate change-related stressors.

In Australia, it is heatwaves and extremely hot weather that present the greatest climate-related threat resulting in hundreds of deaths every year and killing more people than any other natural hazard does [2]. With climate change, heat waves will become more severe in major Australian cities such as Sydney, Melbourne and Adelaide and these events will be intensified by trends such as increasing urbanisation and an ageing population [3]. The urban heat island (UHI) effect—a phenomenon where urban areas exhibit higher temperatures than the surrounding areas and natural settings—is an additional threat that is adversely affecting cooling energy demand, peak electricity demand, concentration of pollutants, outdoor thermal comfort, liveability and health in cities across Australia and around the world [4, 5].

To reduce overheating and protect cities against heatwaves and UHI, several mitigation strategies have been studied and deployed [1]. Amongst them, green infrastructure (GI)—defined as the interconnected network of natural features that bring multiple ecosystem services and benefits to the urban environment—has been identified as a nature-based solution that can contribute to moderating microclimate, sequestrating carbon, controlling stormwater runoff and improving people's health and wellbeing [6, 7]. In particular, urban greenery supports human health and wellbeing through the moderation of urban temperatures (via shading, evapotranspirative cooling and wind modification), improvement of air quality, and the provision of opportunities for physical activities, spaces for social and cultural interactions as well as aesthetic and psychological benefits [8].

There are an increasing number of studies that evaluate the influence and benefits of urban greenery on human health. Numerous reviews focus on the associations between GI characteristics and perceived general and mental health [9–14] but few studies have thoroughly investigated the explicit relationships between urban vegetation and heat-related mortality (HRM), morbidity and vulnerability levels [15–21] and no such studies have been completed for Adelaide. In Australia, previous research in HRM has extensively focused on the correlations between ambient weather conditions (mostly under heatwave conditions) and mortality rates (i.e., by calculating incident rate ratios—IRR) [22–30], with research on the relationship between vegetation cover on the potential reduction of heat-related deaths being more limited [31–35].

This chapter provides a quantitative investigation of the potential influence of urban canopy cover on reducing HRM rates using the Greater Adelaide Metropolitan area as case study. The study employs spatial analytic techniques at the district level (known as local government areas or LGAs in Australia) to analyse the spatial variability and relationships between increased tree coverage, potential reduction of HRM rates and various socio-economic and demographic risk factors. The study is intended to support decision-making by local governments for critical urban greenery interventions and initiatives that can increase liveability and thereby help reduce the health burden and risk of deaths, particularly among vulnerable populations across Adelaide in the future.

2 Background

2.1 Adelaide as Case Study: Climate, Extreme Heat Events and Urban Greenery

Adelaide is the capital city of South Australia and is located in the south-central portion of the Australian continent. It is the fifth-most populous city of the country with a population of 1.2 million [36]. According to the Köppen Climate classification, Adelaide exhibits a Mediterranean climate (Csa) with hot, dry summers and cool winters, a mean annual maximum temperature of 21.6 °C (1955–2019) and moderate rainfall [37] as well as a mean annual rainfall of 437.4 mm [38]. Adelaide suffers a moderate boundary layer of UHI with a maximum urban–rural thermal difference of 6.7 °C during the day [39] and 5.9 °C during the night [40]. It also observed a significant thermal coastal-inland gradient with seashore suburbs up to 5 °C cooler than the easternmost parts of the city [41]. Over the last century, Adelaide has been experiencing a significant increase in the annual number of hot days (>40 °C) rising from 1 to 3(1910–1999) to over 7 days (2000–2019) [42]. Adelaide has also experienced a long duration of heatwaves, with maximum temperatures above 35 °C recorded in the summers of 2008 and 2009, lasting for 15 and 12 days, respectively [27]. In the heatwave of 2019, Adelaide registered the hottest air temperature ever recorded in an Australian capital city with 47.7 °C measured at the West Terrace weather station within the City of Adelaide's LGA [41].

The Greater Adelaide Metropolitan area (1210.3 km^2) encompasses 16 LGAs (as per 2019 statistical subdivision [36]) and it is delimited by the Smithfield Plains and St Kilda-Chapman Creek Aquatic Reserve to the north, the Adelaide Hills to the south and east and the coastal areas adjacent to the St. Vincent Gulf to the west (Fig. 1). The largest forested areas typically concentrate in the hilly suburbs towards the east, north-east and south-east, while vast agricultural land (mostly vineyards) and bushland are commonly found towards the south. A particular feature is the presence of the parklands and associated urban forests surrounding the City of Adelaide LGA. The seashore and northern suburbs are typically characterised by residential areas with limited tree canopy cover interspersed with commercial, mixed-use and small to medium-size industries. The northwest corner is dominated by large industrial zones and marshlands (near Port Adelaide) with few or no tree coverage. The Greater Adelaide Metropolitan area possesses many morphological characteristics and climate conditions that make it an ideal case study to analyse the potential effect of increased tree canopy on urban overheating and consequently on HRM.

2.2 *Heat-Related Mortality and Morbidity in Adelaide*

The association of excess mortality and morbidity with high ambient temperatures and heatwaves has been recorded in Australia's major cities for the past two decades [23, 24, 28, 30, 43–46]. Weather predictions for Adelaide suggest a rise in average summer temperatures of up to 1.3 °C by 2030 and 4.0 °C by 2070, which are likely to increase the frequency, duration and intensity of heatwaves in the following years [47]. These conditions indicate significant impacts on the most vulnerable populations as well as an increased cost and social burden on the healthcare systems and emergency departments across the city [48]. Several quantitative assessments of heat-related health data during extreme heat events experienced in Adelaide between 1984 and 2014 found a moderate surge in morbidity (ambulance call-outs and hospital admissions), particularly for mental and renal diseases [25–27, 29, 49, 50]. Compared to previous events, the heatwave of 2009 has been associated with excess deaths due to the unprecedented intensity and duration of the heatwave [27].

The abovementioned studies have also found that children (<5 years), older (>65 years) and populations with lower social integration and limited access to green spaces and resources are particularly susceptible to extreme heat [48, 51]. Nonetheless, the majority of these studies have been based on time-series analyses of health data aggregated at the city level and have not considered the possible spatial variation of heat-related mortality rates in relation to existing and increased tree canopy covers, population density and socio-economic factors at the district (or LGA) level.

In the present study, the potential reduction in HRM rates across the Greater Adelaide Metropolitan area is estimated from the association between the predicted average peak daily temperature drop and the increment in tree coverage at LGA level. The spatial variability of HRM rates is then compared against various demographic and socio-economic indicators using spatially-explicit exploratory data analyses (described in Sect. 3.3). To our knowledge, this is the first spatial study focusing on the influence of tree canopy coverage on HRM in Adelaide.

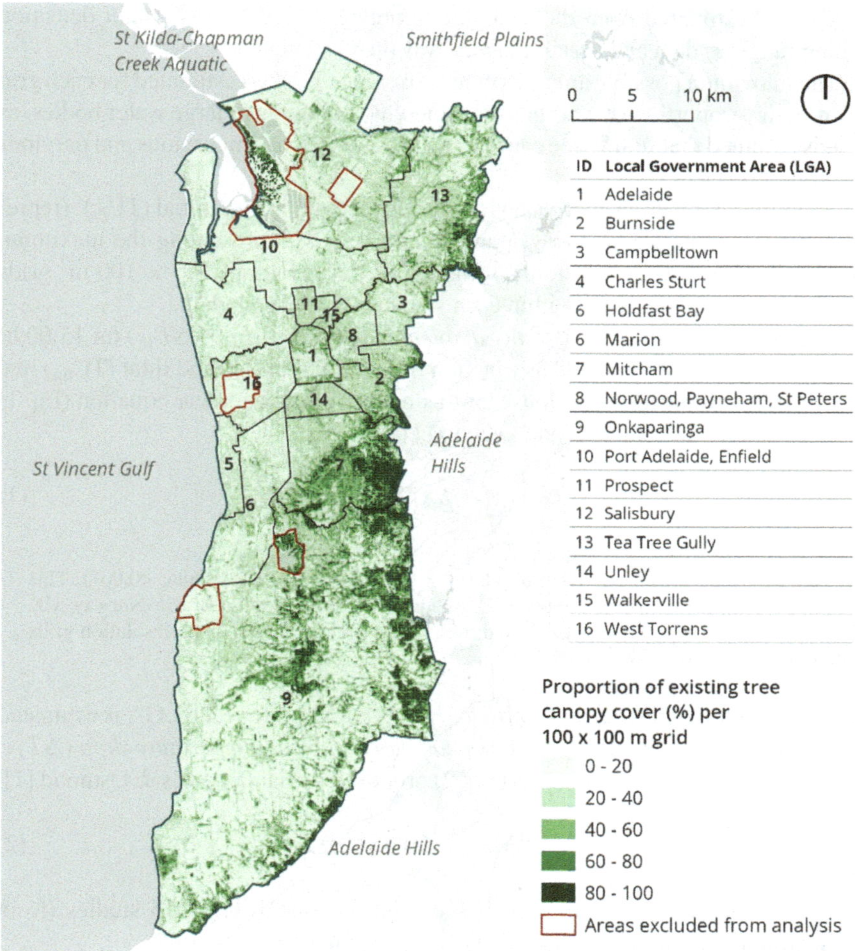

Fig. 1 Distribution of the proportion of existing tree canopy cover estimated per 100 × 100 grid across the Greater Adelaide Metropolitan area encompassing 16 Local government Areas (LGAs) considered for the present study. Note the heavily forested Adelaide Hills in the east (LGA's 2, 7, 9, 13) and the concentrated ring of trees around Adelaide City Centre (LGA 1)

3 Methodology

3.1 The Methodological Approach

The methodological approach implemented in this study involved the following steps:

1. The entire study area was divided into a grid of 100 × 100 m^2 (size of grid as per Stewart and Oke [52] recommendations). The percentage (or fraction) of each grid occupied by building, trees, permeable ground, impermeable ground and

water is estimated from the available remotely-sensed LiDAR (light detection and ranging) data and aerial imagery was then calculated.

2. The maximum possible amount of tree coverage (TC_{fin}) is estimated for each grid from the proportion of land 'not-occupied' by buildings or large water bodies, as it is assumed that future tree canopy can grow and cover impervious and pervious ground surfaces.
3. A simple aggregation approach is applied by averaging the 'initial (TC_{in})' (representing existing tree coverage) and 'final (TC_{fin})' (representing the maximum possible tree coverage in future) fraction of trees from all 100×100 m^2 grids with a centroid that falls within each statistical LGA boundary.
4. The potential *average peak daily ambient temperature drop (ΔT_{15})* (at 15:00 h) for each LGA is calculated as function of the initial (TC_{in}) and final (TC_{fin}) tree coverage estimated in previous steps using the following linear equation (Eq. 1) proposed by Santamouris and Osmond [1]:

$$\Delta T_{15} = a + b \cdot \mathrm{TC}_{in} + c \cdot \mathrm{TC}_{\mathrm{fin}} \tag{1}$$

where $a = -0.0145$, $b = -0.01068$ and $c = 0.0167$ ($R^2 = 0.83$; p-value <0.001). This equation is based on a meta-analysis of 20 studies (from Australia, Brazil, Norway, UK and USA) employing various local- and meso-scale simulation tools with resolution grids ranging between 6×6 m^2 and 2×2 km^2.

5. The *potential percentage decrease of HRM* (HRM_{dec}) for each LGA is estimated as a linear function of the *average peak daily ambient temperature drop* (ΔT_{15}) (from Eq. 1) using the equation (Eq. 2) proposed by Santamouris & Osmond [1]:

$$\mathrm{HRM}_{\mathrm{dec}} = 30.454 \cdot \Delta T_{15} \tag{2}$$

The equation was estimated based on a meta-analysis of 13 studies (from Australia and abroad) with an $R^2 = 0.896$ (p-value $= 2.9 \times 10^{-7}$).

6. HRM_{dec} is used as a proxy for the potential maximum capacity of increased tree canopy cover to reduce the percentage of heat-related deaths in each LGA. To examine the spatial variability of HRM_{dec} between LGAs and identify and prioritise greening interventions in locations with higher heat-vulnerability levels, HRM_{dec} values are spatially compared against selected indicators of social vulnerability (listed in Table 1) using various spatial analysis techniques as described in Sect. 3.3.

Table 1 Local government areas (LGAs) considered in the present study and social vulnerability data (demographic and socio-economic indicators) used to assess the risks factors associated to the spatial distribution of the estimated heat-related mortality across Greater Adelaide

ID	Local government areas *(LGAs)*	Total population *(Tot_Pop)*	Population density (hab/km^2) *(Pop_Dens)*	Total indigenous population[a] *(Tot_Ind)*	% Indigenous population *(Fr _Ind)*	Children (<5 yo) *(Pop_5yo)*	% Children (<5 yo) *(Fr_5yo)*	Older population (>65 yo) *(Pop_65yo)*	% Older population (>65 yo) (Fr_65yo)	IRSD Score[b]
1	Adelaide	25,456	1634.1	375	1.5	573	2.3	3446	13.5	1014
2	Burnside	45,816	1664.4	164	0.4	2002	4.4	10,167	22.2	1081
3	Campbelltown	52,192	2142.8	331	0.6	2809	5.4	10,615	20.3	1012
4	Charles Sturt	118,943	2169.9	1968	1.7	6453	5.4	22,639	19.0	985
5	Holdfast Bay	37,435	2720.8	313	0.8	1458	3.9	9171	24.5	1043
6	Marion	93,448	1678.8	1245	1.3	5567	6	16,693	17.9	1001
7	Mitcham	67,474	892.8	426	0.6	3447	5.1	13,249	19.6	1068
8	Norwood, Payneham, St Peters	37,056	2452.7	263	0.7	1688	4.6	7443	20.1	1029
9	Onkaparinga	172,938	333.6	3175	1.8	10,223	5.9	31,488	18.2	987
10	Port Adelaide - Enfield	127,740	1391.6	3527	2.8	8160	6.4	19,940	15.6	936
11	Prospect	21,520	2760.8	196	0.9	1313	6.1	3070	14.3	1046
12	Salisbury	143,560	897.8	3343	2.3	9828	6.8	20,735	14.4	917
13	Tea Tree Gully	100,261	1052.8	1159	1.2	5436	5.4	19,144	19.1	1031
14	Unley	39,208	2746.4	198	0.5	1847	4.7	7589	19.4	1066
15	Walkerville	8000	2265.6	65	0.8	344	4.3	1755	21.9	1072

(continued)

Table 1 (continued)

ID	Local government areas *(LGAs)*	Total population *(Tot_Pop)*	Population density (hab/km^2) *(Pop_Dens)*	Total indigenous population[a] *(Tot_Ind)*	% Indigenous population *(Fr_Ind)*	Children (<5 yo) *(Pop_5yo)*	% Children (<5 yo) *(Fr_5yo)*	Older population (>65 yo) *(Pop_65yo)*	% Older population (>65 yo) (Fr_65yo)	IRSD Score[b]
16	West Torrens	60,842	1641.9	704	1.2	3165	5.2	10,654	17.5	1002

[a]Aboriginal and Torres Strait Islander Peoples as defined by ABS [36]

[b]A lower index of relative socio-economic disadvantage (IRSD) score indicates areas that are more socially and economically disadvantaged [36]

3.2 Data Acquisition and Processing

3.2.1 Remote Sensing Data

LiDAR data and aerial imagery used in this study were collected, rectified and processed by Aerometrex Ltd. across the Greater Adelaide Metropolitan area. High-resolution four-band (R:G:B:NIR) aerial imagery was captured on 25 September 2018 with a spatial resolution of 75 cm and subsequently down-sampled to 50 cm. LiDAR datasets (with a point cloud density of 8 pts.m^{-2}) were retrieved in two separate flight campaigns between 21 and 24 April 2019 and 30 October to 3 November 2019, respectively. An automatic, semantic classification was conducted by *Aerometrex* researchers to distinguish between different land surface covers including trees (>3 m), permeable and impermeable ground, buildings and water by combining the LiDAR data and aerial imagery. Results of an error matrix show an overall accuracy of 89.5% and Kappa index of 0.78, which are acceptable for the present study.

Raster images corresponding to each land surface cover were created with a spatial resolution of 1 m. A 100 × 100 m^2 grid was generated for the entire study area and the fraction of each land cover was calculated for each grid cell using the zonal statistics tool available in ArcGIS®. The initial tree canopy cover (TC_{in}) corresponds with the portion of each grid occupied by existing trees as extracted from the LiDAR data. The final or maximum proportion of tree coverage (TC_{fin}) per grid was estimated by subtracting the percentage of area occupied by buildings and water as it is assumed that tree canopy can overlay permeable and impermeable ground surfaces in the understorey. Note that the potential application for green roofs and green walls on buildings' envelopes is excluded from this study.

Digital boundaries representing the statistical LGAs for the entire study area were retrieved from the Australian Bureau of Statistics (ABS) (https://www.abs.gov.au/). A list of the 16 LGAs selected for this study is shown in Fig. 1 and summarised in Table 1. TC_{in} and TC_{fin} values were averaged for each LGA boundary using the centroids of grids (Euclidean distance function) as selection method. These values were implemented in Eqs. 1 and 2 using raster calculator to estimate ΔT_{15} and HRM_{dec} for each LGA boundary. It is worth noting that airports (Adelaide and Parafield), water reservoir (Happy Valley), aquatic reserves (i.e., Barker-Inlet St. Kilda) and large infrastructure, industrial, desalination, waste and sewage treatment plants (in Dry Creek, Londsdale and Bolivar) have been excluded from estimations (Fig. 1).

3.2.2 Social Vulnerability Data

Demographic and socio-economic data for 2019 related to heat- and health-vulnerability factors were retrieved for each LGA from the Public Health Information Development Unit (PHIDU) atlases (https://phidu.torrens.edu.au/) developed by the Torrens University Australia. Spatially extensive (total population numbers) and intensive (percentage of population) variables for each LGA are summarised in Table 1. In addition, the Index of Relative Socio-economic Disadvantage (IRSD) provided by ABS on the basis of the 2016 census was also retrieved for each LGA (Table 1). This is a general socio-economic score that measures the relative disadvantage based on income, educational attainment and skilled occupations, where low scores indicate relatively greater disadvantage overall.

3.3 Spatial Analysis

Several spatial econometrics techniques [53] were conducted using GeoDa 1.20.0.8 software [54] to evaluate the spatial and statistical patterns, variability and relationships between the estimated potential percentage decrease of HRM (HRM_{dec}) and the social vulnerability indicators listed in Table 1. Spatial univariate analysis was performed using boxplot maps (or boxmaps) [55] to evaluate the distribution and highlight extreme values (spatial outliers) at the lower and upper end of the scale. Multivariate analysis via scatterplots and bubble charts was performed to assess the association between multiple indicators with the goal of identifying LGAs or locations that require priority attention for future greening interventions. In addition, spatially constrained hierarchical clustering [56] was performed using a standardised (Z) transformation, ward's-linkage method and Euclidean distance function to identify regions or high-level clusters with locational similarity in terms of the reduction of HRM rates and several vulnerability risks factors. A spatial weight matrix was constructed in GeoDa using a 'queen' first-order-contiguity for the clustering analysis [55].

4 The Potential Effect of Increased Tree Canopy Coverage on Heat-Related Mortality in Adelaide

A proof of concept has been applied to the Greater Adelaide Metropolitan area in order to estimate the potential maximum percentage decrease of HRM as a function of the maximum average peak daily ambient temperature drop via the increment in tree canopy coverage at sub-metropolitan level across 16 LGAs based on Eqs. (1) and (2) as proposed by Santamouris and Osmond [1].

4.1 Increment of Tree Canopy Coverage and the Potential Impact in Reducing Heat-Related Mortality Rates

Table 2 and Fig. 2 present a summary of the numerical and spatial distributions (using boxmaps, boxplots and summary statistics) of the existing (TC_{in}) and maximum potential increase (TC_{fin}) of tree coverage, maximum average peak daily ambient temperature drop (ΔT_{15}) and maximum reduction of HRM rates (HRM_{dec}) estimated for 16 LGAs in Adelaide.

Currently available tree coverage in Adelaide ranges between 12.7 and 52.2%. The highest proportion of trees can be found in the eastern council areas of Mitcham (52.2%), Burnside (41.7%) and Tea Tree Gully (32.1%) (adjacent to the Adelaide Hills), while the lowest proportion was found in the north-western councils: Port Adelaide—Enfield (12.7%), Charles Sturt (14.6%) and West Torrens (17.7%) (Fig. 2a).

Table 2 Summary of existing and maximum potential tree coverage, maximum average peak daily ambient temperature drop and maximum reduction of HRM rates estimated per LGA

ID	Local government areas (LGAs)	Existing tree coverage (%)(TC_{in})	Maximum potential tree coverage (%)(TC_{fin})	Maximum average peak daily ambient temperature drop (°C) (ΔT_{15})	Maximum reduction of HRM rates (%) (HRM_{dec})
1	Adelaide	26.1	77.6	1.0	30.5
2	Burnside	41.7	76.7	0.8	25.0
3	Campbelltown	25.9	69.4	0.9	26.4
4	Charles Sturt	14.6	65.2	0.9	28.0
5	Holdfast Bay	18.7	62.8	0.8	25.4
6	Marion	18.4	77.6	1.1	33.0
7	Mitcham	52.2	86.8	0.9	26.7
8	Norwood -Payneham –St. Peters	26.9	63.4	0.8	23.1
9	Onkaparinga	28.3	94.2	1.3	38.2
10	Port Adelaide -Enfield	12.7	73.4	1.1	32.8
11	Prospect	24.0	62.6	0.8	23.6
12	Salisbury	18.9	79.7	1.1	34.0
13	Tea Tree Gully	32.1	81.8	1.0	30.7
14	Unley	30.1	62.5	0.7	21.6
15	Walkerville	30.2	37.4	0.8	24.1
16	West Torrens	17.7	50.4	0.9	28.4

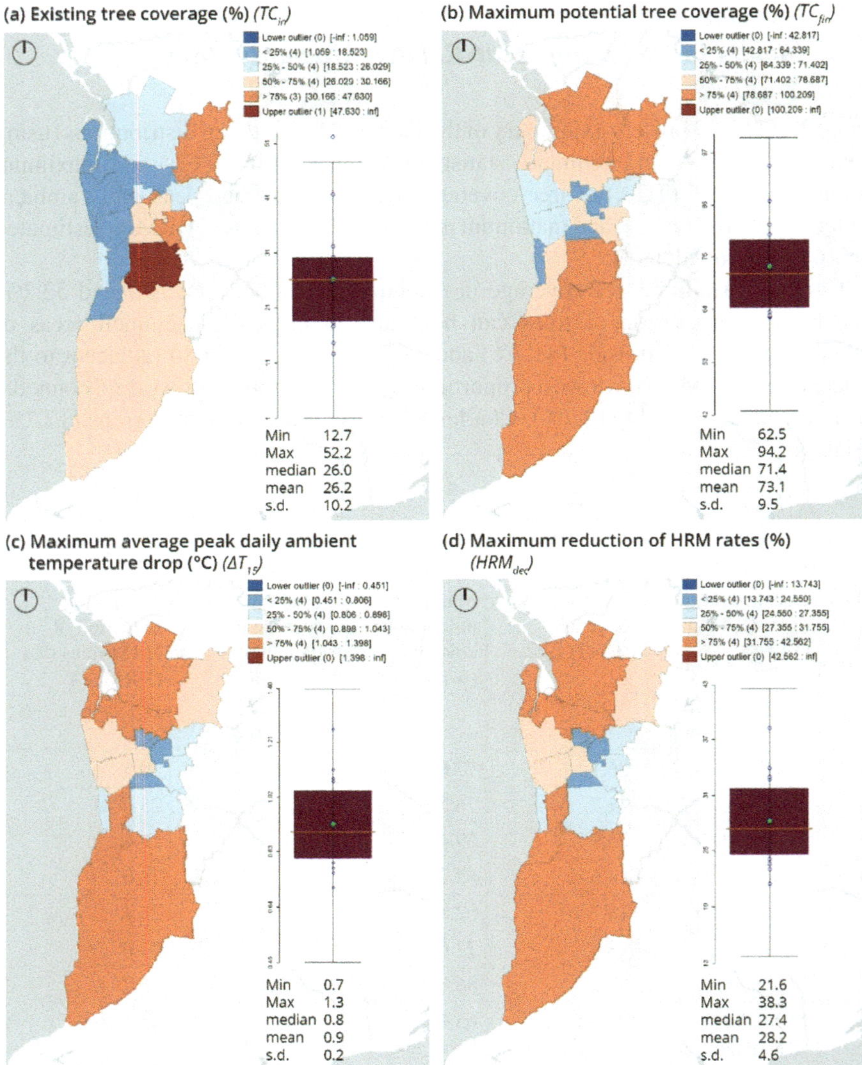

Fig. 2 Boxmaps, boxplots and summary statistics showing the numerical and spatial distribution of the existing **a** and maximum potential **b** proportion of tree canopy coverage; maximum average peak daily ambient temperature drop **c**; and maximum reduction of HRM rates **d** calculated for 16 LGAs of the Greater Adelaide metropolitan area. Note: first quartile (<25%); interquartile range (25–75%); Third quartile (>75%)

The spatial analysis indicates that a maximum tree coverage between 62.5 and 94.2% on the available land that is currently not covered by trees can be potentially achieved for the entire metropolitan area with the highest values calculated for Onkaparinga (94.2%), Mitcham (86.8%) and Tea Tree Gully (81.8%); and the lowest for Unley (62.5%), Prospect (62.6%) and Holdfast Bay (62.8%) (Fig. 2b). It is estimated that a maximum peak daily ambient temperature reduction of 0.7–1.3 °C (and a mean of 0.9 °C) can be attributed to an increment in the urban forest between 32.4% and 65.9% (Fig. 2c). This can be translated into a potential maximum reduction of HRM rates in the range of 21.6–38.2% for the entire metropolitan area. Accordingly, increasing the proportion of urban canopy above 60% would be particularly beneficial to decrease HRM rates in Onkaparinga (by 38.2%), Salisbury (by 34%), Marion (by 33%) and Port Adelaide—Enfield (32.8%). The lowest percentage decrease of HRM was estimated for Unley (by 21.6%), Norwood—Payneham—St Peters (by 23.1%) and Prospect (23.6%) where space constraints reduce the possibility of expanding the urban forest in future (Fig. 2d).

4.2 Spatial Association Between Potential Reduction of Heat-Related Mortality Rates and Socio-Economic Vulnerability Factors

Bivariate analysis shows a moderate correlation between HRM_{Dec} and spatially extensive (total population numbers) socio-economic and vulnerability variables ($R^2 = 0.63$–0.71, $p < 0.001$) (Fig. 3a–d); while low-to-moderate correlation ($R^2 = 0.20$–0.68, $p < 0.001$) is observed between HRM_{Dec} and spatially intensive (percentage of population) variables (Fig. 3e–h). Spatial analysis indicates that higher reduction of HRM rates (>32%) can be achieved by LGAs with a large number of inhabitants and particularly with large sizes of indigenous, children and older populations, namely Onkaparinga, Salisbury and Port Adelaide-Enfield. However, due to their territorial size, the population density in these LGAs is lower than the other councils (300–890 people/km^2 vs 1020–2760 people/km^2). Conversely, lower reduction of HRM rates (<25%) is observed for LGAs (surrounding City of Adelaide) with small number of vulnerable populations (i.e. Unley, Walkerville, Burnside, Prospect, Holdfast Bay and Norwood) within higher population density (>2200 people/km^2).

Figure 4 shows a moderate negative linear association ($R^2 = 0.54$, $p = 0.000$) between HRM_{Dec} and the *IRSD* scores while considering the population density (represented by bubble sizes) and the potential maximum decrease of peak ambient temperatures (ΔT_{15}) (represented by colours). Similar to previous results, higher HRM_{Dec} is observed for LGAs with more socio-economically disadvantaged populations (or low IRSD scores) such as Salisbury, Port Adelaide- Enfield and Onkaparinga where a higher peak ambient temperature drop is expected via the increment in urban greenery. Nonetheless, more disadvantaged areas tend to exhibit lower population densities compared to wealthier councils (i.e., Unley, Norwood, Walkerville) where

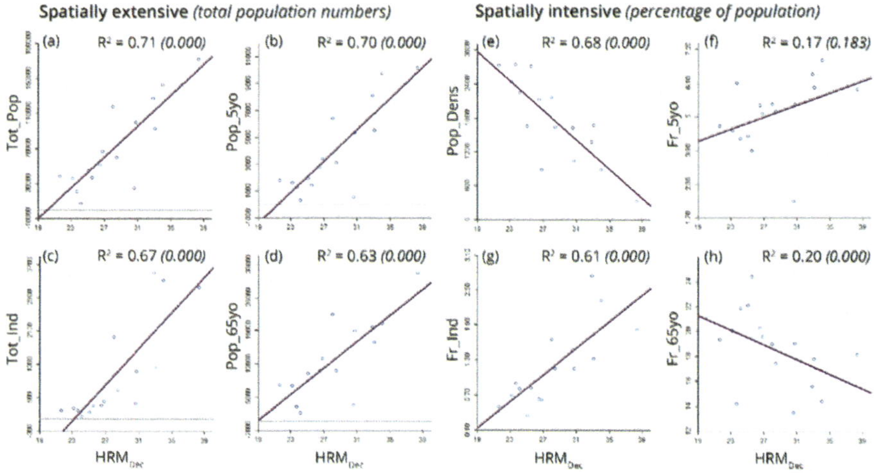

Fig. 3 Bivariate correlation analysis between the percentage decrease of HRM and various spatially extensive **a–d** and intensive **e–h** socio-economic and vulnerability factors estimated for 16 LGAs across Greater Adelaide. Refer to Table 1 for abbreviations

a lower reduction of HRM rates and peak ambient temperatures is expected due to the limited space for increasing tree canopy cover in future.

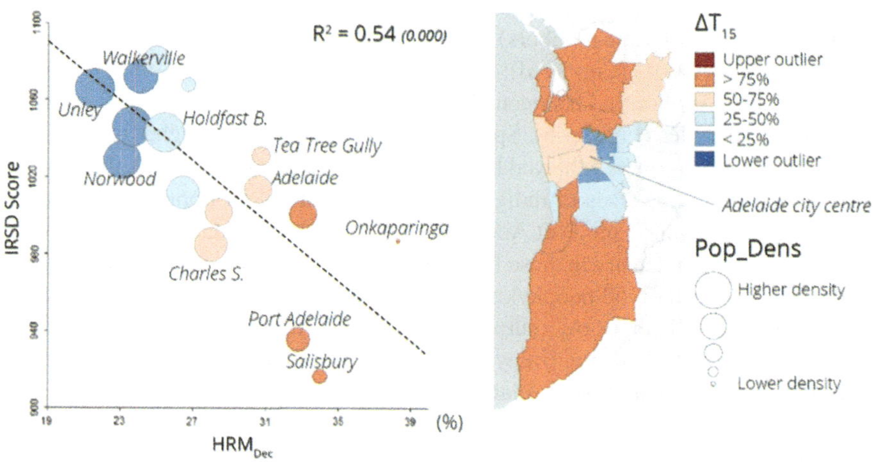

Fig. 4 Bivariate correlation analysis (left) between the percentage decrease of HRM and IRSD scores estimated for 16 LGAs across Greater Adelaide. Bubble sizes represent population density while colour represents the statistical distribution of ΔT_{15} values as per boxmap (right)

4.3 Identification of Urban Clusters Influenced by Increased Tree Canopy Coverage in Adelaide

A spatially constrained hierarchical clustering was performed to identify regions across Greater Adelaide Metropolitan with attribute and spatial similarity by considering the percentage decrease of HRM, population density and IRSD scores. Four clusters or regions were identified as illustrated by the map, dendrogram and cluster centre values in Fig. 5.

Cluster 1 corresponds to seven central-eastern LGAs characterised by lower HRM_{Dec} and higher population density. This region is currently the greenest (and closest to the Adelaide Hills) and the least socially and economically disadvantaged across Greater Adelaide Metropolitan area.

Cluster 2 comprises five central-western LGAs exhibiting relatively low HRM_{Dec}, medium population density with only a relatively small portion of disadvantaged and vulnerable populations. This cluster also possesses the lowest proportion of tree canopy coverage at the moment.

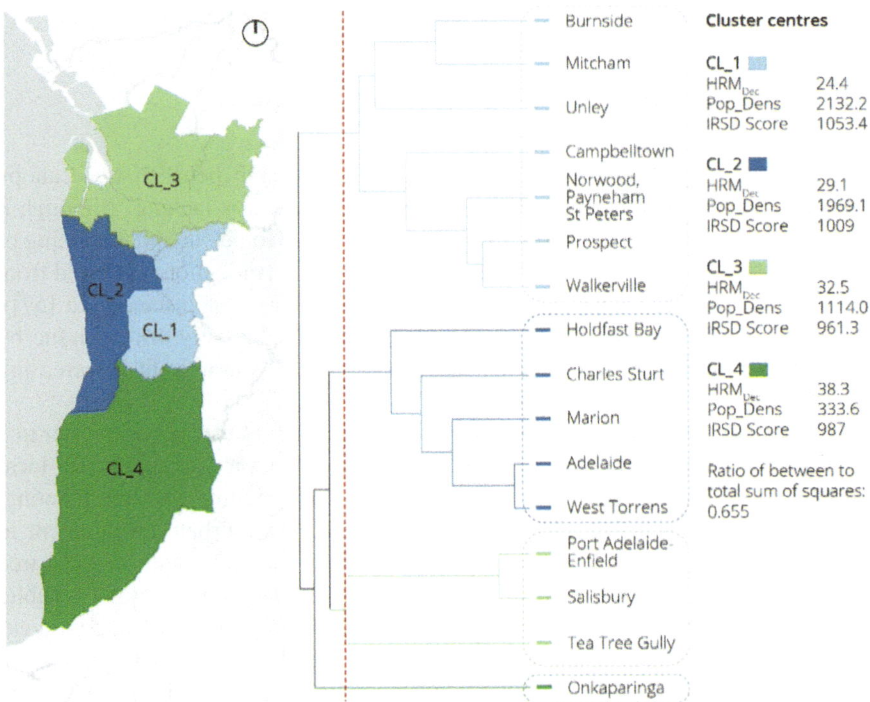

Fig. 5 Results of the spatially constrained hierarchical clustering showing the spatial clusters (left), dendrogram with respective LGAs per cluster and values of percentage decrease of HRM, population density and IRSD score estimated for each cluster centre (right)

Cluster 3 consists of three northern LGAs where it is observed higher HRM_{Dec}, medium population densities and a significant number of socially and economically disadvantaged and vulnerable populations (children, older and indigenous).

Cluster 4 comprises the LGA of Onkaparinga which is characterised by the highest value of HRM_{Dec} due to the vast amount of agricultural land available without tree covers. Although the population density is significantly low, there is a considerable number of vulnerable and disadvantaged populations in this region.

5 Discussion and Future Outlook

In this study, the prediction of HRM has been based on the estimation of potential increase of tree coverage from the current tree coverage, which is then used to estimate the potential drop in average peak daily ambient temperature. Based on a function developed by Santamouris and Osmond [1] derived from meta-analysis of 13 studies from a number of countries, the potential decrease of heat-related mortality for each LGA in the Greater Adelaide Metropolitan area has been estimated.

5.1 The Approach: Challenges and Opportunities

Prior to our study, an evidential basis for the relationship between tree canopy coverage, health outcomes and HRM in South Australia was lacking. Although it is well recognised that trees and green spaces contribute to health and wellbeing of people, studies thus far in South Australia were based on perceptions (inferred from epidemiology surveys for example) rather than hard data (for instance see [57]). Our study has been able to fill this knowledge gap on this very critical issue by providing estimations on the potential reduction of mortality as a result of increasing tree coverage.

Understanding the relationship between the presence (or absence) of trees or tree canopy coverage and health outcome at the LGA level is particularly useful for local governments who are primarily responsible for urban greening and tree planting. This study can then help inform decisions to better shape urban development in each LGA. Such metrics can guide what type of development and how much tree coverage should be aimed for with clear links to health benefits. Improving public health outcomes is particularly urgent in the face of increasing urban temperatures.

The method applied in this study, however, is intended to indicate maximum potential reductions in HRM through tree canopy enhancement. Such an approach is based on the spatial analysis of the existing areas currently "not occupied" by trees and assumed to be all covered by trees in the future scenario. This method may therefore overestimate future practical tree coverage and thus overestimate the potential reduction in HRM, as it does not consider practical limitations of tree planting which may be environmental, cultural or economic in each of the LGAs.

For example, there is a need for careful placement of trees to build new football and recreational fields or to allow solar access to dwellings, school buildings and hospitals. Equally, cultural and economic barriers to tree planting exist in many of the LGAs especially those with socio-economic disadvantages.

This research is therefore intended to support local governments in developing urban greening policies and accessing funding for improved public health outcomes. This applies also to LGAs whose current tree coverage is already high and can assist them in considering a further increase in their tree coverage to maximise the health outcomes in their areas. For instance, the distribution of trees in LGAs such as Mitcham and Burnside is often less than ideal with GI such as footpaths exposed to the harsh sun and lacking in a continuous shading and protective canopy. A useful and practical target is to ensure all footpaths and road carriages are fully shaded by tree canopy, particularly during summer. Many of Adelaide's best and lushest streets in the inner and middle-ring suburbs demonstrate the feasibility of such a target (Fig. 6).

Another challenge encountered with using this remote sensing-based approach is the use of LiDAR and aerial imagery data from 2018 to 2019. It is likely that tree coverage across the Greater Adelaide Metropolitan area has changed since then due to many recent developments, thus the prediction of the existing and future potential tree coverage may also be under or overestimated. The prediction on the percentage reduction in HRM has also not taken into account that there may be changes in the population density in each LGA, particularly due to the impact of the COVID-19

Fig. 6 Example of two streets in Unley and Mitcham council areas with full tree canopy coverage. Left: Northgate St, Unley with exotic species (Platanus acerifolia); Right: Walter Young Ave, Mitcham with native species (Corymbia citriodora). (Photos from authors)

pandemic that resulted in people moving to regional areas or people from outside South Australia moving into Adelaide. Future initiatives that address challenges of longitudinal datasets and accessing continuous remote sensed open data, therefore, require further examination and collaborative development [58].

The method also does not differentiate between public and private land. In estimating potential future tree coverage, it assumes that all available land can be planted with trees. In reality, much of this land is private and so the onus on planting more trees is on the landowner. Active research and programmes that stimulate community greening efforts on private land and verges are needed to transform districts with low IRSD scores. These have been trialled in places such as Sydney which face many of the same convergent challenges of low tree cover, disadvantage and hot landscapes and urban development patterns [59].

Finally, the method assesses tree canopy coverage as a homogenous land cover and does not consider the types nor the health of the trees as well as different characteristics of green patches within the landscape [60]. This is an issue, both in relation to resilience and performance of the tree canopy. It is possible that some trees will not withstand future climatic shocks such as heatwaves or droughts [61]. Likewise, varying tree species have different shading and cooling benefits [62], thus the results may overestimate the impact of the coverage from these trees to reduce HRM in the future. Trees such as Eucalyptus and other local natives have high biodiversity value but often are more limited in their cooling effects due to their characteristically sclerophyllous leaf structure and low evapotranspiration. However, these same qualities increase their health and resilience in conditions that may kill or sicken many introduced species [63, 64]. Such complex variables require further study and research in combination with the metropolitan scale urban remote sensing analytic approaches adopted here.

5.2 The Relationship of Tree Canopy Coverage with Socio-Economic Factors and Public Health Policy

Despite the shortcomings of the method, the study has highlighted and confirmed current perceptions that areas with less tree coverage tend to be those with low IRSD scores. These areas, except Onkaparinga, are where industrial buildings and larger roads (due to heavy vehicles for the industrial activities) dominate a certain portion of the LGA and where the workers of those industries reside (Salisbury and Port Adelaide-Enfield). These areas also tend to have a larger population of older and vulnerable people compared to many other LGAs. Tree planting in such areas can be challenging because of the requirements for maintaining the trees in the long run, which can be costly. These areas also tend to be on the western and northern parts of the city, areas with hotter and drier climates and where the soil is less fertile compared to the central and eastern parts of the metropolitan area, which are closer to the cooler Adelaide Hills and often benefit from the deeper, more fertile soils of the Adelaide Plain.

The study has demonstrated that there is a great potential for reducing HRM in these areas by increasing their tree canopy coverage. The result of the analysis can be used by the local governments to lobby for more funding from the State Government to help them increase the tree coverage in order to reduce HRM. The study also highlights the need for future research and initiatives concerning community and business-based greening efforts on private land-holdings in these specific local socio-economic contexts.

5.3 Relative Effectiveness of Increased GI on HRM Rate in Different LGAs and Implications for Urban Planning and Design

All the various LGAs demonstrate different reductions in HRM with increased tree canopy coverage. It is evident that even LGAs with smaller percentage decreases in HRM stand to benefit from increased tree canopy coverage. Relatively green LGAs such as Unley, Burnside, Norwood, Walkerville and Prospect show that GI enhancement will deliver substantial reductions in HRM. Even for localities within the lowest range, in the Mitcham LGA for example, a dramatic 20% decrease of HRM can potentially be achieved by increasing urban tree canopy coverage.

The four identified clusters in Sect. 4.3 and Fig. 5 require different priorities and strategies regarding GI enhancement. While Cluster 1 (the central-eastern LGAs closer to the Adelaide Hills) may have a relatively high existing GI and less socially and economically disadvantaged population which makes them more resilient to HRM, Cluster 2 (central-western LGAs) suffer significantly from very low tree canopy coverage and can benefit more from urban greening projects. Cluster 3 (northern LGAs) shows a higher decrease in HRM, despite their lower population densities due to a significant number of socially and economically disadvantaged and vulnerable populations. This cluster has significant available land for urban greening and can benefit the most from increased GI relative to its overall and relative populations. Cluster 4 (the LGA of Onkaparinga), however, has a large area and low population. It shows the highest decrease in HRM due to the vast amount of agricultural land available without tree covers, however, its low population density and a considerable number of vulnerable and disadvantaged population in this region requires targeted GI enhancement in population centres. The development of specific urban greening strategies suited to local contexts is therefore required to gain the most public health benefit from tree canopy enhancement.

Current trends in Cluster 1 LGAs include urban consolidation and the removal of trees to accommodate denser urban types. The relationship between urban greenery, canopy coverage and density are not necessarily linear as different medium density types and configurations can accommodate, maintain or increase the same quantity of canopy coverage if designed and planned thoughtfully [65, 66]. Cities with high quantities of urban greenery can also be remarkably dense; the medium-density city of Berlin and other Northern European cities are some of the examples [67].

Current trends for clusters 3 and 4 are similar in that they are dominated by low-density single family dwellings and new suburbs with little shade or canopy coverage. Challenges to improve greening in these areas again include altering the type and pattern of urban development to allow space for substantial tree planting in public spaces and either larger backyards that allow trees or high quantities of public green space and parkland. Unfortunately, current trends of the developments in these areas maximise low-density built space while minimising green space. This needs to be reversed with higher densities of built space with smaller footprints to allow for more green space. Urban development metrics and urban design types can be integrated to prioritise or even mandate such approaches [68]. Cluster 2 containing both the City of Adelaide and coastal and port areas is highly heterogeneous so will require greater levels of master planning to ensure that when urban renewal of post-industrial sites and lower-density estates do take place, it integrates public spaces that support urban forests and continuous canopy. In particular, street redesign plays an important role in urban greening in this area and future street renewal must allow significant verge and median space for urban greening within the street section. This means thinking creatively about transport mode choices and limiting on-street and surface car parking [69, 70].

5.4 Future Work at the Intersection of Public Health and Urban Greening

The potential contribution of diverse GI types, from wetlands to irrigated turf, to urban gardens, to green roofs and green walls is excluded from this study; however, a range of green infrastructures need to be considered to cool cities such as Adelaide and in a wide variety of cultural contexts. For instance, wetland systems and irrigated GI can provide even greater cooling for urban areas than increased tree canopy alone, while large cooling effects can be derived from the greening of industrial roofs which currently intensify UHI effects.

Future research can consider tree species and types that have better cooling capacity to better withstand future climates. Effective policy measures and stewardship to maintain existing trees is also a requirement to ensure the existing tree coverage can sustain to their potential. As much urban land is private, encouraging and incentivising citizens and businesses and other landowners to plant more trees in their private land is an area for further research. Such research is especially needed in the new burgeoning suburbs, which typically lack substantial tree cover for a range of reasons including urban development types favouring closely packed single family dwellings that allow for little greenspace. Compounding to this issue is the fact that trees in such new developments have not had time to establish before the new residents move in.

As Adelaide's localities become greener, there is also a need to consider environmental, economic and social outcomes that might limit the effectiveness and equity of such approaches. Frequently within cities around the world, a process of "green gentrification" [71] occurs where urban greening efforts in disadvantaged neighbourhoods have the effect of increasing property values and displacing populations. Such complex intertwining social and cultural factors associated with urban greening and urban development make identifying trade-offs between public health, economic development, climate mitigation and biodiversity an essential part of tree canopy enhancement and urban greening [72].

6 Conclusions

This study indicates the public health benefits possible from increasing tree canopy coverage in the Greater Adelaide Metropolitan area. Specifically, three distinct findings emerge from the work along with a series of questions that can stimulate future research and actions regarding the intersecting policy areas of public health, urban planning and urban greening.

Our first finding is that there is a clear and urgent need for close alignment of public health policy with urban planning and greening policies for a whole of city's benefit. Remarkably *all* areas considered in the study stand to benefit from a HRM and public health perspective from an enhanced tree canopy and resultant urban cooling. There is a clear case for such findings to be translated into current planning tools such as the new South Australian planning e-platform to mandate conservation of trees and increased tree canopy coverage at the district level. Secondly, we found that there is a solid case for tailoring urban greening and tree canopy enhancement at the local LGA level based on the socio-economic and cultural characteristics of local populations. Participatory engagement with communities is not only significant but also essential if positive greening strategies are to be successful. Finally, given the many dynamics and complex socio-economic and urban greening factors involved in shaping public health outcomes, there is a need to continuously monitor the relationship between these variables to ensure the moving targets of public health and environmental quality is accurately addressed.

Apart from monitoring the spatial quantity and quality of the urban tree canopy coverage, future work requires the development and testing of a range of urban development types and tree species to ensure that the city can become greener over time and stay greener. We must act so that in a hundred years from now the mighty Redgums and shady avenues that characterise so much of Adelaide will be joined by many other trees to greatly expand the urban forest. It is clear that society's health depends upon such initiatives.

Acknowledgements Authors would like to thank the staff and personnel from Aerometrex Ltd., in particular Dr Samuel Holt, who collected, pre-processed and provided the geospatial datasets used in the present study.

References

1. Santamouris M, Osmond P (2020) Increasing green infrastructure in cities: impact on ambient temperature, air quality and heat-related mortality and morbidity. Buildings 10(12):233
2. Coates L, Haynes K, O'Brien J, McAneney J, de Oliveira FD (2014) Exploring 167 years of vulnerability: an examination of extreme heat events in Australia 1844–2010. Environ Sci Policy 42:33–44
3. Coutts AM, Beringer J, Tapper NJ (2007) Impact of increasing urban density on local climate: spatial and temporal variations in the surface energy balance in Melbourne, Australia. J Appl Meteorol Climatol 46(4):477–493
4. Sharifi E, Boland J (2018) Limits of thermal adaptation in cities: outdoor heat-activity dynamics in Sydney Melbourne and Adelaide. Architect Sci Rev 61(4):191–201
5. Vos T, Barber RM, Bell B, Bertozzi-Villa A, Biryukov S, Bolliger I, Charlson F, Davis A, Degenhardt L, Dicker D et al (2015) Global, regional, and national incidence, prevalence, and years lived with disability for 301 acute and chronic diseases and injuries in 188 countries, 1990–2013: a systematic analysis for the global burden of disease study 2013. The Lancet 386(9995):743–800
6. Tzoulas K, Korpela K, Venn S, Yli-Pelkonen V, Kaźmierczak A, Niemela J, James P (2007) Promoting ecosystem and human health in urban areas using Green infrastructure: a literature review. Landsc Urban Plan 81(3):167–178
7. Hawken S, Rahmat H, Sepasgozar SME, Zhang K (2021) The SDGs, ecosystem services and cities: a network analysis of current research innovation for implementing urban sustainability. Sustainability 13(24):14057
8. Roy S, Byrne J, Pickering C (2012) A systematic quantitative review of urban tree benefits, costs, and assessment methods across cities in different climatic zones. Urban Forestry Urban Greening 11(4):351–363
9. Kondo MC, Fluehr JM, McKeon T, Branas CC (2018) Urban green space and its impact on human health. Int J Environ Res Public Health 15:3
10. Gascon M, Triguero-Mas M, Martínez D, Dadvand P, Forns J, Plasència A, Nieuwenhuijsen MJ (2015) Mental health benefits of long-term exposure to residential green and blue spaces: a systematic review. Int J Environ Res Public Health 12(4):4354–4379
11. Lee ACK, Maheswaran R (2011) The health benefits of urban green spaces: a review of the evidence. J Public Health (Oxf) 33(2):212–222
12. Di Nardo F, Saulle R, La Torre G (2010) Green areas and health outcomes. a systematic review of the scientific literature. Italian J Public Health7(4)
13. van den Berg M, Wendel-Vos W, van Poppel M, Kemper H, van Mechelen W, Maas J (2015) Health benefits of green spaces in the living environment: a systematic review of epidemiological studies. Urban Forestry Urban Greening. 14(4):806–816
14. Bowler DE, Buyung-Ali LM, Knight TM, Pullin AS (2010) A systematic review of evidence for the added benefits to health of exposure to natural environments. BMC Public Health 10:456
15. Son J-Y, Lane KJ, Lee J-T, Bell ML (2016) Urban vegetation and heat-related mortality in Seoul Korea. Environment Res 151:728–733
16. Son J-Y, Liu JC, Bell ML (2019) Temperature-related mortality: a systematic review and investigation of effect modifiers. Environ Res Lett 14(7):73004
17. Kalkstein LS, Sheridan S (2003) The impact ofheat island reduction strategies on health-debilitating oppressive air massive in urban areas. U.S. EPA Heat Island Reduction Initiative, Berkeley, CA, USA
18. Mills DM, Kalkstein L (2012) Estimating reduced heat-attributable mortality for an urban revegetation project. J Heat Island Inst Int 7(2):18–24
19. Silva HR, Phelan PE, Golden JS (2010) Modeling effects of urban heat island mitigation strategies on heat-related morbidity: a case study for Phoenix, Arizona, USA. Int J Biometeorol 54(1):13–22

20. Murage P, Kovats S, Sarran C, Taylor J, McInnes R, Hajat S (2020) What individual and neighbourhood-level factors increase the risk of heat-related mortality? a case-crossover study of over 185,000 deaths in London using high-resolution climate datasets. Environ Int 134:105292
21. Burkart K, Meier F, Schneider A, Breitner S, Canário P, Alcoforado MJ, Scherer D, Endlicher W (2016) Modification of Heat-related mortality in an elderly urban population by vegetation (urban green) and proximity to water (urban blue): evidence from Lisbon Portugal. Environment Health Perspect 124(7):927–934
22. Guest CS, Willson K, Woodward AJ, Hennessy K, Kalkstein LS, Skinner C, McMichael AJ (1999) Climate and mortality in Australia: retrospective study, 1979–1990, and predicted impacts in five major cities in 2030. Clim Res 13:1–15
23. Nicholls N, Skinner C, Loughnan M, Tapper N (2008) A simple heat alert system for Melbourne, Australia. Int J Biometeorol 52(5):375–384
24. Vaneckova P, Hart MA, Beggs PJ, Dear RJ de (2008) Synoptic analysis of heat-related mortality in Sydney, Australia, 1993–2001. Int J Biometeorol 52(6): 439–51
25. Herbst J, Mason K, Byard RW, Gilbert JD, Charlwood C, Heath KJ, Winskog C, Langlois NEI (2014) Heat-related deaths in Adelaide, South Australia: review of the literature and case findings - an Australian perspective. J Forensic Leg Med 22:73–78
26. Nitschke M, Tucker GR, Bi P (2007) Morbidity and mortality during heatwaves in metropolitan Adelaide. Medical J Australia 187,11/12
27. Nitschke M, Tucker GR, Hansen AL, Williams S, Zhang Y, Bi P (2011) Impact of two recent extreme heat episodes on morbidity and mortality in Adelaide, South Australia. a case-series analysis. Environmental Health 10(42)
28. Bi P, Williams S, Loughnan M, Lloyd G, Hansen A, Kjellstrom T, Dear K, Saniotis A (2011) The effects of extreme heat on human mortality and morbidity in Australia: implications for public health. Asia-Pacific J Public Health 23 (2_suppl):27S–36
29. Williams S, Nitschke M, Sullivan T, Tucker GR, Weinstein P, Pisaniello DL, Parton KA, Bi P (2012) Heat and health in Adelaide, South Australia: assessment of heat thresholds and temperature relationships. Sci Total Environ 414:126–133
30. Williams S, Nitschke M, Weinstein P, Pisaniello DL, Parton KA, Bi P (2012) The impact of summer temperatures and heatwaves on mortality and morbidity in Perth, Australia 1994–2008. Environ Int 40:33–38
31. Santamouris M, Paolini R, Haddad S, Synnefa A, Garshasbi S, Hatvani-Kovacs G, Gobakis K, Yenneti K, Vasilakopoulou K, Feng J, Gao K, Papangelis G, Dandou A, Methymaki G, Portalakis P, Tombrou M (2020) Heat mitigation technologies can improve sustainability in cities. An holistic experimental and numerical impact assessment of urban overheating and related heat mitigation strategies on energy consumption, indoor comfort, vulnerability and heat-related mortality and morbidity in cities. Energy Build 217:110002
32. Chen D, Wang X, Thatcher M, Barnett G, Kachenko A, Prince R (2014) Urban vegetation for reducing heat related mortality. Environ Pollut (Barking, Essex: 1987). 192:275–84
33. Haddad S, Barker A, Yang J, Kumar DIM, Garshasbi S, Paolini R, Santamouris M (2020) On the potential of building adaptation measures to counterbalance the impact of climatic change in the tropics. Energy Build 229:110494
34. Yenneti K, Ding L, Prasad D, Ulpiani G, Paolini R, Haddad S, Santamouris M (2020) Urban overheating and cooling potential in australia: an evidence-based review. Climate 8(11):126
35. Vaneckova P, Beggs PJ, Jacobson CR (1982) Spatial analysis of heat-related mortality among the elderly between 1993 and 2004 in Sydney Australia. Soc Sci Med 70(2):293–304
36. Australian Bureau of Statistics (2021). 2016 Census QuickStats. https://quickstats.censusdata.abs.gov.au/census_services/getproduct/census/2016/quickstat/UCL401001?opendocument. Accessed 5 Nov 2021
37. Kottek M, Grieser J, Beck C, Rudolf B, Rubel F (2006) World map of the Köppen-Geiger climate classification updated. metz. 15(3):259–63
38. Bureau of Meteorology. 2021. *Climate statistics for Australian locations*. http://www.bom.gov.au/climate/averages/tables/cw_023034.shtml. Accessed 5 Nov 2021

39. Erell E, Williamson T (2007) Intra-urban differences in canopy layer air temperature at a mid-latitude city. Int J Climatol 27(9):1243–1255
40. Soltani A, Sharifi E (2017) Daily variation of urban heat island effect and its correlations to urban greenery: a case study of Adelaide. Frontiers Architect Res 6(4):529–538
41. Ossola A, Jenerette GD, McGrath A, Chow W, Hughes L, Leishman MR (2021) Small vegetated patches greatly reduce urban surface temperature during a summer heatwave in Adelaide Australia. Landscape Urban Plan 209:104046
42. CSIRO (2018) Australian climate observations reference network—surface air temperature dataset. http://www.bom.gov.au/metadata/catalogue/19115/ANZCW0503900450
43. Khalaj B, Lloyd G, Sheppeard V, Dear K (2010) The health impacts of heat waves in five regions of New South Wales, Australia: a case-only analysis. Int Arch Occup Environ Health 83(7):833–842
44. Loughnan ME, Nicholls N, Tapper NJ (2010) When the heat is on: threshold temperatures for AMI admissions to hospital in Melbourne Australia. Appl Geogr 30(1):63–69
45. Li M, Gu S, Bi P, Yang J, Liu Q (2015) Heat waves and morbidity: current knowledge and further direction-a comprehensive literature review. Int J Environ Res Public Health 12(5):5256–5283
46. Sadeghi M, Chaston T, Hanigan I, Dear R de, Santamouris M, Jalaludin B, Morgan GG (2021) The health benefits of greening strategies to cool urban environments—a heat health impact method. Build Environ 108546
47. Suppiah R, Preston B, Whetton PH et al (2006) Climate change under enhanced greenhouse conditions in South Australia. Melbourne, CSIRO Marine and Atmospheric Research
48. Wondmagegn BY, Xiang J, Dear K, Williams S, Hansen A, Pisaniello D, Nitschke M, Nairn J, Scalley B, Varghese BM, Xiao A, Le Jian, Tong M, Bambrick H, Karnon J, Bi P (2021) Impact of heatwave intensity using excess heat factor on emergency department presentations and related healthcare costs in Adelaide, South Australia. Sci Total Environ 781:146815
49. Hansen A, Bi P, Nitschke M, Ryan P, Pisaniello D, Tucker G (2008) The effect of heat waves on mental health in a temperate Australian city. Environ Health Perspect 116(10):1369–1375
50. Hansen AL, Bi P, Ryan P, Nitschke M, Pisaniello D, Tucker G (2008) The effect of heat waves on hospital admissions for renal disease in a temperate city of Australia. Int J Epidemiol 37(6):1359–1365
51. Wondmagegn BY, Xiang J, Dear K, Williams S, Hansen A, Pisaniello D, Nitschke M, Nairn J, Scalley B, Xiao A, Le Jian, Tong M, Bambrick H, Karnon J, Bi P (2021) Increasing impacts of temperature on hospital admissions, length of stay, and related healthcare costs in the context of climate change in Adelaide, South Australia. Sci Total Environ 773:145656
52. Stewart ID, Oke TR (2012) Local climate zones for urban temperature studies. Bull Am Meteor Soc 93(12):1879–1900
53. Anselin L, Rey SJ (2014) Modern spatial econometrics in practice. GeoDaSpace and PySAL. GeoDa Press, Chicago, IL, A guide to GeoDa
54. Anselin L (2021) DeoDa. University of Chicago, Center for Spatial Data Science
55. Anselin L (1994) Exploratory spatial data analysis and geographic information systems
56. Murray AT, Grubesic TH (2002) Identifying non-hierarchical spatial clusters. Int J Ind EngTheory Appl Pract 9(1):86–95
57. Sugiyama T, Leslie E, Giles-Corti B, Owen N (2008) Associations of neighbourhood greenness with physical and mental health: do walking, social coherence and local social interaction explain the relationships? J Epidemiol Community Health 62(5):e9
58. Hawken S, Han H, Pettit C (2020) Open cities | Open data. Springer, Singapore
59. Gallagher E (2014) Beyond "Green" streets—mitigating climate change through residential streetscape design, University of Sydney
60. Malkinson D, Kopel D, Wittenberg L (2018) From rural-urban gradients to patch—matrix frameworks: plant diversity patterns in urban landscapes. Landsc Urban Plan 169:260–268
61. Burley H, Beaumont LJ, Ossola A, Baumgartner JB, Gallagher R, Laffan S, Esperon-Rodriguez M, Manea A, Leishman MR (2019) Substantial declines in urban tree habitat predicted under climate change. Sci Total Environ 685:451–462

62. Rahman MA, Armson D, Ennos AR (2015) A comparison of the growth and cooling effectiveness of five commonly planted urban tree species. Urban Ecosyst. 18(2):371–389
63. Ossola A, Hoeppner MJ, Burley HM, Gallagher RV, Beaumont LJ, Leishman MR (2020) The Global urban tree inventory: a database of the diverse tree flora that inhabits the world's cities. Global Ecol. Biogeogr. 29(11):1907–1914
64. Chen X, Zhao P, Hu Y, Ouyang L, Zhu L, Ni G (2019) Canopy transpiration and its cooling effect of three urban tree species in a subtropical city—Guangzhou China. Urban Forestry Urban Greening 43:126368
65. Chen S, Haase D, Xue B, Wellmann T, Qureshi S (2021) Integrating quantity and quality to assess urban green space improvement in the compact city. Land 10(12):1367
66. Haaland C, van den Bosch CK (2015) Challenges and strategies for urban green-space planning in cities undergoing densification: a review. Urban Forestry Urban Greening 14(4):760–771
67. Kabisch N, Strohbach M, Haase D, Kronenberg J (2016) Urban green space availability in European cities. Ecol Ind 70:586–596
68. Uytenhaak R (2008) Cities full of space. Qualities of density. O1O, Rotterdam
69. Silvera Seamans G (2013) Mainstreaming the environmental benefits of street trees. Urban Forestry Urban Greening. 12(1):2–11
70. Coutts AM, White EC, Tapper NJ, Beringer J, Livesley SJ (2016) Temperature and human thermal comfort effects of street trees across three contrasting street canyon environments. Theor Appl Climatol 124(1–2):55–68
71. Curran W, Hamilton T (2018) Just green enough: urban development and environmental gentrification
72. Bodnaruk EW, Kroll CN, Yang Y, Hirabayashi S, Nowak DJ, Endreny TA (2017) Where to plant urban trees? a spatially explicit methodology to explore ecosystem service tradeoffs. Landsc Urban Plan 157:457–467

Chapter 14
Impact of Urban Overheating and Heat-Related Mortality in Hong Kong

Junyi Hua, Yuan Shi, Chao Ren, Kevin Ka-Lun Lau, and Edward Yan Yung Ng

Abstract Climate change and the urban heat Island (UHI) effect jointly drive urban overheating, which increase mortality risks, in the high-density context of Hong Kong that is experiencing a rapid growth of extreme heat events. This chapter summarizes recent studies on mapping of urban overheating and estimation of heat mortality risks in Hong Kong. Both the UHI effect and daytime and nighttime extreme heat events have been found to be spatially variant and related to urban morphological factors. Exposure to higher UHI intensities, extreme heat events, particularly hot nights, and lower air ventilation could increase mortality risks. High-risk communities were identified by integrated extreme heat hazards and vulnerable population exposure. Research findings can provide informative references for heat-health impacts, UHI studies, urban planning, and health actions to achieve a livable, healthy, and climate-resilient high-density city.

The original version of this chapter was revised: The affiliation of Hua Junyi has been changed. The correction to this chapter is available at https://doi.org/10.1007/978-981-19-4707-0_18

C. Ren
Faculty of Architecture, The University of Hong Kong, Pokfulam, Hong Kong, China
e-mail: renchao@hku.hk

Y. Shi · K. K.-L. Lau
Institute of Future Cities, The Chinese University of Hong Kong, Shatin, Hong Kong, China
e-mail: shiyuan@cuhk.edu.hk

K. K.-L. Lau
e-mail: kevinlau@cuhk.edu.hk

E. Y. Y. Ng
School of Architecture, The Chinese University of Hong Kong, Shatin, Hong Kong, China
e-mail: edwardng@cuhk.edu.hk

J. Hua (✉)
School of International Affairs and Public Administration, Ocean University of China, Qingdao, China
e-mail: huajunyi@ouc.edu.cn

N. Aghamohammadi and M. Santamouris (eds.), *Urban Overheating: Heat Mitigation and the Impact on Health*, Advances in Sustainability Science and Technology,
https://doi.org/10.1007/978-981-19-4707-0_14

Keywords Urban overheating · Heat mortality · Heat wave · Extreme heat event · Hot night · Urban heat island · Urban morphology · Air ventilation · High-density city · Adaptation and mitigation

1 Introduction

Hong Kong, located in South China, as one of the most densely populated cities in the world, is suffering from urban overheating that increases mortality risks [1]. In a subtropical climate, Hong Kong has hot and humid summers, and climate change has driven a warming trend and more frequent and intense heat waves. It has been predicted that Hong Kong will see more heat extremes in the rest of this century [2]. The urban overheating is aggravated by the urban heat island (UHI) effect in the high-rise, high-density urban areas of this almost urbanized city, which accommodates approximately 7.5 million residents on the approx. 270 km^2 built-up land that accounts for only about one-fourth of the total territory [3] and leads to the living density up to over 55,000 people per km^2 in the metro area. Moreover, narrow and poor living conditions and the aging society call for emphasis on social vulnerability to heat stress [4]. All these facts have fueled local investigations of the hot conditions and related health risks taking into account various aspects including extreme heat events, diurnal variations, socioeconomic vulnerability, and built environments for heat adaptation and mitigation.

1.1 The Effects of Urban Morphology in High-Density Cities

Urban morphology affects urban climate, particularly in high-density cities [5]. Higher-density urban areas generally show greater UHI intensities [6]. In the summer months, the UHI intensities can be up to 5 °C in some urban areas of Hong Kong [7]. Local studies have investigated the effects of urban morphology indicators on air temperature [8] and the wind environment [9] in the high-density urban environment of Hong Kong. In specific, Chen and colleagues [8] identified a negative relationship between the sky view factor (SVF) and the summer daytime intra-urban temperature differences, and Ng and colleagues [9] quantified the negative association of the frontal area density (FAD), that describes surface permeability, with air ventilation. High-rise, high-density built environment hinders heat losses between buildings, which is aggravated by poor ventilation conditions.

1.2 Hot Nights

Nighttime hot weather and its adverse health impacts have caught more and more attention. The UHI effect, especially in high-density cities, slows down urban areas' cooling at night. It has been found that high temperatures in the nighttime are associated with greater mortality risks in urban areas [10–12]. Exposure to nighttime heat may hinder human bodies' relief from daytime heat exposure if hot nights follow hot days, which may increase mortality risks [11]. Lee, Kim [13] identified elevated mortality risks of tropical night events in Seoul and Hong Kong and suggested that tropical nights could be regarded as a heat-health indicator for high-density cities in South-East Asia. According to the Hong Kong Observatory, the annual number of hot nights increased from 8.73 for 1961–1990 to 23.60 for 1991–2020, more quickly than that of very hot days from 13.37 to 17.47[14]. Due to the rising trend of hot nights in Hong Kong driven by global warming and the intensified UHI effect [2, 15], local studies have focused on the health impacts of not only single hot nights but also consecutive hot nights and even combinations with hot days [16, 17].

1.3 Local Indices of Extreme Heat

Owing to Hong Kong's worsening overheating situation, heat indices have been developed for local heat stress measurement. The Hong Kong Observatory and the Chinese University of Hong Kong have collaboratively developed the Hong Kong Heat Index (HKHI),[1] which combines the data of dry bulb temperature, natural wet bulb temperature, and globe temperature derived from five weather stations and therefore takes into account of heat stress-related factors, e.g., air temperature, humidity, wind, and solar radiation, based on the relationship with non-accidental hospitalization [18]. A threshold of 30.5 approximate to the 90th percentile of the HKHI associated with excess hospitalization was determined as one criterion of issuing the very hot weather warning. However, the HKHI, dependent on the measurement data from single weather stations, does not well account for the UHI effect and the spatial pattern of hot weather. Currently, the index is mainly used for warning purpose.

Given the attention to hot nights, two temperature-related indices, the numbers of very hot days (VHDs)[2] and hot nights (HNs),[3] have also been locally defined to measure daytime and nocturnal heat extremes, separately [15]. Very hot days refer to days with a maximum temperature of 33 °C or above, and hot nights mean that the daily minimum temperature reaches at least 28 °C. Not only used for very hot weather warnings, the two indices have also been widely adopted to study diurnal and spatial variations of extreme heat and heat-related health effects in the complex high-density urban environment in Hong Kong by using temperature records from

[1] https://www.hko.gov.hk/en/wxinfo/ts/display_element_hkhi.htm.

[2] https://www.hko.gov.hk/en/cis/statistic/hngtday_statistic.htm.

[3] https://www.hko.gov.hk/en/cis/statistic/hngtday_statistic.htm.

the HKO weather station network [4, 16, 17, 19–21]. Related studies are introduced in detail in the following sections.

2 Overheating Under the High-Density City Context

The subtropical climate in South China brings Hong Kong hot and humid summers, and the city is experiencing a long-term warming trend under global climate change with a greater rate during 1991–2020 [22]. Compound heat waves embracing both daytime and nighttime extreme heat were dominant ones in China over the half past century [23]. In Hong Kong, the last three decades have witnessed notably rising occurrences of locally defined VHDs and HNs, especially the latter as mentioned before [14]. Figure 1 shows that during 2000–2018, the growth rates have been further boosted to 1.23 and 1.02 days per year, respectively, and the rising trend of the days in the representative extreme heat (EH) events, which were featured as two consecutive VHDs with three HNs (2D3N) in a recent study [17], was 0.64 per year. According to the HKO's projection [2], Hong Kong will see significantly increasing extreme high temperature occurrences in this century. It has been also found that the UHI effect is intensified during extreme heat events in Hong Kong [19].

A comprehensive understanding of the spatiotemporal patterns and features of overheating under the high-density urban context of Hong Kong could inform local decision-making for heat mitigation and health risk reduction. Spatial and temporal variations in overheating in Hong Kong in terms of UHI and extreme heat events have been recently observed by local researchers [19, 20, 24]. Such investigations combined air temperature data, collected from the local weather station network, with land use regression (LUR), a common parametric spatial modeling method, to map the fine-scale distribution of heat conditions. Despite high spatial resolutions, remotely sensed land surface temperature, widely used for UHI analysis [25] and

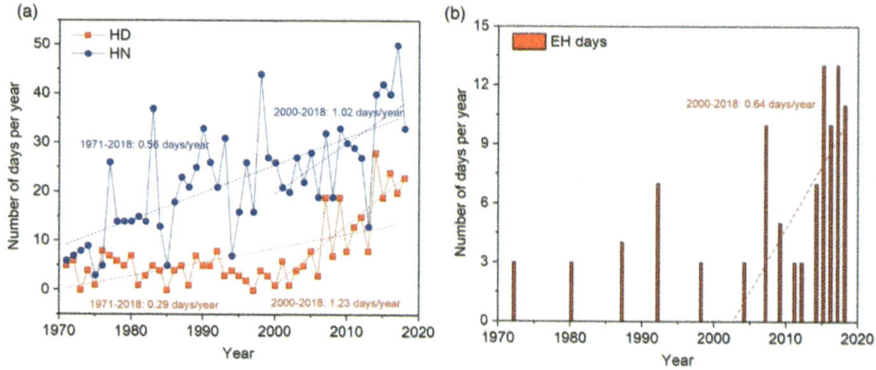

Fig. 1 Annual statistics of **a** very hot day (HD) and hot night (HN); **b** EH event days in Hong Kong. *Source* [19]

heat-health risk assessment [26], is different from the ambient air temperature that people perceive. The LUR technique is able to overcome the limitation of low-resolution station-based meteorological records.

Shi et al., initiatively attempted to adopt the LUR modeling to estimate the spatiotemporal pattern of the UHI effect in Hong Kong by taking urban morphology into account [24]. Annual and seasonal distributions of air temperature for the years 2013–2016 derived from the network of 42 automatic weather stations (AWSs) maintained by the Hong Kong Observatory (HKO) were mapped over the city to investigate the daytime and nighttime UHI effect through the LUR on a series of meteorological and geospatial predictors. The locations of the AWSs and the fine-scale summer nighttime UHI mapping are presented in Fig. 2. Temporally, background wind speed and cloudiness were correlated with UHI. Spatially, land use and land cover and urban surface morphology were found to be the main determinants of the UHI distribution. Specifically, residential land use reflecting anthropogenic heat emission, especially in summer, and fractional vegetation cover improving cooling positively and negatively affect UHI, respectively. Compared with the built-up areas in the New Territories, the downtown areas in the north of the Hong Kong Island and the Kowloon Peninsula, featured with high building volume density (BVD) and SVF, suffered from higher temperature, particularly in the nighttime.

Based on the initial study by Shi et al. [24], Shi et al. [20] extended the LUR modeling to the spatial mapping of extreme hot weather conditions in Hong Kong. Two indicators were developed from VHDs and HNs—very hot day hours (VHDHs, the total number) and hot night hours (HNHs), which refer to the total number of hours at least 33 °C and 28 °C in the daytime and nighttime, respectively, and calculated from the HKO weather stations' records for June–August from 2011 to

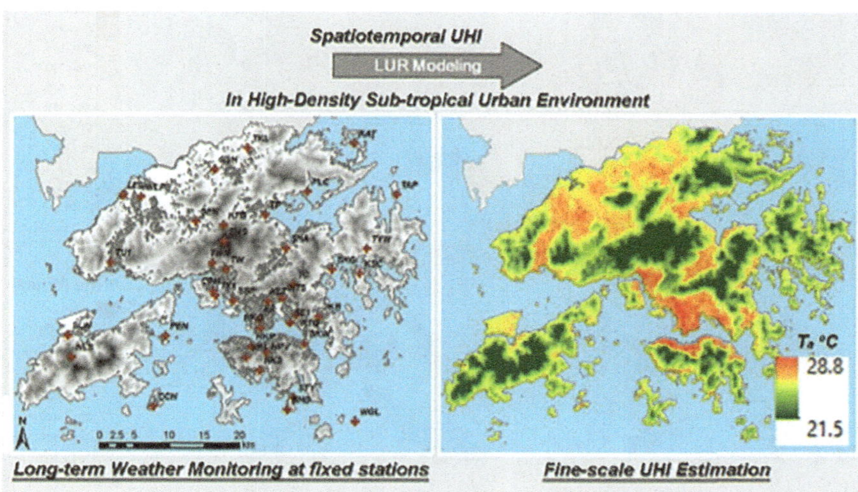

Fig. 2 Locations of long-term weather monitoring stations and spatial mapping of UHI. *Source* [24]

2015. This study found that the five-year average annual VHDHs and HNHs were spatially variant and the distributions were also driven by urban surface morphology. Fine-scale and community-level mapping results, shown in Fig. 3, were produced separately for building energy consumption analysis and public health measures. From Fig. 3, the daytime extreme heat distribution is significantly influenced by traffic heat emission from roads, and the nighttime condition shows a more obvious contrast between built-up areas and non-built-up areas. Downtown areas and new towns were faced with significantly greater nocturnal extreme heat due to slow heat loss in built-up areas, which was embodied in less open space but lower morphological permeability (higher FAD) in the modeling. This study contributes to not only a spatial understanding of extreme hot weather conditions in Hong Kong but also the development of the hourly based extreme heat indicators that are applicable and transferable to other places.

Considering the synergies between UHIs and heat waves [27, 28], Ren et al. [19] identified that VHD- and HN-based extreme heat events in the light of Wang et al.'s [17] study could enhance both the UHI and urban cool island (UCI) effects. Selected weather stations were categorized into urban, urban oasis, suburban, and rural classes for (1) estimating the diurnal and inter-annual variations in UHI/UHI intensities, (2) mapping city-wide UHI/UCI degree hours (UHIdh and UCIdh) [29], and (3) exploring the association of wind with UHI. All related results were contrasted between days with and without extreme heat in the period of 2000–2018. Comparatively, when extreme heat events occurred, enhanced daytime UCI and nighttime UHI were identified. However, in the urban context, the annual gaps in both UHI and

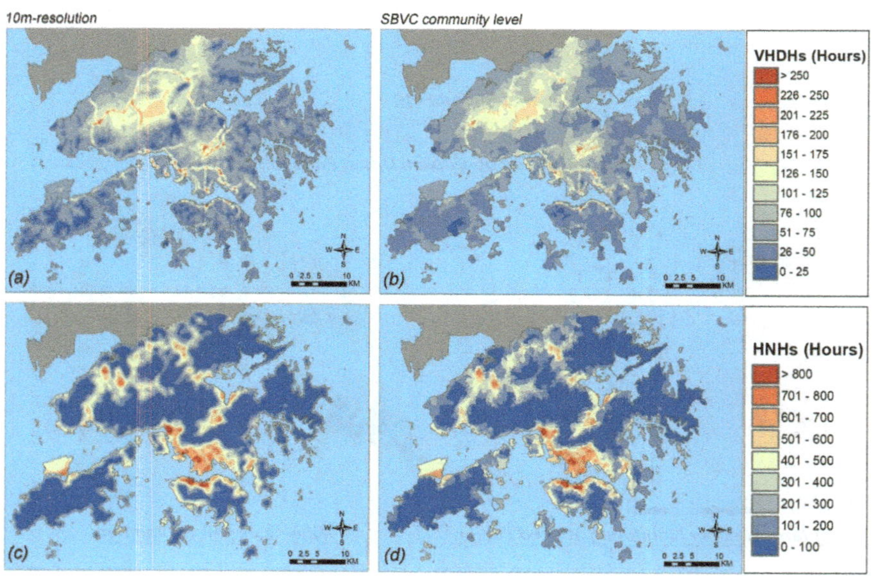

Fig. 3 The 10 m-resolution LUR mappings of the VHDHs (**a**) and HNHs (**c**) and the resultant spatial maps of the VHDHs (**b**) and HNHs (**d**) at the community level of Hong Kong. Source [20]

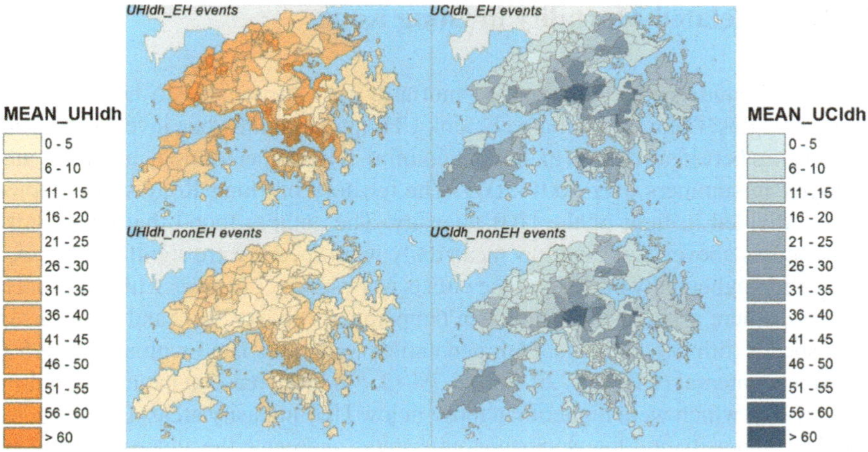

Fig. 4 The spatial maps of UCIdh and UHIdh in Hong Kong under EH and non-EH events at the TPU level. Source [19]

UCI intensities have turned to be narrow between periods with and without extreme heat in the current warming trend. Following the mapping studies introduced before, this study employed LUR to map the spatial patterns of UHIdh and UCIdh at the tertiary planning unit (TPU) level. Results show that during extreme heat events, the UHI effect was pushed up, and it was more notable in high-density downtown areas (Fig. 4), which is consistent with the two foregoing studies' findings. The wind environment, specifically, windward/leeward index (WLI), was found to be one of the explanatory spatial predictors, which also contained elevation, whose effect was also confirmed by the foregoing studies, vegetation coverage, and building compactness, of UHIdh and UCIdh in extreme heat days. Temporally, air ventilation could better reduce the UHI effect in the nighttime during extreme heat. Overall, wind conditions play an important role in heat mitigation in cities.

Overall, the overheating in Hong Kong under a high-density urban context, synergistically contributed by the UHI effect and extreme heat events, has been found to be temporally increasing and spatially variant. The city is experiencing an accelerating growth of both very hot days and hot nights. Meanwhile, the complex hilly terrain and compact urban form make the built-up areas, especially, downtown areas, bear much higher heat stress, particularly in the nighttime. Lower building permeability and lower vegetation coverage jointly weaken urban cooling [9, 30]. The wind environment in terms of wind speed and direction has been found to be influential to the UHI intensity. In addition, anthropogenic heat emission also makes a large contribution to the UHI effect. All the findings are informative of both mitigation and adaptation planning and measures to reduce heat-related health impacts.

3 Heat-Related Mortality in Hong Kong

A series of studies have investigated mortality risk attributable to heat in Hong Kong. Goggins, Chan [31] initiatively took UHI into account when investigating the relationship between short-term thermal comfort-related meteorological factors and mortality in summers over 2001–2009. The territory of Hong Kong was dichotomously divided in light of the UHI intensity. The authors found that mean daily temperatures above 29 °C and low mean daily wind speeds were significantly associated with higher daily mortality. The effects were stronger in areas with a high UHI intensity, where a 1°C increase in mean temperatures above 29 °C and lower mean wind speeds (5th percentile vs. 95th percentile) were associated with a 4.1% (95% confidence interval (CI) 0.7–7.6) and a 5.7% (2.7–8.9) increase in natural mortality, respectively, which were higher than those in low UHI intensity areas. Furthermore, high UHI intensity and low-socioeconomic status aggregate the adverse impacts of high temperature and low wind speed, as shown in Fig. 5.

Not limited to short-term meteorological factors, prolonged exposure to extreme heat can cause increased mortality risks [32, 33]. Ho, Lau [16] defined six types of prolonged extreme heat events based on the occurrence of VHDs and HNs within days between 2007 and 2014 and compared the temperature-mortality relationships of the different types by using a time-stratified analysis. They found up to 7.99% (95% CI 7.64–8.35) increase in all-cause mortality lagged for 0–3 days for 1 °C rise in daily minimum air temperature (T_{min}) after five consecutive NHs. Under the circumstances of non-consecutive heat, after seven days including at least five VHDs and five HNs, immediate excess mortality lagged for 0–1 day (15.61%, 14.52–16.70) for 1 °C rise in T_{min} was found in contrast to mortality lagged for 2–3 days (−2.00%, −2.83 to −1.17). This study uncovered variations in the temperature-mortality association across different types of prolonged extreme heat events that are comprised of consecutive or intermittent VHDs and/or NHs.

Given the temperature-mortality relationships in prolonged extreme heat events, a further study focused on the relative risk of mortality attributed to exposure to prolonged extreme heat events [17]. By extracting VHDs and HNs from a ten-year period of 2006–2015, Wang et al. identified various types of extreme heat events in terms of lengths of consecutive days, occurrence intervals of non-heat between

	Temp>29°C. (per 1°C. increase) % Increase (95% CI)	P-value	Mean Wind Speed (5th %ile vs. 95th %ile) % Increase (95% CI)	P-value
Hot/Low SES	5.6 (0.5, 10.9)	.028	8.0 (3.6, 12.7)	.0003
Cool/Low SES	2.6 (−1.9, 7.3)	.26	3.3 (−1.0, 7.7)	.13
Hot/High SES	3.0 (−1.7, 8.0)	.22	2.6 (−1.6, 7.1)	.23
Cool/High SES	−1.2 (−5.2, 3.1)	.59	−3.2 (−7.4, 0.8)	.12

doi:10.1371/journal.pone.0038551.t003

Fig. 5 Results of generalized additive models for all natural causes mortality in Hong Kong June–September, 2001–2009 stratified by SES and UHI of TPU of residence. Source [31]

VHDs or HNs, and combination of VHDs and HNs. By using Poisson generalized additive models controlling for long-term and short-term trends and allowing lag effects, the relative risk was estimated for the entire population and different gender and age groups. Apart from one single HN, five or more consecutive VHDs or HNs and 2D3N showed notably greater mortality risks. The so-called 2D3N with an increased risk of 5.32% (95% CI 1.83–8.93) for the entire population, the effect of which could significantly lag for five days, was considered representative in Hong Kong. In addition, across the significant event types mentioned above, female and elderly people were found to have higher risks than their counterparts and therefore considered to be vulnerable groups.

As we discussed before, besides air temperature, the wind environment is also associated with urban heat conditions and people's thermal comfort. In a very recent study, Wang, Goggins [34] quantified the long-term relationship between mortality and air ventilation, which was mirrored by the urban morphology indicator, FAD, for different demographic groups and death causes at the TPU level in Hong Kong. The indicator has been used to measure urban permeability and air ventilation [9]. High FAD in high building density contexts means low air ventilation, which could hinder heat dissipation in summers, within building blocks. Results show a 10% (95% CI 2–19) increase in all-cause mortality and a 21% (2–45) increase in death caused by asthma during 2008–2014 significantly associated with an interquartile increase in FAD when all seasons were accounted for.

The studies mentioned above revealed significantly higher mortality risks associated with Hong Kong's adverse thermal conditions in terms of the UHI intensity, air temperature, the wind environment, and extreme heat events. Not limited to high temperature, air ventilation and prolonged extreme heat events also tip the scales. Thanks to the local measurement of extreme heat events, specific patterns of persistent hot weathers with greater mortality risks in Hong Kong were identified, though similar effects have been recognized in other places [35, 36]. Besides, the mortality effect of air ventilation was quantified in a pioneering study using FAD as a proxy.

4 Spatiotemporal Pattern of Extreme Heat-Health Risk

After the investigations of the spatiotemporal patterns and mortality effects of overheating, an overall spatiotemporal assessment of the extreme heat-related health risk can inform urban planning and health actions for risk reduction [37]. Regarding the spatial pattern of the heat-health risk, both extreme heat conditions and the exposure of **vulnerable** population matter. Hence, Hua and his colleagues [4] developed an assessment accounting for three risk-related components—hazard, exposure, and vulnerability—based on the Crichton's Risk Triangle [38] as well as the Intergovernmental Panel on Climate Change's (IPCC) risk conception [39]. The annual cumulative VHDHs and HNHs, population density, and demo-socioeconomic vulnerability factors were integrated to estimate the extreme heat risk. The daytime and nighttime heat risks were separately classified and mapped for about 150 communities represented by the Large Tertiary Planning Unit (LTPU) for 2006, 2011, and 2016. Spatially variant patterns of heat risks were found (Fig. 6), and high-risk spots were identified for daytime and nighttime with underlying determinants behind them. Overall, high-risk spots were mainly distributed in the core urban areas of Kowloon and the northern coast of Hong Kong Island. Despite more high-risk spots in the nighttime, several old communities in Kowloon kept at high risk all day within the ten years. In addition, some new towns in the New Territories showed a rising trend in the heat risk. Targeted adaptation strategies are expected to be arranged for high-risk communities featured with specific driving factors.

5 Discussion

5.1 Implications to Heat-Health Impacts

Although cold weather has been found associated with higher mortality risks than hot weather in Hong Kong [40], given the warming trend, the health risks of hot weather should not be looked down. The higher risk of cold weather is possibly because Hong Kong residents have acclimatized to the warm climate and are more physically sensitive to cold weather [41, 42], especially senior people. Liu, Hansen [40] found a reversed J-shaped temperature-mortality relationship in Hong Kong. According to this study, relative to the model-specific minimum mortality temperatures, moderately high temperature (the 90th percentile, 29.53 °C) was related to higher relative risks and more deaths of injuries and mental and behavioral disorders with respective attributable fractions of 4.74% (95% CI 1.01–7.93) and 4.49% (0.54–7.84) than extremely high temperature (the 99th percentile, 30.60 °C). After losses caused by moderately hot weather, measures are generally taken to protect people from extremely hot weather. However, when adverse conditions are superimposed or extended as described in the studies in Sect. 3, such as high temperature or low wind speed in areas with high UHI intensities, prolonged extreme heat events, consecutive VHDs and/or HNs, or long-term poor ventilation, mortality risks can be pushed up, especially for vulnerable population groups.

Hot nights deserve special attention. Due to the recent trend of increasing VHDs and HNs in Hong Kong, there will be more consecutive occurrences of HNs, some of which are accompanied by intermittent VHNs, within a period of days [2, 17]. The findings from recent studies mentioned in Sect. 3 support the cumulative effects of extreme heat events with more hot nights on mortality in Hong Kong [16, 17]. Future studies on heat-health effects are expected to take long-term and short-term acclimatization to hot nights into account [43], given that the warming trend will last and the occurrences of VHDs and HNs have been extending to May and September in Hong Kong's long summers.

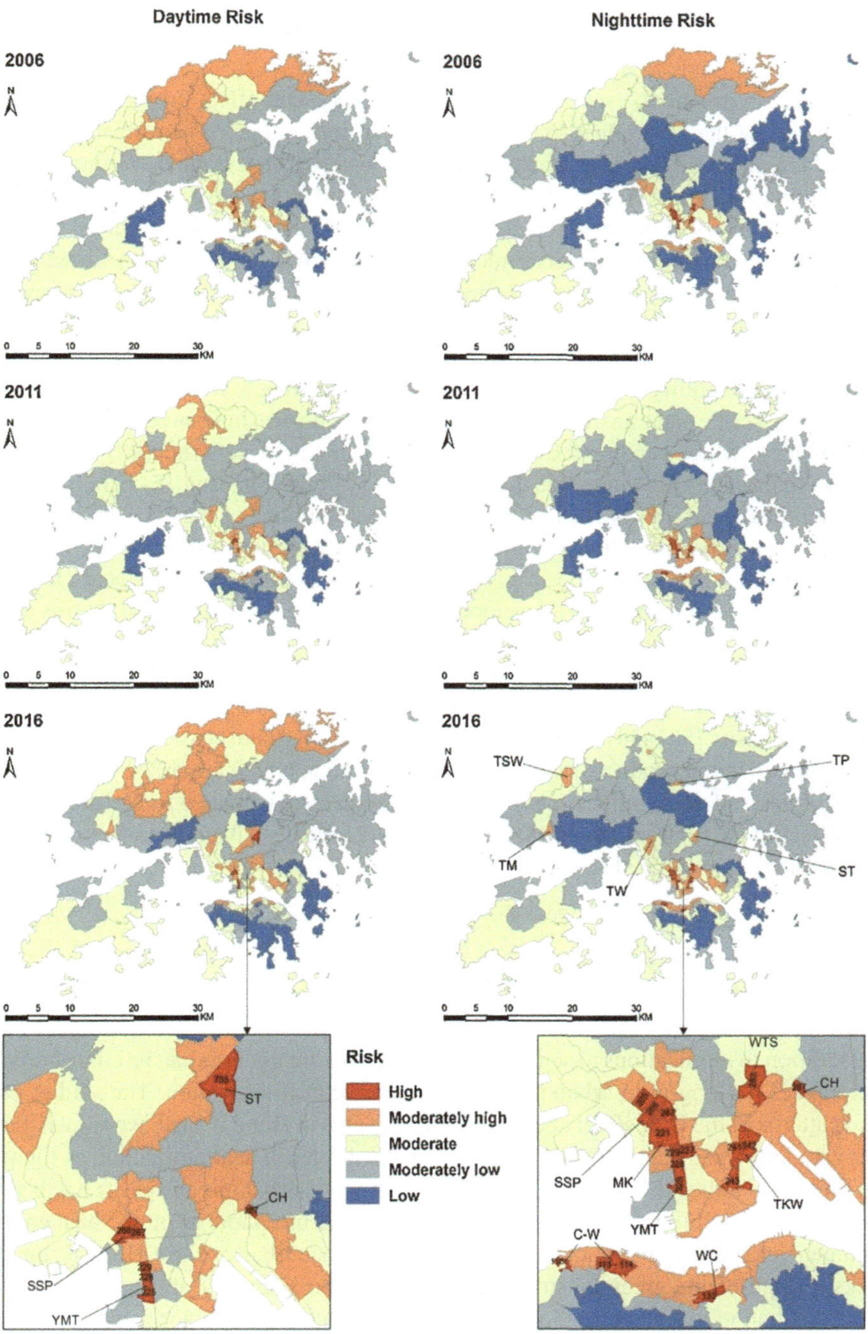

Fig. 6 Daytime and nighttime extreme heat risk maps of Hong Kong for 2006, 2011, and 2016. Source [4]

5.2 Implications to Urban Heat Island Studies

The built environment, especially urban morphology, can affect urban climates [5, 44] including the UHI effect [6, 45, 46] and therefore influence heat-related health risks [47]. In Hong Kong's complex and high-density urban context, various urban morphology parameters, e.g., BVD, SVF, FAD, and WLI, have been found to be influential factors of the spatial pattern of the UHI effect and heat extremes [19, 20, 24, 48] through pathways towards microclimate conditions [49], such as air temperature, humidity, wind direction and speed, and solar radiation, which further impact people's thermal comfort [50, 51]. Increasing attention has recently been given to air ventilation, which is shaped by urban morphology, specifically, the FAD measuring building permeability [9] and linked with heat-related health mortality as reported. Hence, fine-scale and transferable extraction of three-dimensional urban morphology information, which can be used to investigate urban climates, particularly the urban wind environment, is suggested by Ren, Cai [52] and Xu, Ren [5]. The anthropogenic heat emission is found to be another determinant of the UHI intensity in Hong Kong. Urban morphology influentially modifies air flow and therefore impacts anthropogenic heat dispersion in high-density urban areas [53]. The local climate zone (LCZ) classification [44] developed to categorize urban surface structures and covers, with urban morphology information inputs, has been increasingly adopted for studies on urban climates, especially, thermal conditions [54, 55]. The built types of LCZs featured with different morphological properties vary in the anthropogenic heat flux density [56] and air ventilation [57]. Chao Ren's team has developed various LCZ classification products for Hong Kong and China, which are based on different approaches and data sources [58–61]. Future studies are expected to be focused on the effect of air ventilation on urban climate at local scales taken into account urban morphology that can be holistically described by LCZss.

5.3 Implications to Urban Planning

In addition to implications to scientific research, existing findings can be contributive references for long-term urban planning to improve heat mitigation. The studies on the patterns of the UHI effect and extreme heat events in Hong Kong have shown that air ventilation and vegetation cover play important roles in heat mitigation, as found in previous studies [30, 62, 63]. The cooling effect of urban green spaces has been well documented [64, 65]. Given the effects of urban morphology on the heat environment and limited green space supply in the compact urban areas of Hong Kong, it asks for optimizing greening strategies to create thermal comfortable environments taking urban morphology into account [66, 67]. Constructing breezeways in urban areas can help wind take away not only heat but also air pollutants [9]. The effects of urban morphology on air ventilation have been long studied for improving the urban heat environment in Hong Kong's urban planning practice. A Feasibility Study for

Establishment of Air Ventilation Assessment (AVA) System [68] was driven by the 2003 severe acute respiratory syndrome (SARS) to ameliorate air ventilation for public health [69]. Air ventilation design and assessment were then involved into the Hong Kong Planning Standards and Guidelines for planning practitioners [70] and the Sustainable Building Design Guidelines [71] and have also been implemented in other Chinese cities [72]. Directed by Edward Ng, a professional team investigated the feasibility of developing standards for wind environments by using the tool of urban climatic maps with substantial work in Hong Kong [73–75]. The recent HKGBC Guidebook on Urban Microclimate Study released by the Hong Kong Green Building Council [76] illustrates a series of practical strategies from aspects of wind, thermal radiation, temperature, and precipitation for microclimate design in the outdoor environment. Specifically, the guidebook recommends increasing ventilation by, e.g., increasing permeability of blocks and buildings, and different methods using vegetation for cooling, such as providing tree canopies and installing green walls.

5.4 Implications to the Energy Sector

Urban overheating contributes to the energy consumption increase in Hong Kong. In a subtropical climate, Hong Kong faces high cooling demand in the summer season. When extreme heat events occur, it turns to be worse. Morakinyo, Ren [77] estimated that extreme heat events contributed to an 80–140% rise in cooling degree days during 2011–2015. Regarding future consumption, considering shared socioeconomic pathways, Liu, Zeng [78] predicted that relative to 2018, Hong Kong's residential and commercial sectors' electricity demands would separately grow by 89.40% and 54.34% in the 2090s under the representative concentration pathway 8.5. Urban overheating and consequent increased energy consumption challenge the reliability of power supply and elevate the risk of power outages. Power outages happening in heat waves increase heat exposure and therefore heat-related mortality and morbidity risks in a short time, for example, those occurred in New York City in 2003 and 2006 [79, 80]. Therefore, power supply infrastructure in densely populated communities that are exposed to more heat should be well maintained.

5.5 Implications to Health Actions

Existing findings of the spatial and temporal patterns of UHIs, extreme heat conditions, and the extreme heat risk and the mortality effects can serve as scientific references for relevant decision-makers and stakeholders at different levels in taking right health actions in the right place at the right time for the right persons. For dealing with extreme heat events, communities need targeted responses, e.g., opening and increasing cooling centers in hot areas, taking special care of susceptible groups in vulnerable communities, and making preparedness of the public health system in high-risk areas. Considering the exacerbated effect of prolonged extreme heat events, prompt heat alerts and action responses are of higher significance. Including warnings and advisories for very hot weather, the HKO has provided a series of climate services [81], especially information services for senior citizens with a collaboration with the Senior Citizen Home Safety Association [82]. With the UHI effect and extreme heat events increasing, cross-level and cross-sector collaborations will be badly needed to improve health actions.

6 Summary

Hong Kong suffers from the intense UHI effect and increasing extreme heat events, both of which have been found to be spatially variant and strongly related to urban morphology in the high-density urban context. The hot conditions in terms of UHI intensities, daytime and nighttime heat extremes, and air ventilation are driving factors of mortality risks. Spatially integrating extreme heat hazards and exposure of vulnerable populations helped identify high-risk communities. Research findings provide informative reference for local strategies on urban planning and climate responses. The findings are evidence for improving the urban microclimate and green and blue infrastructure through optimizing urban planning and design to strengthen the urban fabric for climate adaptation required by the "Hong Kong's Climate Action Plan 2030+" [83] and further achieve the goal for a livable compact city as promised in the "Hong Kong 2030+: Towards a Planning Vision and Strategy Transcending 2030" [84].

References

1. Ebi KL et al (2021) Hot weather and heat extremes: health risks. The Lancet 398(10301):698–708
2. Lee T-C, Chan K-Y, Ginn W-L (2011) Projection of extreme temperatures in Hong Kong in the 21st century. Acta Meteor Sin 25(1):1–20
3. PlanD (2020) Land utilization in Hong Kong 2019. Available from: https://www.pland.gov.hk/pland_en/info_serv/statistic/landu.html
4. Hua J et al (2021) Spatiotemporal assessment of extreme heat risk for high-density cities: a case study of Hong Kong from 2006 to 2016. Sustain Cities Soc 64:102507
5. Xu Y et al (2017) Urban morphology detection and computation for urban climate research. Landsc Urban Plan 167:212–224
6. Li Y et al (2020) On the influence of density and morphology on the Urban Heat Island intensity. Nat Commun 11(1):2647
7. PlanD (2012) Urban climatic map and standards for wind environment—feasibility study. Planning Department, HKSAR Government
8. Chen L et al (2012) Sky view factor analysis of street canyons and its implications for daytime intra-urban air temperature differentials in high-rise, high-density urban areas of Hong Kong: a GIS-based simulation approach. Int J Climatol 32(1):121–136

9. Ng E et al (2011) Improving the wind environment in high-density cities by understanding urban morphology and surface roughness: a study in Hong Kong. Landsc Urban Plan 101(1):59–74
10. Laaidi K et al (2012) The impact of heat Islands on mortality in Paris during the August 2003 heat wave. Environ Health Perspect 120(2):254–259
11. Murage P, Hajat S, Kovats RS (2017) Effect of night-time temperatures on cause and age-specific mortality in London. Environ Epidemiol 1(2):e005
12. Royé D (2017) The effects of hot nights on mortality in Barcelona Spain. Int J Biometeorol 61(12):2127–2140
13. Lee J et al (2019) Is 'tropical night' an effective Heat-Health Indicator of Heatwave for South-East Asian cities? Environ Epidemiol 3:203
14. HKO (2021) Climate change in Hong Kong—extreme weather events. Available from: https://www.hko.gov.hk/en/climate_change/obs_hk_extreme_weather.htm
15. Wong MC, Mok HY, Lee TC (2011) Observed changes in extreme weather indices in Hong Kong. Int J Climatol 31(15):2300–2311
16. Ho HC et al (2017) Characterizing prolonged heat effects on mortality in a sub-tropical high-density city Hong Kong. Int J Biometeorol 61(11):1935–1944
17. Wang D et al (2019) The impact of extremely hot weather events on all-cause mortality in a highly urbanized and densely populated subtropical city: a 10-year time-series study (2006–2015). Sci Total Environ 690:923–931
18. Lee KL et al (2015) The development of the Hong Kong Heat Index for enhancing the heat stress information service of the Hong Kong Observatory. Int J Biometeorol 60(7):1029–1039
19. Ren C et al (2021) Investigating the urban heat and cool island effects during extreme heat events in high-density cities: a case study of Hong Kong from 2000 to 2018. Int J Climatol 41(15):6736–6754
20. Shi Y et al (2019) Assessing spatial variability of extreme hot weather conditions in Hong Kong: a land use regression approach. Environ Res 171:403–415
21. Zhang X et al (2019) Towards a comprehensive heat-health warning system of Hong Kong: an evidence-based multi-stage approach. Environ Epidemiol 3:463
22. HKO (2021) Climate change in Hong Kong—temperature. Available from: https://www.hko.gov.hk/en/climate_change/obs_hk_temp.htm
23. An N, Zuo Z (2021) Changing structures of summertime heatwaves over China during 1961–2017. Sci China Earth Sci 64(8):1242–1253
24. Shi Y, Katzschner L, Ng E (2018) Modelling the fine-scale spatiotemporal pattern of urban heat island effect using land use regression approach in a megacity. Sci Total Environ 618:891–904
25. Tomlinson CJ et al (2011) Remote sensing land surface temperature for meteorology and climatology: a review. Meteorol Appl 18(3):296–306
26. Cheng W et al (2021) Approaches for identifying heat-vulnerable populations and locations: A systematic review. Sci Total Environ 799
27. Li D, Bou-Zeid E (2013) Synergistic Interactions between Urban Heat Islands and Heat Waves: the impact in cities is larger than the sum of its parts. J Appl Meteorol Climatol 52(9):2051–2064
28. Ramamurthy P, Bou-Zeid E (2017) Heatwaves and urban heat islands: a comparative analysis of multiple cities. J Geophys Res: Atmos 122(1):168–178
29. Yang X et al (2017) The urban cool island phenomenon in a high-rise high-density city and its mechanisms. Int J Climatol 37(2):890–904
30. Ng E et al (2012) A study on the cooling effects of greening in a high-density city: an experience from Hong Kong. Build Environ 47:256–271
31. Goggins WB et al (2012) Effect modification of the association between short-term meteorological factors and mortality by urban heat islands in Hong Kong. PLoS ONE 7(6):e38551
32. Anderson BG, Bell ML (2011) Heat Waves in the United States: mortality risk during heat waves and effect modification by heat wave characteristics in 43 U.S. Communities. Environ Health Perspect 119(2):210–218
33. Gasparrini A, Armstrong B (2011) The impact of heat waves on mortality. Epidemiology 22(1):68–73

34. Wang P et al (2021) Long-term association between urban air ventilation and mortality in Hong Kong. Environ Res 197:111000
35. Yin Q, Wang J (2017) The association between consecutive days' heat wave and cardiovascular disease mortality in Beijing, China. BMC Public Health 17(1):223
36. Rocklöv J, Ebi K, Forsberg B (2011) Mortality related to temperature and persistent extreme temperatures: a study of cause-specific and age-stratified mortality. Occup Environ Med 68(7):531
37. Wolf T, McGregor G (2013) The development of a heat wave vulnerability index for London United Kingdom. Weather Clim Extremes 1:59–68
38. Crichton D (1999) The risk triangle. In: Ingleton J (ed) Natural disaster management. Tudor Rose Holdings Limited, Leicester, UK, pp 102–103
39. IPCC, Climate Change 2014: Impacts, adaptation and vulnerability. Part A: global and sectoral aspects. contribution of working group ii to the fifth assessment report of the intergovernmental panel on climate change. 2014, Cambridge, UK and New York, NY, USA, Cambridge University Press
40. Liu J et al (2020) Cause-specific mortality attributable to cold and hot ambient temperatures in Hong Kong: a time-series study, 2006–2016. Sustain Cities Soc 102131
41. Anderson BG, Bell ML (2009) Weather-related mortality: how heat, cold, and heat waves affect mortality in the United States. Epidemiology 20(2):205–213
42. Yi W, Chan APC (2015) Effects of temperature on mortality in Hong Kong: a time series analysis. Int J Biometeorol 59(7):927–936
43. Medina-Ramón M, Schwartz J (2007) Temperature, temperature extremes, and mortality: a study of acclimatisation and effect modification in 50 US cities. Occup Environ Med 64(12):827
44. Stewart ID, Oke TR (2012) Local climate zones for urban temperature studies. Bull Am Meteor Soc 93(12):1879–1900
45. Steeneveld GJ et al (2011) Quantifying urban heat island effects and human comfort for cities of variable size and urban morphology in the Netherlands. J Geophys Res: Atmos 116(D20)
46. Touchaei AG, Wang Y (2015) Characterizing urban heat island in Montreal (Canada)—Effect of urban morphology. Sustain Cities Soc 19:395–402
47. Ellena M, Breil M, Soriani S (2020) The heat-health nexus in the urban context: a systematic literature review exploring the socio-economic vulnerabilities and built environment characteristics. Urban Clim 34:100676
48. Yuan C et al (2016) A modelling-mapping approach for fine-scale assessment of pedestrian-level wind in high-density cities. Build Environ 97:152–165
49. Tong S et al (2017) Impact of urban morphology on microclimate and thermal comfort in northern China. Sol Energy 155:212–223
50. Huang C-H, Tsai H-H, Chen H-C (2020) Influence of weather factors on thermal comfort in subtropical urban environments. Sustainability 12(5):2001
51. Rupp RF, Vásquez NG, Lamberts R (2015) A review of human thermal comfort in the built environment. Energy Build 105:178–205
52. Ren C et al (2020) Developing a rapid method for 3-dimensional urban morphology extraction using open-source data. Sustain Cities Soc 53
53. Yuan C et al (2020) Mitigating intensity of urban heat island by better understanding on urban morphology and anthropogenic heat dispersion. Build Environ 176
54. Lehnert M et al (2021) Mapping local climate zones and their applications in european urban environments: a systematic literature review and future development trends. ISPRS Int J Geo-Inf 10(4)
55. Xue J et al (2020) Applications of local climate zone classification scheme to improve urban sustainability: a bibliometric review. Sustainability 12(19)
56. Chen G et al (2021) Future "local climate zone" spatial change simulation in Greater Bay Area under the shared socioeconomic pathways and ecological control line. Build Environ 203
57. Yang J et al (2019) Local climate zone ventilation and urban land surface temperatures: Towards a performance-based and wind-sensitive planning proposal in megacities. Sustain Cities Soc 47

58. Chung LCH, Xie J, Ren C (2021) Improved machine-learning mapping of local climate zones in metropolitan areas using composite earth observation data in Google Earth Engine. Build Environ 107879
59. Ren C et al (2019) Assessment of local climate zone classification maps of cities in China and feasible refinements. Sci Rep 9(1):18848
60. Wang R et al (2018) Mapping the local climate zones of urban areas by GIS-based and WUDAPT methods: a case study of Hong Kong. Urban Clim 24:567–576
61. Zheng Y et al (2018) GIS-based mapping of local climate zone in the high-density city of Hong Kong. Urban Climate 24:419–448
62. Wong MS et al (2010) A simple method for designation of urban ventilation corridors and its application to urban heat island analysis. Build Environ 45(8):1880–1889
63. Ng E, Cheng V (2012) Urban human thermal comfort in hot and humid Hong Kong. Energy and Buildings 55:51–65
64. Bowler DE et al (2010) Urban greening to cool towns and cities: a systematic review of the empirical evidence. Landsc Urban Plan 97(3):147–155
65. Gunawardena KR, Wells MJ, Kershaw T (2017) Utilising green and bluespace to mitigate urban heat island intensity. Sci Total Environ 584–585:1040–1055
66. Tan Z, Lau KK-L, Ng E (2017) Planning strategies for roadside tree planting and outdoor comfort enhancement in subtropical high-density urban areas. Build Environ 120:93–109
67. Tan Z, Lau KK-L, Ng E (2016) Urban tree design approaches for mitigating daytime urban heat island effects in a high-density urban environment. Energy Build 114:265–274
68. Ng E (2007) Feasibility study for establishment of air ventilation assessment system (AVAS). J HKIP 22:39–45
69. Ng E (2009) Policies and technical guidelines for urban planning of high-density cities—air ventilation assessment (AVA) of Hong Kong. Build Environ 44(7):1478–1488
70. PlanD (2020) Chapter 11 urban design guidelines, in Hong Kong planning standards and guidelines. 2020, Planning Department, HKSAR Government
71. BD (2016) Sustainable building design guidelines. 2016, Building Department, HKSAR Government
72. Ren C et al (2018) Creating breathing cities by adopting urban ventilation assessment and wind corridor plan—The implementation in Chinese cities. J Wind Eng Ind Aerodyn 182:170–188
73. Ng E, Ren C (2015) The urban climatic map: a methodology for sustainable urban planning. Routledge
74. Ng E, Ren C, Katzschner L (2009) Urban climatic mapping in Hong Kong. J Heat Island Inst Int 7(2):55–64
75. Ren C, Ng EY, Katzschner L (2011) Urban climatic map studies: a review. Int J Clim 31(15):2213–2233
76. HKGBC (2017) HKGBC guidebook on urban microclimate study. Hong Kong Green Building Council Limited
77. Morakinyo TE et al (2019) Estimates of the impact of extreme heat events on cooling energy demand in Hong Kong. Renewable Energy 142:73–84
78. Liu S et al (2021) Predicting long-term monthly electricity demand under future climatic and socioeconomic changes using data-driven methods: A case study of Hong Kong. Sustain Cities Soc 70
79. Dominianni C et al (2018) Health impacts of citywide and localized power outages in New York City. Environ Health Perspect 126(6):067003
80. Lin S et al (2011) Health impact in New York City during the Northeastern blackout of 2003. Public Health Rep 126(3):384–393
81. Lee T-C, Wong W-K, Tam K-H (2018) Urban-focused weather and climate services in Hong Kong. Geosci Lett 5(1):18
82. Lee TC, Leung I (2016) Protecting the elderly from heat and cold stress in Hong Kong: using climate information and client-friendly communication technology, case 3B, in Shumake-Guillemot J, Fernandez-Montoya L (eds) Climate services for health: improving public health decision-making in a new climate. WHO/WMO, Geneva

83. Environment Bureau (2017) Hong Kong's Climate Action Plan 2030+. Environment Bureau, HKSAR Government
84. PlanD (2016) Hong Kong 2030+: planning and urban design for a liveable high-density city. Planning Department, The Government of the Hong Kong Special Administrative Region

Chapter 15
The Trend of Heat-Related Mortality in European Cities

Fabrizio Ascione, Rosa Francesca De Masi, Margherita Mastellone, Mattheos Santamouris, Francesco Tariello, and Giuseppe Peter Vanoli

Abstract Climate change is one of the main issues affecting the earth, and many effects appear in our cities, with serious consequences for our survival. The temperature growth acts synergically to the urban heat island in generating urban overheating during extreme climatic events. It is also well-known that urban overheating can widely affect our wellbeing compromising our health. Indeed, additional energy demands, due to growing urban temperatures, have an impact on environmental quality and contribute to increased vulnerability and energy poverty of cities. Disproportionate use of plants to satisfy high energy demands of buildings during heat waves causes a large increase of pollutant emission, consequently exacerbating health, and respiratory problems; at the same time, the need for healthy and comfortable indoor environments, even to face health problems, results in more intense use of

F. Ascione (✉)
Department of Industrial Engineering, DII—Università degli Studi Di Napoli Federico II, Napoli 80125, Italy
e-mail: fabrizio.ascione@unina.it

R. F. De Masi
Department of Engineering—DING, Università degli Studi del Sannio, Benevento 82100, Italy
e-mail: rfdemasi@unisannio.it

M. Mastellone
Department of Architecture, DIARC, Università degli Studi di Napoli Federico II, Napoli 80134, Italy
e-mail: margherita.mastellone@unina.it

M. Santamouris
Faculty of Built Environment, University of New South Wales, Sydney, NSW 2052, Australia
e-mail: m.santamouris@unsw.edu.au

F. Tariello
Department of Agricultural, Environmental and Food Sciences, Università degli Studi del Molise, Campobasso 86100, Italy
e-mail: francesco.tariello@unimol.it

G. P. Vanoli
Department of Medicine and Health Sciences—Vincenzo Tiberio, Università degli Studi del Molise, Campobasso 86100, Italy
e-mail: giuseppe.vanoli@unimol.it

N. Aghamohammadi and M. Santamouris (eds.), *Urban Overheating: Heat Mitigation and the Impact on Health*, Advances in Sustainability Science and Technology,
https://doi.org/10.1007/978-981-19-4707-0_15

air-conditioning systems and consequently in peaks in energy demand. The synergic relation between climate change and urban overheating, energy, health, and pollution, was deepened in this chapter, starting from a large scale up to European cities. The serious consequences of global warming on an economic and environmental level were evaluated, and particular attention was paid to the close correlation between the increase in mortality rate and the earth's temperature rise.

Keywords Global overheating · Heat waves · Heat-related mortality · Urban heat islands · European cities

1 Intro: European Data, Cities, Climates, and Variation

Europe is one of the less populated continents of the World, apart from Oceania and the almost inhabited Antarctica, having a population of about 723 million persons (≈ 470 million in the EU Member States). On the other hand, Europe has a very high density of inhabitants (≈ 71 persons/km^2), being second only to Asia (100.1 persons/km^2). In Table 1 and Fig. 1, the main data and information about continents' population and size are reported. Here, we are using a common classification in which 6 continents of the World are identified, according to a historical and etymological criterion. This approach is widely used in Europe, and it evaluates a continent in a modern way, according to the current language; finally, the following six areas are the ones identified also by the United Nations Geoschema: Africa, Asia, Europe, America, Oceania, and Antarctica.

Even if access to energy is not available in the same way all around the World, however, through the artificial lighting of earth, the energy density of some world regions is quite evident. In particular, as evident in [1] Europe, Eastern U.S., and Japan show a quite high and uniform wide lighting, immediately noticeable. Of course, this is a measure of energy intensity, and it definitively underlines how Europe, together with other world areas, is characterized by a very high anthropogenic activity, population, density of persons, and energy usage. It is quite evident that the whole

Table 1 Size, population and occupancy density in the world

	Surface (km^2)	Population (–)	Density (pers/km^2)	Percentage of world area (%)	Percentage of world population (%)
Asia	44,579,000	4,462,676,731	100.1	29.7	59.2
Europa	10,180,000	723,000,000	71.0	6.8	9.6
Africa	30,221,532	1314,000,000	43.5	20.1	17.4
America	42,549,000	1,001,463,142	23.5	28.4	13.3
Oceania	8,525,989	40,117,432	4.7	5.7	0.5
Antarctica	14,000,000	1000	0.0	9.3	0.0

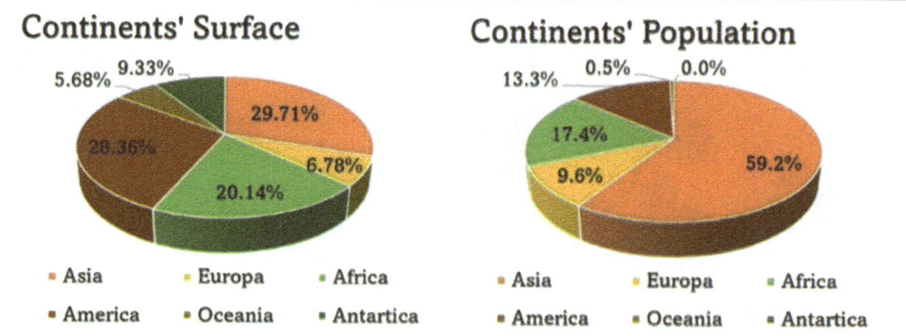

Fig. 1 Size and estimated population of the World's continents

continent, mainly on the coast lines of the Mediterranean area and North Sea, is overcrowded.

About the European climates, according to Peel et al. [2], in Europe there are 4 dominant climate types, on the basis of the Köppen-Geiger climate classification, namely the cold climate "*D*" ($\approx$ 44.4% of land), followed by arid "*B*" ($\approx$ 36.3%), temperate climate *C* ($\approx$ 17.0%), and polar one "*E*" ($\approx$ 2.3%). In the cited study, the European boundaries are, on the east side, the Urals, while, on the south, it includes the Arabian Peninsula and the Middle East countries, and thus, this explains the very significant "arid" quota BWh.

In Fig. 2, current and expected future climates of Europe (2071–2100) [3] are provided, concerning the measured rise of temperatures. By comparing Fig. 2a, b, the progressive increase of temperature can be seen that will determine a climate change, from current types to warmer ones, along the next years and decades. For instance, it becomes visible how many warm and mild areas of southern Europe would become tropical, arid, or desertic, along next years, according to the projections. Analogously, for the climatic conditions of the northern area of Europe, progressive and generalized overheating will carry on the transformation from the current conditions of cool continental climate to temperate ones.

About the climate change in Europe, it is quite accepted a generalized overheating also of the old continent. "Europe is getting warmer" [4]: This is a very interesting study of 2018. More in deep, the average temperature growth in the period 1900–2017 has been about 1 °C, even if in Italy, Spain, and Eastern European regions, an increase in the range 1.2–1.5 °C has been registered. In particular, in [4] it is shown that, with reference to 2003, about 70,000 extra deaths have been recorded because of the heat waves.

In the same study, it is inferred that, during 2017, the so-called Lucifer heatwave provided a significant excess of death in the Mediterranean regions, with temperatures above 40 °C in Spain, Italy, and Balkan countries. This study is performed by investigating data from the European Centre for Medium-Range Weather Forecasts (ECMWF). Really, summer temperatures, in July and August, above the thresholds

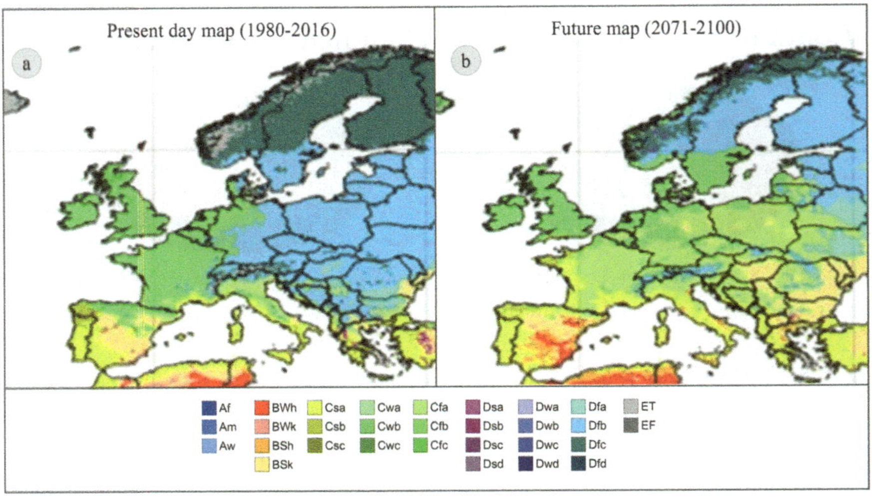

Fig. 2 Present **a** and future **b** climates of Europe, according to the Köppen-Geiger scheme. Extraction from the world maps of Beck et al. [3]. Color scheme adopted by Peel et al. [2]. License copyright link: http://creativecommons.org/licenses/by/4.0/

of 40 °C are quite common today, during the summer heatwave, which happens every year during July and August, so that these extremes are not exceptional.

At the global level, already about 20 years ago, a milestone review of Basu and Samet [5] related high temperatures of the environment to a clear increase in mortalities. The numbers are very impressive: More than 400 persons die, yearly, in the U.S. only (where the study is centered). The mortality increase due to heatwaves is primarily dependent on the high ambient temperatures.

According to the Web site https://www.onedegreewarmer.eu/list [6], 2018 has been, in more than 200 of the 558 investigated cities, the warmest experienced year, while, in the majority of these, this year was within the five warmest ones. The data and statistics have been defined, starting from the ECMWF values (more than 100,000,000 of measurements), and these are updated at spring 2019.

All implications of high temperature heat waves are really unpredictable. Direct implications are, obviously, heat-related mortalities but, as cited in [4], also quite unpredictable secondary effects can be noted, as the increment of violence and aggression [7]. These are results from a study performed in 1987 in the United States.

In Europe, evaluating the impact of extreme weather and other climate-related events on the economic sector, the Member States have lost around 499 billion EUR in the period between 1980 and 2019. Indeed, Fig. 3 (left side) reports the total amount of billion euro lost per country, and Fig. 3 (right side) shows the losses per capita in euro, due to the extreme climatic events for the EU members and UK. It is evident that Switzerland, Luxemburg, and Denmark have seen the highest losses per

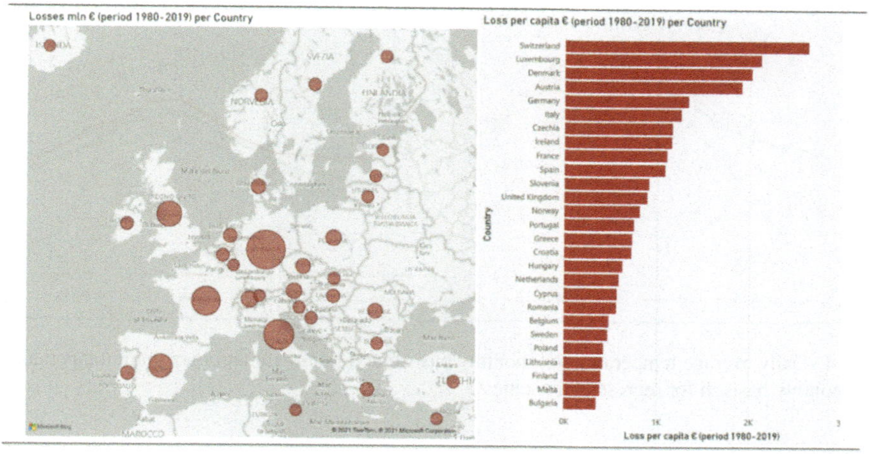

Fig. 3 Total amount of billion euro lost (on the left side) and losses per capita in euro, per country, in the period 1980–2019 due to extreme climatic events (on the right side). Data from European Environment Agency [8]

capita in the reference period (1980–2019) but Germany, Italy, and France have had the greatest economic losses [8].

With reference to the temperature rise, according to the data of [6], eight European cities, in the twenty-first century compared to the values of the twentieth century, have increased the average temperature by 1.5 °C or more. These are:

- Næstved, in Denmark, with an overall increase of temperate of 1.67 °C
- Helsinborg (Sweden, + 1.54 °C)
- Copenhagen (Denmark, + 1.54 °C)
- Malmö (Sweden, + 1.54 °C)
- Banja Luka (Bosnia and Herzegovina, + 1.51 °C)
- Linares (Spain, + 1.51 °C)
- Roskilde (Denmark, + 1.50 °C)
- Córdoba, Spain, + 1.50 °C).

Really, if the temperature increase threshold is 1 °C, then the cities that have recorded increased temperatures, in the twenty-first century compared to the previous one, are 317 on 558 [4].

As expressed by EDJNet [6], some comparisons and analyses required corrections, and, when datasets were not directly comparable, a reconciliation method has been necessary to compare the values. On average, the European capital city with the highest annual temperatures is La Valletta (+23.0 °C), followed by Athens (+22.3°), while the ones with the coolest ones are Helsinki (9 °C) and Moscow (+9.6 °C). Of course, by dimensions, Moscow and Athens are more representative.

By including in the analyses also two other cities, and thus Rome and Berlin, both with more than 2 million inhabitants, the average monthly temperatures and the extreme ones are inferred in Fig. 4.

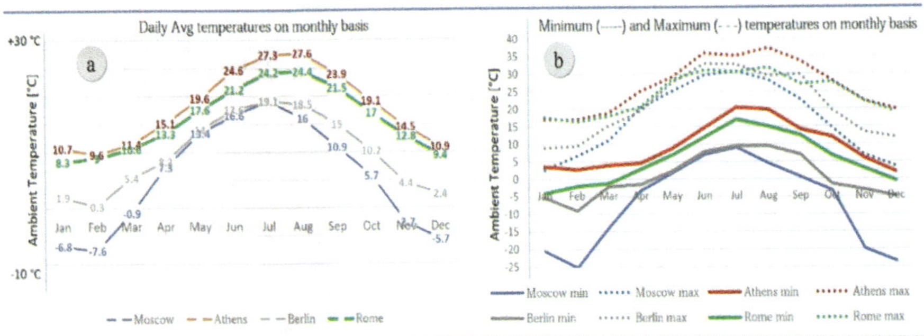

Fig. 4 Daily average temperature on monthly basis **a** and minimum and maximum temperatures on monthly basis, **b** for representative cities

It should be noted that data of Fig. 4 have been derived by the available IWEC file ".stat" from the database of Energy Plus [9]: these are the international weather data for energy calculations, developed by ASHRAE [10]. All around Europe, in the presence of the common heatwaves in summer, such maximum daily temperatures can rise also by about 10 °C with respect to the average values, and such rise is much more accentuated in the urban environments compared to the backcountry, because of the so-called phenomenon of urban heat islands, exacerbated by the city pollutions and heat, including the condensation heat from mechanical cooling systems [11], more and more spread all around the world, in both developed and underdeveloped countries.

Finally, "Europe is getting warmer" [4]. A such temperature increase is changing the climate, are determining collateral effects, and thus the risks for the survive of traditional species, the appearance of new animal and plant species, the alteration of ecosystems, several direct effects, and thus the intensification of extreme events, not only heatwaves, but also storms, hurricanes and violent and sudden weather phenomena.

At the end of October 2021, for the first time in Europe, Sicily, South Italy, was interested by an extremely violent storm, a so-called Medicane (i.e., Mediterranean hurricane), with winds with velocities higher than 100 km/h, destructions, and enormous damages for the nature and the built environments, besides losses in terms of human and animal lives. The same country, in the north-east part (i.e., Alpes and Dolomites), at the end of October 2018, has been interested in a devastating windstorm, with around 42,500 ha. of forest almost destroyed.

During the last summer, and thus August 2021 in Germany, torrential rains in Rhineland-Palatinate and North Rhine Westphalia caused more than 80 victims and hundreds of lost persons. Tens of deaths have been registered also in Belgium.

All around Europe, as it can be seen in Fig. 5, there is a progressive heating of the environment. The heating degree days are under a progressive reduction, while the cooling needs are increasing. This trend, progressively increasing, must be stopped. The necessary reduction and braking of climate change have been recognized as the

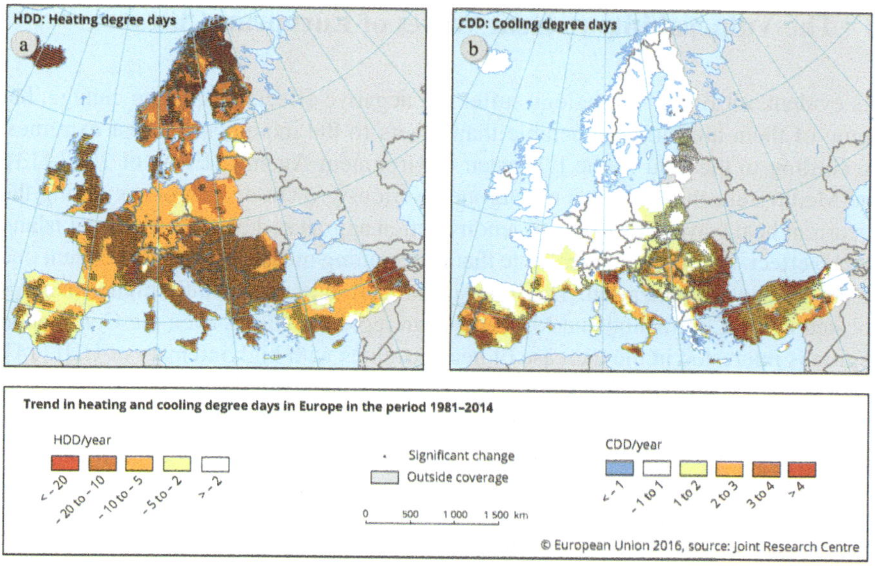

Fig. 5 Trend of European heating **a** and cooling, **b** degree day (1981–2014) by European Environment Agency (EEA) [12]

future, mandatory, and challenge to face, and this is recently (i.e., autumn 2021) established at both the G20 of Rome (Italy) and the COP26, conference of Party in Glasgow.

With reference to Fig. 5 (provided by [12]), the trend of HDD (Heating Degree Day) and CDD (Cooling Degree Day) for the period 1981–2014 is reported.

Heating Degree Day (HDD) is an indicator to describe how cold is a location. Higher the number of HDDs, higher is the thermal energy need for heating. Cooling Degree Day (CDD), analogously, is also a weather-based index, used for estimating how intense is the need for the space cooling of a building. Higher the number of CDDs, higher the cooling energy need. In Fig. 5, the progressive variation in the number of degree days can be seen: more in deep, the following phenomena are evidenced:

- an undebatable reduction in the heating necessity (the number of HDD is decreasing, mainly in the Mediterranean areas and in the cooler regions of Scandinavia),
- the number of cooling degree days is, conversely, increasing, so that the installation of mechanical cooling is, at today, more and more frequent.

Finally, we are going toward a very dangerous direction, in which the increasing necessity of space heating is interesting also climates that, some years ago, did not require cooling. The EU climate is changing, with progressive heating, and thus, urgent measures to face its impacts are mandatory, starting from buildings.

2 The Vulnerability to Heat Waves of European Cities

As evident, all European regions suffer the negative effects of climate change, but some of them are more vulnerable than others to the increment in heat extremes. According to the data of the European Environment Agency (EEA) of 2017 [13], the Mediterranean region has undergone a serious growth of heat peaks during the recurrent heat waves. This phenomenon is often aggravated by the urban heat island (UHI) effect, an emerging challenge that is becoming more intense and frequent due to the growing urbanization. The intensity of UHI is predominantly higher in cities located in the Continental and Mediterranean regions. Indeed, even the Continental region is interested in the increase of heat extremes with high intensity of UHI [14]. Recent studies have underlined the risk of heat waves and the interaction between heat waves and UHI in some European cities.

Interesting research was conducted by Geletič et al. [15] to show the seasonal variability of the surface UHIs in three European cities: Prague, Brno, and Novi Sad. The cities have different urban areas, structures, and topography. By considering the land surface temperatures for the period 2013–2018, high intensity of UHIs was revealed in densely built-up and industrial zones, while a lower intensity in sparsely built city outskirts. The results of the study showed the highest differences in the surface temperatures between pairs of local climatic zones during the summer, in Brno (86%), then in Prague (88.4%), and finally, in Novi Sad (75.5%). The role of vegetation in reducing the intensity of UHI was underlined but it was also evidenced the importance in evaluating the building height, the density of structures, and materials to fully understand the UHI variability. Focusing on the same topics, but on a larger scale, Nastra et al. [16] studied the relationship between UHI and the extent, shape, and distribution of green urban infrastructures. By considering different European cities, several space green planning traditions were compared in terms of UHI magnitude in a ten-year period. The study revealed that:

- Cities' longitude has a stronger impact on the UHI magnitude than latitude.
- Higher is the share of agricultural land uses, lower is the magnitude of UHI in Nordic and British cities. A contrary effect verifies in the Mediterranean cities.
- Higher is the forest proportion, lower is the magnitude of UHI in new member and Mediterranean cities. Conversely, fragmented forests have a negative effect on UHI magnitude.

According to the UHI maps reported in [14], France is interested in high intensity of UHIs, and several studies were focused on the relationship between land use and UHI phenomenon in France cities. Indeed, for a large set of French cities, Gardes et al. [17] analyze the correlation between UHI intensity and urban morphology and geographical factors. A statistical relationship between six predictors (the population density, the distance to coast, the elevation differences, local climatic zone, the distance to the agglomeration center, and the climatic region) and the UHI intensity was derived. In particular, the analysis pointed out that the UHI intensity increases when the number of inhabitants rises and when the distance from the coast is higher.

More in general, Semi-Continental and Mountain climatic regions have a higher intensity of UHI than Pure Atlantic and Pure Mediterranean climatic regions. For the specific case of Paris, Lemonsu et al. [18] evaluated the impact of urban expansion scenarios on UHI and heat stress. The authors identified five urban expansion scenarios assuming different local land-use regulations: a "spread out city," which follows the historical urban development of Paris over the twentieth century; a "compact city" with strict urban policies to contrast urban sprawl and protect natural areas; and a "green city," with no land-use regulation but with numerous parks to the extent of 10, 30, and 50% of all built surfaces. Through a future projection of the regional climate, the expansion city scenarios were compared with a reference scenario ("Paris today") and it was showed how the heat stress of Paris today is higher than the possible future city developments.

For the case of Italy, by considering the increasing occurrence of extreme phenomena of climate change, and thus the frequency of UHI and heat waves, 10 Italian peninsular metropolitan cities were studied, to evaluate the change in the urban microclimate in relation to impervious and tree cover surfaces [19]. Low tree areas and growing impervious density affect negatively surface UHI, conversely to the effect of the sea, and thus the proximity to the coast. It was revealed that Turin (a metropolitan city in Italy) has the most intense surface UHI, and for an increment of 10% of highly impervious surfaces and low tree coverage in the metropolitan center, the surface UHI grows of 4.0 °C. By considering that more than 30% of the Italian population lives in metropolitan cities and that the Mediterranean region is interested in the highest increment in heat extremes, the study remarks the necessity of UHI mitigation strategies. A similar investigation but for different cities—Prague and Bucharest, respectively, located in central and south-eastern Europe—was carried out to evidence the influence of the atmospheric circulation on the diffusion of UHI [20]. The temperatures of pairs of urban and peri-urban stations were compared, and the results remarked a higher intensity of UHI in Bucharest than in Prague. By considering the trend of UHI intensities for the period 1981–2016, statistically, a significant increasing trend was revealed for Bucharest, worse than in Prague, for the future 100 years. For both cities, UHIs are more intense under anticyclonic situations with southern winds. Even in Athens, the correlation between heat island and heat waves was found and the nocturnal intensity of UHI was recorded exceptionally amplified under extremely hot weather [21].

Obviously, extreme heat phenomena are not limited to European cities. In Phoenix metropolis, Chow et al. [22] evaluated the spatial distribution of vulnerability to heat extremes in the period 1990–2000. This city has one of the largest and fastest growth of population size and land areas in Central Arizona, and it is characterized by a sub-tropical climate with high average temperatures and low precipitations. So, during the summer, the residents are exposed to high risk for extreme heat (temperature over the 43 °C during the daytime) even increased by the UHI phenomenon. These conditions have an impact on the mortality rate; indeed, in Arizona, more than 250 deaths were caused by heat exposure during the period 1993–2002. To evaluate the heat vulnerability of the city of Phoenix, the authors [22] developed an index (VItotal) based on seven measures, equally weighted, about physical heat exposure (i.e., mean minimum and maximum temperatures, mean normalized difference vegetation index) and adaptive capacity (e.g., age, economic conditions, ethnic status) of the population and determined the portion of different racial and ethnic minority populations which lived in conditions of a critical vulnerability. Indeed, the population capacity to face hazards was connected to the income level, racial and ethnic status, age, gender, housing tenure, and migration status. To map the VItotal, the Phoenix metropolis was divided into four areas, by considering the central business direct as a reference point. The results for the period 1990–2000 underline the differences in the heat vulnerability according to time, space, and demographic segments of the population. Particularly, Hispanic population results as the most exposed to heat extremes. Thus, it is underlined the necessity of a policy intervention in the urban design to face both heat extremes and social disparities. The social disparities due to climatic conditions in Phoenix, Arizona, were also pointed out by Harlan et al. [23]. The study shows that, according to biophysical and social data, affluent whites predominately live in more vegetated neighborhoods, less climatically stressed than those of low-income Latinos. The correlation between microclimate, vegetation, land use, and socio-economic context was investigated, and it was evidenced how the social attributes vary with biophysical environments. More in general, land use, the presence of vegetation, and environmental quality, which affect human health and wellbeing, were all considered important aspects to understand the urban socioecological systems.

Reid et al. [24] evidence how vulnerability to heatwaves can be marked by specific population and community characteristics. The authors identified different vulnerability factors in the United States: six are referred to the demography (i.e., poverty level, age, education level, ethnicity, and living conditions), two to the availability of air-conditioning systems, and two others to the land cover and percentage of diabetes. A United States map was developed by identifying the heat vulnerability of the different nations, to reduce and prevent heat-related morbidity and mortality. The analysis showed the correlation between heat vulnerability and the use of air-conditioning systems; indeed, the higher is the air-conditioning prevalence, the lower is the heat vulnerability. In addition, the highest urban heat phenomena were evidenced in the downtowns of metropolitan areas, which are also social vulnerability areas. According to the authors, reducing social inequalities could be a step to reduce health outcomes, and to reach this scope, public health professionals could prepare climate change adaptation plans just starting from vulnerability maps.

Other main causes of urban overheating, according to previous literature, besides the extreme climatic events, are as follows: (a) the lack of vegetation in cities, which contribute to reducing temperatures through evapotranspiration and shading effect; (b) the heat emissions from human metabolism, building energy use, industrial activities, and vehicles; (c) the construction materials which alter the heat absorption, conduction, and storage; and (d) the building geometry which affects the temperature and radiative exchange in the urban canopy. Considering the increasing UHIs, possible mitigation strategies have been proposed over the past decades and can be divided into nature-based solutions and engineered solutions. The first ones include green infrastructures, green roofs, and walls, and also blue infrastructures and water bodies, suitable to reduce temperatures and increase latent heat fluxes. On the other hand, common engineered strategies mainly reduce the absorption of solar radiation by pavements and building surfaces and include cool roofs and pavements, and cool facades [25]. About that, Kolokotsa et al. [26] investigated the potential of green and cool roofs in mitigating UHI in European climates. The solar reflectance, thermal mass, and insulation level of the cool roof were examined as well as the type of vegetation and irrigation of green roofs to evaluate their effects in microclimate mitigation. The study confirmed the benefits of these solutions in improving the urban environment and simultaneously reducing the energy demand. Indeed, for the cities of Crete (Greece), Rome (Italy), and London (England), cool roofs with albedo higher than 0.6 involve a maximum sensible heat reduction for the whole summer, close to 178 kW h/m^2, 162 kW h/m^2, and 112 kW h/m^2, respectively. Green roofs with a LAI (Leaf Area Index) higher than 1 involve a maximum summer sensible heat reduction of 147 kW h/m^2 in Crete, 128 kW h/m^2 in Rome, and 70 kW h/m^2 in London.

Many studies demonstrated how climate change acts synergically to the UHI in generating urban overheating during extreme climatic events, and it is also well-known that urban overheating widely affects our wellbeing. It compromises our health, causes high energy demands, has an impact on environmental quality, and contributes to increasing energy poverty and vulnerability (Fig. 6) [27].

According to an investigation of Santamouris [27], the additional energy demand required during heat waves causes a prolonged plant operation that increases significantly pollutant emission and consequently exacerbates health and respiratory problems; in addition, high energy demands involve the construction of additional power

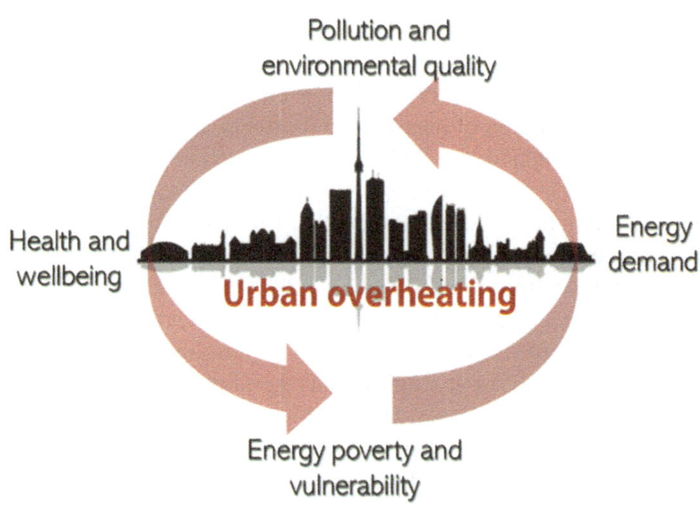

Fig. 6 Urban overheating and the main effects

plants, with a consequent increase in the energy prices and a significant increase in the cities' vulnerability; and the need of healthy and comfortable indoor environments, even to face health issues and criticalities, results in more intense use of air-conditioning systems and consequently in peaks in energy demand. The synergic relation between urban overheating and energy, health, pollution, and city vulnerability evidence the need for an interdisciplinary approach to studying possible mitigation and adaptation strategies for the development of present and future cities.

3 Heat-Attributable Mortality in Europe

Heat-related mortality is the phenomenon of increase in death number that accompanies extremely warm periods (i.e., heatwaves) exacerbated by urban heat island effects and poor health conditions.

Particularly, susceptible people are those suffering from cardiovascular, respiratory, or mental diseases as emerge in the epidemiological studies. Therefore, heat and health correlations with mortality are worthy to be deeply investigated. Data collected from 26 studies in the review paper of Moghadamnia et al. [28] highlights that there is an average increase of 1.3% in the risk of heat-related mortality connected to cardiovascular diseases. This increase is less marked among men (1.1%) than in women (1.4%). Some other factors can directly or indirectly affect heat and cardiovascular mortality. Among this, educational level seems to be not significant, whereas a lower acclimatization capacity of people living in colder areas results in a 0.07% increase in mortality for each degree of increase in latitude. The elderlies are particularly sensitive to the increase in temperature, and their mortality due to heat increases by 8%, while mortality due to low temperatures only grows by 6%. A final outcome concerns the effect of heat on deaths, and it continues for almost a week after an extremely hot period.

The work of Gasparrini et al. of 2021 [29] aimed to find people at high-risk levels studying the correlation of heat-related mortality and health issue during the period 1990–2006 in England and Wales. Results were particularized for 4 ranges of age (0–64, 65–74, 75–84, $\geq$ 85 years). Comprehensively, it was found that heat is responsible for more than 23 thousand deaths, 1.03% of the total ones during the summer period. The maximum increase in mortality when the ambient temperature exceeds regional temperature thresholds (20.9–24.7 °C) took place for respiratory diseases (+4.1% for each degree of increase), while both cardiovascular issues and no-cardiorespiratory problems increased by 1.8%. In absolute terms, cardiovascular-associated extra deaths (8005) were higher than respiratory-attributable ones (5841). The risk for cardiovascular diseases grows with age, whereas it is quite constant in people affected by ills to the respiratory apparatus. On a population of about 30 million inhabitants living in 15 European cities, the heat-attributable average daily deaths correlated to cardiovascular diseases varied from 2.6 in Ljubljana to 61.0 in London, respectively, while the average mortality for respiratory illnesses ranges from 0.4 to 23.7, in the same two cities that account for the minimum and maximum population among the analyzed ones according to the assessments of 2008 reported in [30]. On average, in the cities of the Mediterranean area, the growth factor of heat-related mortality with a 1-degree increase over the apparent temperature thresholds (it is reported below in this section for all the cities, and takes into account ambient temperature and humidity level) was always higher than in the cities of north-continental Europe concerning both cardiovascular and respiratory causes, respectively, 3.70% versus 2.44% and 6.71% versus 6.10%. These variations raise with the age of the subjects analyzed. By considering the dead people divided into three age groups (15–64, 65–74, and ≥ 75 years), the maximum increment in percentage change was observed in Mediterranean cities and, for respiratory problems, it rose from 1.54% in the 15–64 age group to 8.1% in people of 75 years or more.

The relationship between mental illnesses and heat-related mortality is a less investigated topic, but there is well evidence that a certain connection exists between them. The increase of heat-related mortality in the people category suffering from psychosis, dementia, alcohol, and drug abuse was investigated in [31], by demonstrating an overall increase of 4.9% in the number of deaths concerning 2% growth in the general population. This increment was especially evident in subjects aged less than 65 years (10% increase instead of 4% of the elderly), even if more than 82% of the dead have more than 65 years. The treatment of mental disturbs with some medicines—like hypnotics, anxiolytics, and antipsychotics—analyzing heat-related deaths in England during the period 1998–2007 showed a noticeable increase in mortality of 7–8% for each 1 °C increase in temperature, even more significant in patients at the first diagnosis. The reasons for these data must be sought in the state of loneliness and abandonment in which these people often are and in the alterations that the substances and also therapeutic treatments induce in their physiological responses. The authors of [32] used the case-crossover approach to estimate the effect of high temperature on mortality. They analyzed data of 4 Italian cities (Turin, Milan, Bologna, and Rome) disaggregating results based on age, gender, marital status, area-based income, previous and current hospitalization, cause and period of hospitalization, and place of death. Concerning these last factors, the risk of dying during a day with a mean apparent temperature of 30 °C compared with a day with 20 °C of apparent temperature revealed interesting observations. A first remark worthy of note concerns the heat-related mortality difference between hospitalized and non-hospitalized people (in the previous 2 years); it is higher in the second category. More in detail, heat-related mortality is lower in the patients that have left hospitals in the previous 4 weeks. Among the hospitalized people, the mortality risk increase resulted higher in the patients with long hospitalization periods (more than 59 days). Concerning the patients' diseases, the most susceptible subjects are those suffering from conduction disorder of the heart, depression, psychoses, disorders of the thyroid gland, and cerebrovascular diseases.

In recent years, numerous studies have highlighted the growth of heat-related mortality in the most populated areas of the globe, where overheating is more marked, and the average age of the population is higher.

According to [5], some people groups are affected more than others by heat-related mortality, and these are the elder persons, infants, and low-income populations (due, for instance, to other influencing factors, starting from the not availability of air-conditioning systems). Other influencing factors of risk and safety, besides the availability (or not) of microclimatic control and mechanical cooling, are the availability of transportation, living alone or with other persons, mental illness, and living on the upper floors of multistory buildings. The review study is very wide, by focusing on studies of the frequency of heatwaves in the U.S., the combination of heat-related mortality also when evaluated for persons with cardiovascular and respiratory pathologies, heat-related mortality in health care and medical facilities, and temperature-related mortality. Even if it is quite evident the high temperature of the ambient is a risk factor (i.e., there is a positive and net association between thermal levels and mortality), the effects of air pollutants in combination with temperatures are not so evident. The authors noted that, besides the mortality, even other deep damages to people's health are related to the exposure to heatwaves, and thus skin and eye pathologies, increment of respiratory and cardiovascular diseases, and vector-borne and water-borne transmissions of pathogens. One further very interesting finding is that heat waves are very dangerous in early summer and spring, more than heat waves occurring in the late summer.

3.1 Non-optimal temperature and heat-related mortality

On the basis of the data elaborated from the temperature recorded in 750 locations of 43 countries, about five million deaths per year have been estimated worldwide due to non-optimal temperature conditions in the period 2000–2019. On a continental level (see Fig. 7a), the greatest mortality incidence occurs in Africa (11.77%) and then in Europe (10.27%). In absolute terms, the greatest number of deaths is counted in Asia (over 2.6 million deaths), 51.49% of the total, while, in Europe, there are over eight hundred thousand deaths per year (835,897), 7.70% of the global one [33].

Globally, the deaths for non-optimal temperature account for 9.43% of the total mortality showing a reduction in the period 2000–2019 of 0.3 percentage points. These excess death rates have tended to reduce especially in Africa and America; in Oceania, a reduction took place in the first decade of the new millennium, but it was then followed by a significant increase in the last period (an overall increase of 1.79 percentage point). In Europe (as it can be seen in Fig. 7b), an oscillating trend appeared in the analyzed years, with a net increase of 0.37 percentage points between the period 2000–2003 and 2016–2019 [33].

The global reduction in the incidence of non-optimal temperature-attributable deaths all over the world is mainly driven by the reduction in cold-related mortality, which passed from 8.70% of the period 2000–2003 to 8.19% of the years 2016–2019. Heat-related deaths, instead, increased globally by 0.21 percentage points. On average, the annual deaths recorded in the world due to excessively high temperatures have been estimated to be 489,075; 45.81% of these occurred in Asia and 36.54%

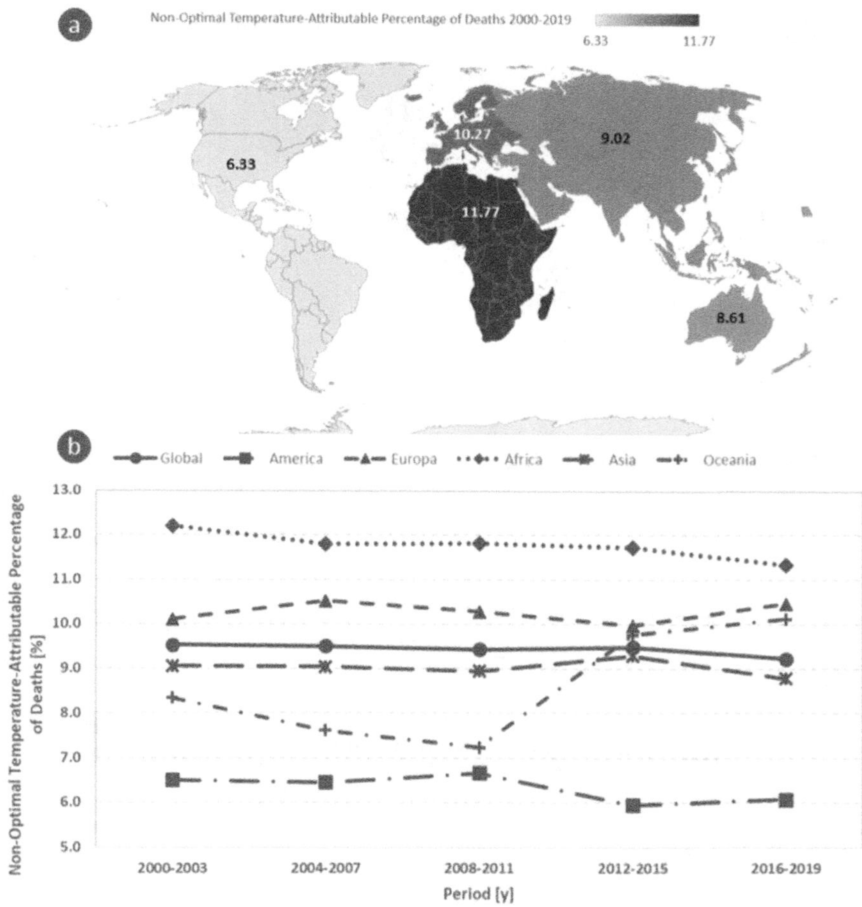

Fig. 7 Non-optimal temperature-related percentage of deaths. Data elaborated from supplementary material of [33]

in Europe. Europe was the continent with the highest percentage of excess deaths of the total ones due to hot temperatures (2.19% as reported in Fig. 8a). Europe and Oceania had also the fastest growing trend of all. Europe passed from 1.99 to 2.63%, while Oceania's rate increased by 0.81 percentage points from 1.22% during the period 2000–2019 (Fig. 8b) [33].

By reading the data reported above, it appears that there is not a geographical uniform distribution of non-optimal attributable mortality, and heat-related death rate is high in Europe, whereas the incidence of cold-related mortality is high in Africa.

In addition, it is evident that low temperatures are still the most significant cause of deaths, but cold-related mortality is experiencing a reduction, whereas heat-attributable one has a certain increase trend following the climate extremization.

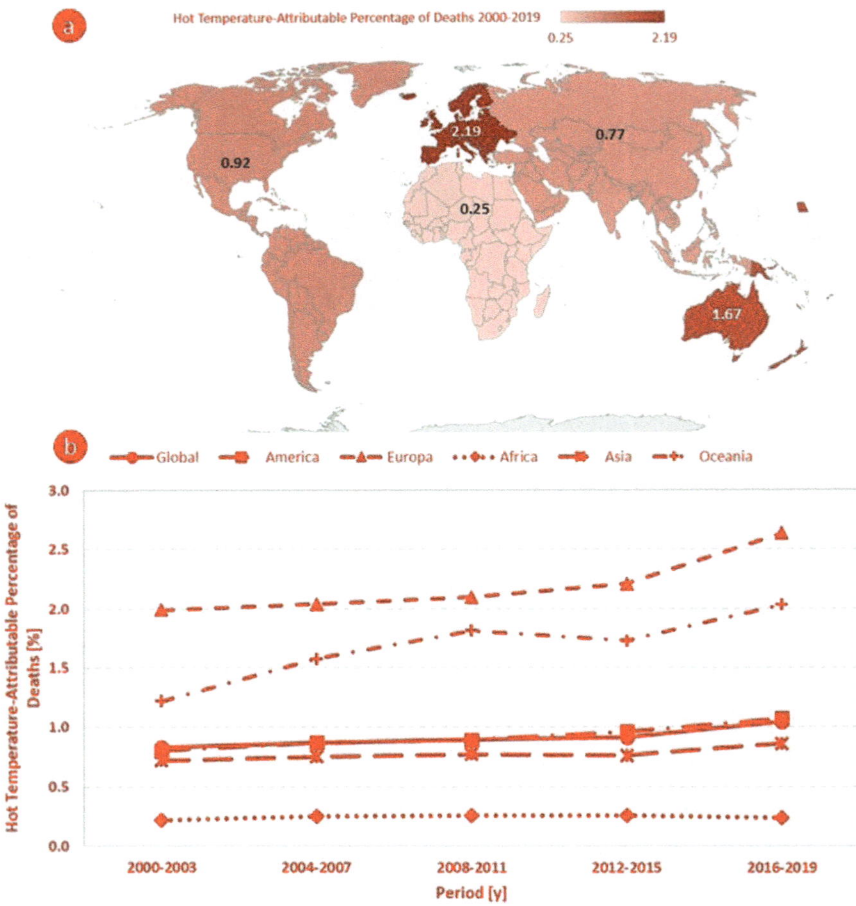

Fig. 8 Heat-related percentage of deaths. Data elaborated from supplementary material of [33]

In 2011, Anderson and Bell [34], regarding many communities in the U.S., evaluated the mortality risk during the days interested by heatwaves and the effect modifications by the same heat anomalies. More in detail, the authors studied heat waves in 43 cities, and thus the periods in which the temperatures are, for a period equal or longer than 2 days, equal or higher than the 95th percentile, concerning the same community, and the 5 months starting in May and finishing at the end of September. The method is based on the use of Bayesian hierarchical modeling, applied to generate global effects for the whole community, also by taking into account the regional and national levels. Three main aspects have been investigated: the role of the period in which the heatwaves occur, the intensity of the heat wave, and the duration. Results are surprising. As in Basu and Samet [5], Anderson and Bell [34] found that the first heat waves (in spring, for instance) are the most dangerous, with increased mortality of + 5% compared with the non-heat wave days, while the later heat waves have an increment of "only" + 2.65%. In addition, at the national level, the mortality increase during heatwaves is around 3.73%. It is very interesting the role of every 1°F (i.e., + 0.55 °C) increase of temperature on the increase in mortality (+2.49%), as well as the role of every 1-day prolongation of the heat wave (+0.38% of deaths).

3.2 *Heat-related mortality in European countries*

Europe is the area where the greatest increase in climate pressure on health is observed, both in terms of average and extreme climatic conditions. The spatial distribution of temperatures is not uniform even at a European level. In addition, even the social and economic vulnerability of society to climate anomalies is not uniform, and thus, their impact on mortality is variable. Finally, the heat-related deaths depend on the three previous factors.

Regarding climate anomalies, in 2004, Johnson et al. [35] measured what happens during the heatwave of August 2003, that interested the whole of Europe, with largely increased mortality in Spain, France, Italy, and other EU countries. In England and Wales, very high temperatures were recorded, with values higher than 38 °C, with high temperatures driven upon England and Wales because of the penetration of hot dry continental air from the Iberian Peninsula: the so-called Spanish plume. Very high temperatures have been measured during the diurnal hours, but also that at the night, with values averagely above 6–8 °C compared to the seasonal ones. An increase in mortality has been measured in the 10 days between 4 and 13 August 2003, the days interested by the heatwaves. The mortalities were compared to the ones of the last five years, in the same periods. The results are net: in general, an increment in mortality of about 17% has been measured (+2091 deaths) and the highest incidence has been measured in the southern half of England, including the area of London, with an increase of + 42%. The highest increase in mortality interested the age group of persons over 75 years, and, for this age sector, in the region of London, the excess mortality was + 59%. At the national level, the mortality of persons over 75 years was + 23%. Infants, children, and sick adults are other people groups exposed at a very high level of risk. In London, also temperatures of 37.9 °C were registered, with night values of 26–27 °C. The peak of mortality (+1600 persons) was registered two days after the heatwave, while, during the peak of ambient temperatures (last days of the heatwave durations), the peak of admissions in the hospital was verified. Already in 2004, a worrying risk underlined by Johnson et al. [35] was the evidence of the increase of intensity and frequency of extreme phenomena. Now, almost twenty years later, we have experienced it, also in form of extreme hurricanes, floods, and windstorms.

The 2003 heatwave has been unforgettable, given the intensity and the long duration. By keeping this in mind, Fouillet et al. [36] in 2007 evaluated impacts of the last heatwaves, and thus the one of July 2006, that interested many regions of Europe. In particular, the authors verified if the countermeasures applied by French Institutions, starting from 2003, have had positive effects. These countermeasures consisted in reducing the risk of ambient air temperatures, by developing a new national system of real-time surveillance, monitoring of health data, the definition of some recommendations for preventing and facing heat-related diseases, the spread of air-conditioning equipment for buildings hosting fragile persons, and thus hospitals and retirement homes, plans, and census of vulnerable persons. The definition and application of such countermeasures were monitored by the main healthcare institutions of France, and a general intensification of control and prevention, by exploiting also the potential communication of media, was established. Implementation of such measures started during 2004, and the authors evaluated the beneficial effects by analyzing the mortality data due to the extreme heatwave of July 2006, whose duration was very long, i.e., 18 days, from 11 to 28 July. For many days, temperatures reached 35 °C during the day with night values above 20 °C. The anomalies in temperatures have been registered in more than 70 French locations. The climate conditions were very hard but, thanks to the applied set of countermeasures, starting from a deep communication, the measured mortality was lower compared to the expectation from models; indeed, the related deaths were 2065 (the increment in mortality has been + 9%), against the predicted 6452 deaths, and thus, this is a clear signal of the importance of reduction of the people vulnerability, through both social and technological measures.

On the basis of the work of Martinez-Solanas et al. [37], the percentage of deaths attributable to non-optimal temperature extrapolated for the period 1998–2012 from the total deaths data of 147 regions in 16 European countries (representing a population of about 420 million people in urban e rural zones) accounts to 7.17%.

The distribution of this percentage across the 16 countries is widely heterogeneous, as it can be seen in Fig. 9a, b, passing from 9.87% of Italy to 4.85% of Germany. Furthermore, mortality for non-optimal temperature is mainly due to cold climatic conditions rather than to warm temperature (blue bars are longer than red ones in Fig. 9a), and heat-attributable deaths are on average lower than the cold-related ones by a factor of 10. The authors of [37] also elaborated scenarios projecting the anomalies in the non-optimal temperature-related percentage of total deaths to the middle and the end of the century, by considering three different greenhouse gas emissions conditions (RCP2.6, RCP6.0, and RCP8.5). In most parts of the considered countries, a reduction is expected in the anomalies by the middle of the century (2035–2064) for all three RCP scenarios considered. Exceptions are Italy and Spain (see Fig. 9c). This circumstance derives from the reduction of cold-related deaths that is only partially balanced by the heat-attributable growth. The longer-term projection (2070–2099), reported in Fig. 9d, shows that a net increase is expected to take place in most parts of the Counties under the effect of warmer conditions, particularly in the scenario with no or low mitigation strategies (RCP6.0 or RCP8.5).

The distribution of anomalies in the heat-related mortality with regard to the reference period (1976–2005) does not show a uniform distribution in Europe as depicted in Fig. 10a, b. The countries of the Mediterranean area—Spain, Italy, and Croatia—and Portugal highlight the highest percentage: 1.01, 1.33, 0.89, and 0.83%, respectively. A common behavior characterizes the projection scenarios in the 16 countries analyzed. An increase is forecasted everywhere and in correspondence of

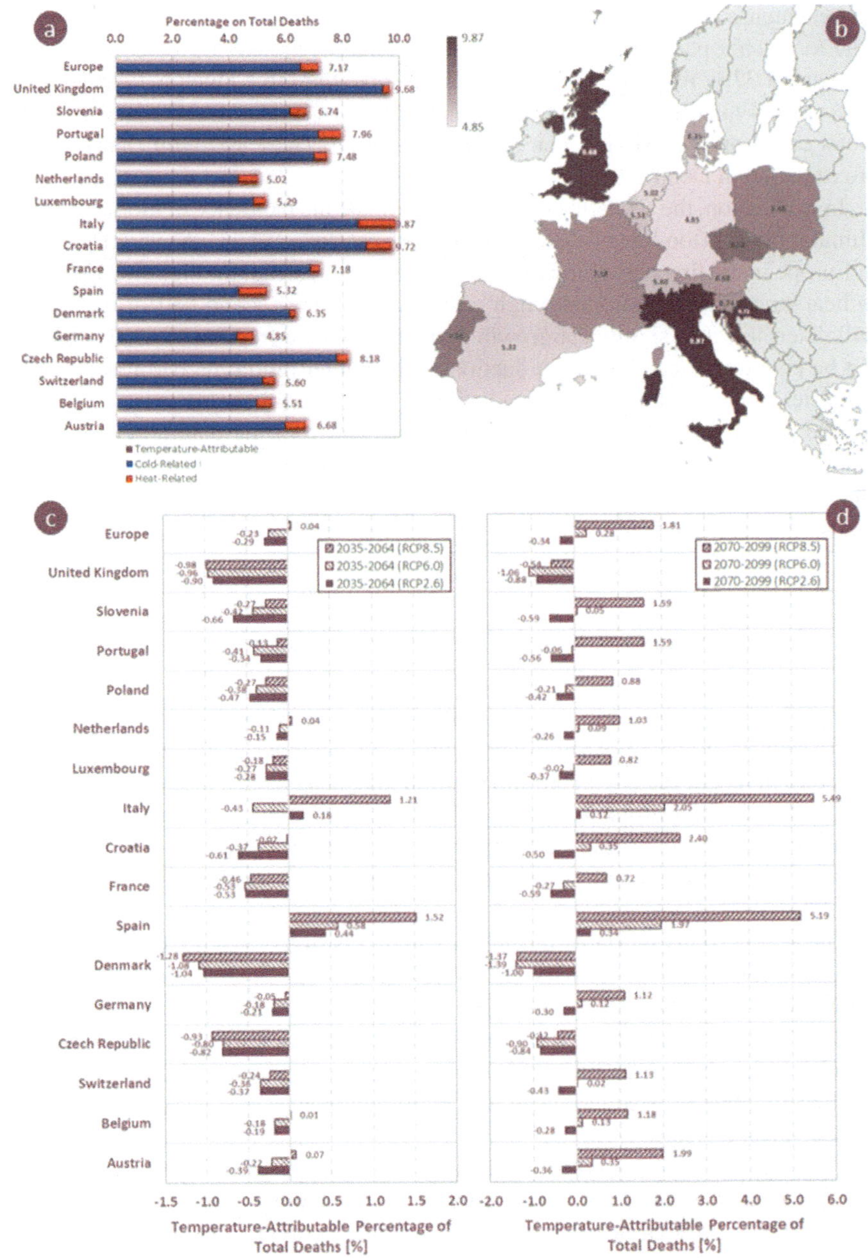

Fig. 9 Temperature-related mortality in European countries and its projection for different greenhouse gas emission scenarios. Data elaborated from supplementary material of [37]

higher intense GHG emissions and long-term perspectives. Italy and Spain will be able to reach a percentage increase of the anomalies in heat-attributable deaths of 9.09 and 7.37%, respectively. Globally in Europe, the same index is foreseen to be 1.78 and 4.49 in the middle and in the final part of the twenty-first century, when the worst emission scenario is taken into consideration. The least significant increases are expected in Denmark.

In conclusion, the scenarios proposed to estimate the incidence of mortality due to climatic factors (too low or extremely high temperatures) show an overall reduction in Europe in the first part of the twenty-first century; instead, the significant increase in heat stress due to extremely high temperature in the second half of the century will determine a so significant growth in the attributable fraction of deaths that will not be counterbalanced by the reduction of the winter mortality, particularly in the countries of the Mediterranean area and with high emission scenarios.

The importance of mitigation policies emerges from the data analyzed before mainly in the countries of the Mediterranean area to attenuate the effect of climate change on population health. In the next section, the most suitable actions proposed in the EU to mitigate climate change will be described.

3.3 Local studies on heat-related mortality in Europe

Alongside these large-scale studies, there are a series of analyses developed on a local scale (at the level of small areas, cities, or neighborhoods) presented in the literature in recent years with the aim of evaluating heat-related mortality and its risk factors throughout Europe. A milestone in the analysis of heat-attributable mortality in European cities consists in the study proposed by Baccini et al. in 2008 [30], which considers 15 important cities in Europe (about 30 million inhabitants) as reported in Fig. 11.

The average daily mortality during the summer period (1 April–30 September) of the years 1990–2000 for natural causes varied from a maximum of 149 in London to a minimum of 6.3 in Ljubljana. Based on the findings extrapolated by the association between high apparent temperature and excess mortality in warm season, a significant correlation has been found; in particular, an average increase of 3.12% and 1.84% resulted in cities of Mediterranean and north-continental area (boxes with dashed contour), respectively.

These growths refer to a 1 °C increase of the apparent temperature over the thresholds, 29.4 °C and 23.3 °C, respectively. The maximum temperature thresholds, above 30 °C, appeared in Athens (32.7 °C), Milan (31.8 °C), and Rome (30.3 °C), whereas the minimum ones were in Stockholm (21.7 °C) and Zurich (21.8 °C). In correspondence with the highest temperatures, the highest rates of mortality growth were recorded, and Athens (5.54%), Rome (5.25%), and Milan (4.29%) showed the maximum risk of deaths for hot weather conditions.

The same critical period was also examined by Kovats et al. [38] which identified the relationship between the hot weather and heatwaves on emergency admissions in

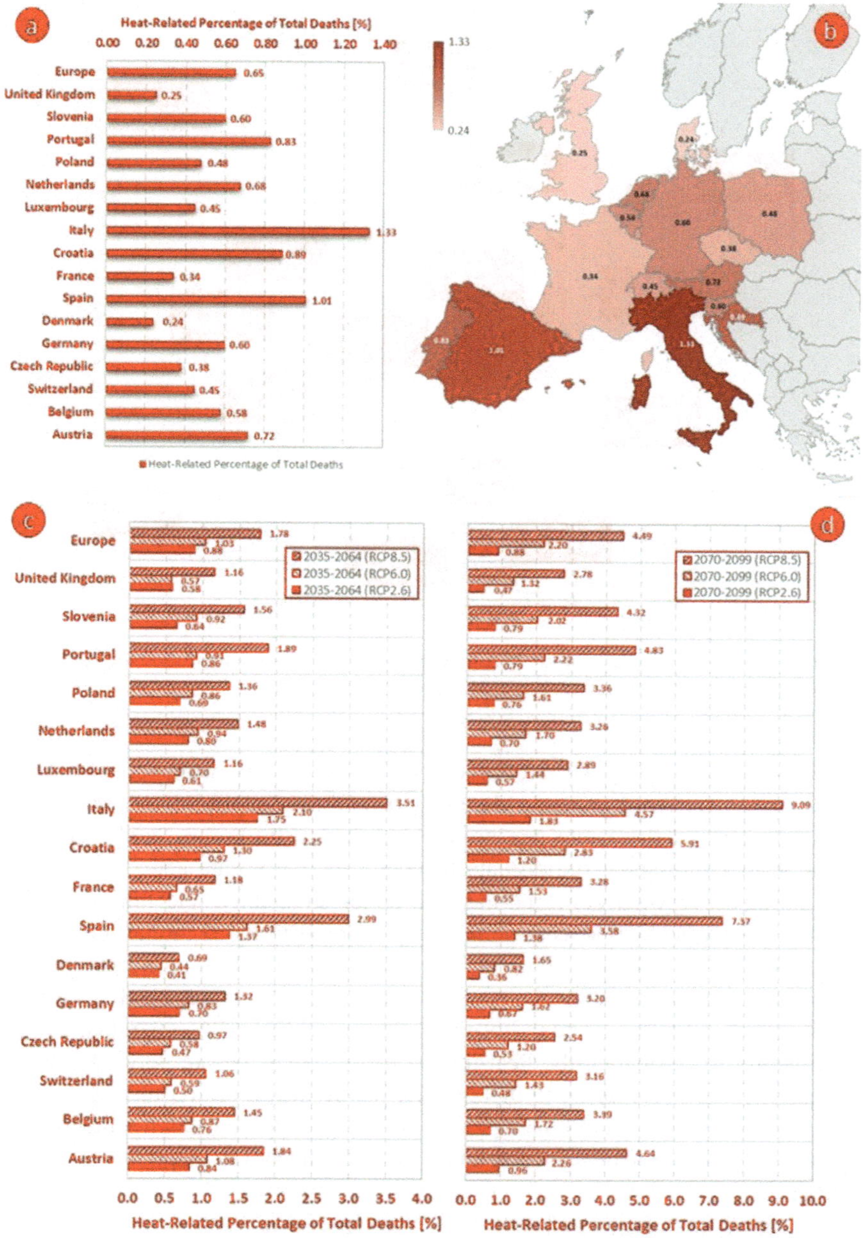

Fig. 10 Heat-related mortality in European countries and its projection for different greenhouse gas emission scenarios. Data elaborated from supplementary material of [37]

Fig. 11 Apparent temperature thresholds and percent variations in natural mortality per a 1 °C increase over apparent temperature threshold. Figure elaborated from [30]

Greater London hospitals. They referred to the data of the Department of Health, for the period between April 1994 and March 2000, for the hospital admissions in Greater London, organized them into age groups and diagnostic groups, and considered Heathrow meteorological station data to obtain the daily data about temperature and humidity. For what concerns the data about air quality, and thus the ozone and airborne particulate concentrations, they referred to the National Air Quality Monitoring Network. To investigate the relationship between hospital admissions and daily mean temperatures, a linear model was adopted. It was pointed out that mortality increases during hot periods, but hospital admissions did not rise significantly and consequently, so probably, the heat-related deaths occur before persons came to medical attention. According to the authors, public health efforts must be paid to vulnerable people who could have a rapid decline if exposed to high temperatures, and thus the youngest and the elderly and the people affected by respiratory and renal diseases.

Here below, instead, a brief collection of the most interesting works about heat-related mortality published in the last two years is reported. For the municipalities of the Madrid region with more than 10,000 inhabitants, López-Bueno et al. [39] considered urban and rural populations, dividing them into 4 areas (Urban Metropolitan Center, Rural Northern Mountains, Rural Center, and Southern Rural). These authors, by analyzing heat-related mortality, found that the percentiles, corresponding to the threshold for the maximum temperature above which a heatwave can be assumed, are lower in Urban Metropolitan Center and Southern Rural zones. The high risk of mortality is reinforced by the deprivation index (that takes into account working and education conditions) and housing indicators unfavorable. A scarce vulnerability appears, instead, in the Rural Center, in part explained by the lowest percentage of the population over 64 years. Another interesting aspect pointed out is the adaptation to heat, and there are fewer extreme heat events associated with increments in mortality in population that daily lives with higher temperatures.

For the city of Madrid, Abadie and Polanco-Martinez [40] studied the increase of possible heatwaves through different climate models and with different emission scenarios in order to derive the probability distribution of mortality and risk correlated to them. For an emission scenario RCP8.5, it has been estimated for 2100 a mean excess of 3.6 °C over the critical value of the temperature of 38 °C with a corresponding number of deaths equal to 1614. The correlation between temperature distribution and mortality due to cardiorespiratory problems in the elderly population has been estimated at the neighborhood level in the city of Lisbon by Morais et al. [41]. They found that the proximity to an area with high vegetation cover reduces heat-related cardiorespiratory mortality, as well as it is lower in the areas with a high number of people with complete tertiary education. An opposite correlation appears, instead, where building conditions are worse. The correlation between heat-related mortality for cardiovascular issues and socio-economic aspects, pollution conditions, and aircraft noise has been highlighted for Zurich (Switzerland) in [42]. With high temperature, especially deaths for hypertension and myocardial infarction in women increase. The risk of mortality grows in older people with a low level of education, living alone and in low-quality buildings. Heat-related heart failure mortality shows a minor rise trend with $PM_{2.5}$ and NO_2 concentration and no correlation with excessive aircraft noise. Similarly, in the Southern Finland region, $PM_{2.5}$, ozone, or both did not modify the heatwave mortality rate due to non-accidental, cardiovascular, or respiratory causes as found by Kollanus et al. [43]. They identify in age (>65 years) and chronic diseases (respiratory diseases, renal diseases, mental and behavioral disorders, diseases of the nervous system, and cardiovascular diseases) as significant factors of heat-attributable mortality in Finland. Furthermore, mortality risk was observed higher in healthcare facilities (especially social structures), and it increases by 13.1% in long-term inpatients. The heat-related mortality in Brussels and Antwerp has been investigated by De Troeyer et al. in [44]. They found that 1585 and 411 deaths could be associated with high temperatures in the two cities, respectively, in the period 2002–2011. Temperature thresholds over which a significant increase in the ordinary mortality rate occurs are 25.2 °C in the capital and 22.8 °C in Antwerp. Every increase of 1 °C above these critical values determines a percentage growth in mortality of 3.1% and 4.9%, respectively. Assuming as an example the city of Thessaloniki in Greece, Kouis et al. [45] assessed the impact on the health of increasing temperatures considering adaptation phenomena like the use of air-conditioning systems. The heat-related and air pollution cardiorespiratory mortality has been foreseen in different scenarios with a projection of the excess deaths in 2080–2099 ranging from 2.4 to 433.7 without adaptation. In the case of adaptation, these values vary depending on the mix of primary energy sources used to produce electricity for air-conditioners. Technological adaptation to heat according to the authors will mitigate direct heat-related mortality but will be counterbalanced by a higher indirect mortality rate due to pollution emissions. Arbuthnott et al. [45, 45] examined the heat-related mortality and years of life lost in the biggest conurbations of England (Greater London, Greater Manchester, and West Midland), by analyzing data referred to the period 1996–2013. Years of life lost are a measure of the premature death of people with respect to the average life expectancy. They found that the mortality risk increases of 3.9, 2.0, and 2.5% for each 1 °C enhance over temperature threshold (18.9, 16.8, and 17.6 °C) in London, Manchester, and West Midland Conurbations, respectively, while the increases in years of life lost are 3.0%, 1.5% and 2.4%, respectively. The incidence of the risk and life reduction is higher for females than male in Greater London and Manchester, whereas the opposite situation appears in West Midland. The highest susceptibility of women was found also in the previously cited [43]. A parameter based on life expectancy at the time of each death was also evaluated in the Czech Republic [47], where it was observed that 1994 and 2015 were the years with the maximum cumulative years of life lost: 8971 and 7850, respectively. These two years were also the most impacted years by heatwaves, in terms of duration and intensity dividing the considered period into two halves. The frequency of heatwaves longer than 10 days increases in the period 2006–2017; four of the top 5 significant events occurred in this time interval correspondingly; and heat-wave-related excess deaths and years of life lost equal to 2909 and 33,053, respectively, compared to 2190 and 24,501 in the first half (1994–2005). Murage et al. [48] evaluated heat-related mortality link with the natural environment, built environment, and socio-economic factors in Greater London on the basis of 185,397 deaths recorded in the period 2007–2016. It was observed that temperature and hence mortality risk are higher where the level of vegetation cover, building quality, and the economic possibilities are lower. Vegetation demonstrates to be the most influencing factor on mortality. Huber et al. [49] estimated temperature-related mortality in the 12 major German cities, starting from death data recorded in the period 1993–2015. Actual data highlights a more important incidence of cold-related mortality (attributable fractions of 5.49%) with respect to the heat-attributable deaths (0.81%). Projection in scenarios with an increase of + 3, + 4, and + 5 °C of the global mean temperature pointed out an opposite situation. Correspondingly, heat-related mortality is expected to rise 2.8-fold, 5.1-fold, and 7.2-fold, respectively. The maximum heat-attributable fractions of mortality in the observed period were computed in Frankfurt (1.09%), Dusseldorf, and Cologne (1.08%), and in the worst projection (+5 °C global warming), the most impacted city remains Frankfurt (8.06%) followed by Cologne (7.61%) and Leipzig (7.17%). The capital Berlin passed from 1.00% to 6.50%.

4 Conclusion

Heat-related mortality is a growing issue in Europe and worldwide. It fundamentally depends on global warming and is exacerbated by UHI. Actually, cold-related deaths account for larger percentages today, but heat-related mortality is expected to grow rapidly especially in pessimistic scenarios that consider high greenhouse gas emission perspectives.

Geographical differences highlight it in large-scale analyses, while local evaluations point out the incidence of secondary factors, like socio-economic conditions, building quality, gender, educational level, chronic diseases, and vegetation cover on heat-attributable mortality.

Mitigation strategies of overheating (cool roof, green areas, outdoor vegetations, etc.) seem to be strategic in preventing the escalation of heat-attributable deaths; therefore, these should be considered in the planning tools of the local and national governments to subvert the unpleasant forecasts that appear at a not-too-distant horizon.

References

1. Earth's city light—1994–1995 Data from the Defense Meteorological Satellite Program (DMSP) Operational Linescan System (OLS). https://visibleearth.nasa.gov/images/55167/earths-city-lights
2. Peel MC, Finlayson BL, McMahon TA (2007) Updated world map of the Köppen-Geiger climate classification. Hydrol Earth Syst Sci 11:1633–1644. https://doi.org/10.5194/hess-11-1633-2007
3. Beck H, Zimmermann N, McVicar T et al (2018) Present and future Köppen-Geiger climate classification maps at 1-km resolution. Sci Data 5:180214. Available online at: https://doi.org/10.1038/sdata.2018.214. Accessed Dec 2021
4. Kayser-Bril N (2018) Europe is getting warmer. European Data Journalism Network. Available online at https://www.europeandatajournalism.eu/eng/News/Data-news/Europe-is-getting-warmer-and-it-s-not-looking-like-it-s-going-to-cool-down-anytime-soon. Accessed Dec 2021
5. Basu R, Samet JM (2002) Relation between elevated ambient temperature and mortality: a review of the epidemiologic evidence. Epidemiol Rev 24(2):190–202
6. EDJNet—The European Data Journalism Network. Available online at: https://www.onedegreewarmer.eu/. Accessed Dec 2021
7. Anderson CA (1987) Temperature and aggression: effects on quarterly, yearly, and city rates of violent and nonviolent crime. J Pers Soc Psychol 52(6):1161–1173. https://doi.org/10.1037/0022-3514.52.6.1161
8. European Environment Agency, the impacts of extreme weather and climate related events in the EEA member countries and the UK (1980–2019). Available online at: https://www.eea.europa.eu/data-and-maps/daviz/impacts-of-extreme-weather-and-2#tab-chart_1. Accessed Dec 2021
9. EnergyPlus 8.9.0. Available online at: https://github.com/NREL/EnergyPlus/releases/tag/v8.9.0. Accessed Dec 2021
10. ASHRAE International weather for energy calculations (IWEC weather files) users' manual and CD-ROM. American Society of Heating, Refrigerating and Air-Conditioning Engineers, Atlanta, GA, USA

11. Ascione F (2017) Energy conservation and renewable technologies for buildings to face the impact of the climate change and minimize the use of cooling. Solar Energy 154. ISSN 34–100 0038 092X https://doi.org/10.1016/j.solener.2017.01.022
12. Trend in heating and cooling degree days (1981–2014) European Environment Agency. Available online at: https://www.eea.europa.eu/data-and-maps/figures/trend-in-heating-and-cooling. Accessed Jan 2022
13. European Environment Agency (EEA) (2017) https://www.eea.europa.eu/highlights/climate-change-poses-increasingly-severe
14. ArcGIS map—Urban Heat Island intensity. Available online at: https://www.arcgis.com/home/webmap/viewer.html?url=%3A%2F%2Fclimate.discomap.eea.europa.eu%2Farcgis%2Frest%2Fservices%2FUAMV%2FUrban_Heat_Island_Intensity%2FMapServer&source=sd. Accessed Dec 2021
15. Geletič J, Lehnert M, Savić S, Milošević D (2019) Inter-/intra-zonal seasonal variability of the surface urban heat island based on local climate zones in three central European cities. Build Environ 156:21–32
16. Nastran M, Kobal M, Eler K (2019) Urban heat islands in relation to green land use in European cities. Urban Urban Greening 37:33–41
17. Gardes T, Schoetter R, Hidalgo J, Long N, Marquès E, Masson V (2020) Statistical prediction of the nocturnal urban heat island intensity based on urban morphology and geographical factors-an investigation based on numerical model results for a large ensemble of French cities. Sci Total Environ 737:139253
18. Lemonsu A, Viguie V, Daniel M, Masson V (2015) Vulnerability to heat waves: impact of urban expansion scenarios on urban heat island and heat stress in Paris (France). Urban Clim 14:586–605
19. Morabito M, Crisci A, Guerri G, Messeri A, Congedo L, Munafò M (2021) Surface urban heat islands in Italian metropolitan cities: tree cover and impervious surface influences. Sci Total Environ 751:142334
20. Zak M, Nita IA, Dumitrescu A, Cheval S (2020) Influence of synoptic scale atmospheric circulation on the development of urban heat island in Prague and Bucharest. Urban Clim 34:100681
21. Founda D, Pierros F, Petrakis M, Zerefos C (2015) Interdecadal variations and trends of the Urban Heat Island in Athens (Greece) and its response to heat waves. Atmos Res 161:1–13
22. Chow WT, Chuang WC, Gober P (2012) Vulnerability to extreme heat in metropolitan phoenix: spatial, temporal, and demographic dimensions. Prof Geogr 64(2):286–302
23. Harlan SL, Brazel AJ, Jenerette GD, Jones NS, Larsen L, Prashad L, Stefanov WL (2007) In the shade of affluence: the inequitable distribution of the urban heat island. In: Equity and the environment. Emerald Group Publishing Limited
24. Reid CE, O'neill MS, Gronlund CJ, Brines SJ, Brown DG, Diez-Roux AV, Schwartz J (2009) Mapping community determinants of heat vulnerability. Environ Health Perspect 117(11):1730–1736
25. Wang C, Wang ZH, Kaloush KE, Shacat J (2021) Perceptions of urban heat island mitigation and implementation strategies: survey and gap analysis. Sustain Cities Soc 66:102687
26. Kolokotsa D, Santamouris M, Zerefos SC (2013) Green and cool roofs' urban heat island mitigation potential in European climates for office buildings under free floating conditions. Sol Energy 95:118–130
27. Santamouris M (2020) Recent progress on urban overheating and heat island research. Integrated assessment of the energy, environmental, vulnerability and health impact. Synergies with the global climate change. Energy Build 207:109482. https://doi.org/10.1016/j.enbuild.2019.109482
28. Moghadamnia MT, Ardalan A, Mesdaghinia A, Keshtkar A, Naddafi K, Yekaninejad MS (2017) Ambient temperature and cardiovascular mortality: a systematic review and meta-analysis. PeerJ 5:e3574. https://doi.org/10.7717/peerj.3574
29. Gasparrini A, Armstrong B, Kovats S, Wilkinson P (2012) The effect of high temperatures on cause-specific mortality in England and Wales. Occup Environ Med 69:56–61. https://doi.org/10.1136/oem.2010.059782

30. Baccini M, Biggeri A, Accetta G, Kosatsky T, Katsouyanni K, Analitis A, Anderson HR, Bisanti L, D'Ippoliti D, Danova J, Forsberg B, Medina S, Paldy A, Rabczenko D, Schindler C, Michelozzi P (2008) Epidemiology 19(5):711–719. https://doi.org/10.1097/EDE.0b013e318176bfcd
31. Page LA, Hajat S, Kovats RS, Howard LM (2012) Temperature-related deaths in people with psychosis, dementia and substance misuse. Br J Psychiatry 200(6):485–490. https://doi.org/10.1192/bjp.bp.111.100404
32. Stafoggia M, Forastiere F, Agostini D, Biggeri A, Bisanti L, Cadum E, de Caranci N, Donato F, De Lisio S, De Maria M, Michelozzi P, Miglio R, Pandolfi P, Picciotto S, Rognoni M, Russo A, Scarnato C, Perucci CA (2006) Vulnerability to heat-related mortality: a multicity, population-based, case-crossover analysis. Epidemiology 17(3):315–323
33. Zhao Q, Guo Y, Ye T et al (2021) Global, regional, and national burden of mortality associated with non-optimal ambient temperatures from 2000 to 2019: a three-stage modelling study. Lancet Planet Health 5(7):e415–e425. https://doi.org/10.1016/S2542-5196(21)00081-4
34. Anderson GB, Bell ML (2011) Heat waves in the United States: mortality risk during heat waves and effect modification by heat wave characteristics in 43 US communities. Environ Health Perspect 119(2):210–218
35. Johnson H, Kovats S, McGregor G, Stedman J, Gibbs M, Walton H, Cook L (2004) The impact of the 2003 heat wave on mortality and hospital admissions in England. Epidemiology 15(4):S126
36. Fouillet A, Rey G, Wagner V, Laaidi K, Empereur-Bissonnet P, Tertre Le A, Hémon D (2008) Has the impact of heat waves on mortality changed in France since the European heat wave of summer 2003? A study of the 2006 heat wave. Int J Epidemiol 37(2):309–317
37. Martínez-Solanas È, Quijal-Zamorano M, Achebak H, Petrova D, Robine JM, Herrmann FR, Rodó X (2021) Projections of temperature-attributable mortality in Europe: a time series analysis of 147 contiguous regions in 16 countries. Lancet Planet Health 5(7):e446–e454. https://doi.org/10.1016/S2542-5196(21)00150-9
38. Kovats RS, Hajat S, Wilkinson P (2004) Contrasting patterns of mortality and hospital admissions during hot weather and heat waves in Greater London. UK. Occup Environ Med 61(11):893–898
39. López-Bueno JA, Navas-Martína MA, Linares C, Mirón IJ, Luna MY, Sánchez-Martínez G, Culqui D, Díaz J (2021) Analysis of the impact of heat waves on daily mortality in urban and rural areas in Madrid. Environ Res 195:110892. https://doi.org/10.1016/j.envres.2021.110892
40. Abadie LM, Polanco-Martínez JM (2022) Sensitivities of heat-wave mortality projections: moving towards stochastic model assumptions. Environ Res 204:111895. https://doi.org/10.1016/j.envres.2021.111895
41. Morais L, Lopes A, Nogueira P (2021) Human health outcomes at the neighbourhood scale implications: elderly's heat-related cardiorespiratory mortality and its influencing factors. Sci Total Environ 760:144036. https://doi.org/10.1016/j.scitotenv.2020.144036
42. Saucy A, Ragettli MS, Vienneau D, de Hoogh K, Tangermann L, Schäffer B, Wunderli JM, Probst-Hensch N, Röösli M (2021) The role of extreme temperature in cause-specific acute cardiovascular mortality in Switzerland: a case-crossover study. Sci Total Environ 790:147958. https://doi.org/10.1016/j.scitotenv.2021.147958
43. Kollanus V, Tiittanen P, Lanki T (2021) Mortality risk related to heatwaves in Finland—factors affecting vulnerability. Environ Res 201:111503. https://doi.org/10.1016/j.envres.2021.111503
44. De Troeyer K, Bauwelinck M, Aerts R, Profer D, Berckmans J, Delcloo A, Hamdi R, Van Schaeybroeck B, Hooyberghs H, Lauwaet D, Demoury C, Van Nieuwenhuyse A (2020) Heat related mortality in the two largest Belgian urban areas: a time series analysis. Environ Res 188:109848. https://doi.org/10.1016/j.envres.2020.109848
45. Kouis P, Psistaki K, Giallouros G, Michanikou A, Kakkoura MG, Stylianou KS, Papatheodorou SI, Paschalidou AK (2021) Heat-related mortality under climate change and the impact of adaptation through air conditioning: a case study from Thessaloniki. Greece. Environ Res 199:111285. https://doi.org/10.1016/j.envres.2021.111285

46. Arbuthnott K, Hajat S, Heaviside C, Vardoulakis S (2020) Years of life lost and mortality due to heat and cold in the three largest English cities. Environ Int 144:105966. https://doi.org/10.1016/j.envint.2020.105966
47. Urban A, Kyselý J, Plavcová E, Hanzlíková H, Štěpánek P (2020) Temporal changes in years of life lost associated with heat waves in the Czech Republic. Sci Total Environ 716:137093. https://doi.org/10.1016/j.scitotenv.2020.137093
48. Murage P, Kovats S, Sarran C, Taylor J, McInnes R, Hajat S (2020) What individual and neighbourhood-level factors increase the risk of heat-related mortality? A case-crossover study of over 185,000 deaths in London using high-resolution climate datasets. Environ Int 134:105292. https://doi.org/10.1016/j.envint.2019.105292
49. Huber V, Krummenauer L, Peña-Ortiz C, Lange S, Gasparrini A, Vicedo-Cabrera AV, Garcia-Herrera R, Frieler K (2020) Temperature-related excess mortality in German cities at 2 °C and higher degrees of global warming. Environ Res 186:109447. https://doi.org/10.1016/j.envres.2020.109447

Chapter 16
The Trend of Heat-Related Mortality in Spain

Laura Carlosena and Alicia Carlosena

Abstract Rapid urbanization, increasing, and aging population combined with rising temperatures and extreme weather events present severe worldwide challenges for the near future. Spain is not an exception, characterized by one of the highest life expectancies in Europe and a very high urban density. Moreover, its diverse climate, ranging from arid to wet, hot, and cold regions, is broadly affected by heatwaves during the summertime. In addition, the Spanish built environment presents some overheating challenges. Therefore, although there is ongoing research on heat-related mortality in Spain, there is a strong need for a systematic overview. This study aims to overview the current state of heat mortality trends from the last 25 years and describe current population trends, climate, heatwaves, and the Spanish built environment to provide a context-specific overview. In addition, this chapter presents the most relevant outcomes in research using the Scopus database, and 27 published papers on heat-related mortality are reviewed and discussed. Moreover, data of heat-specific deaths will be presented for the first time, concluding that 62% of the deceased were over 65 years old, and 58% were men. Finally, the already implemented mitigation strategies in several Spanish cities are discussed. To conclude, we found a strong link between overheating and heatwaves in the number of deaths but a decline in the number of heat-related deaths reported in research. A standardized research methodology is needed to adequately classify and determine heat-specific deaths with indicators such as location, population size, and threshold temperature. In addition, although heat mitigation strategies have been heterogeneously implemented in several Spanish cities, there is scant monitoring to quantify their effectiveness. Future research should address these knowledge gaps to find the most suitable strategies for every climate and urban condition.

L. Carlosena (✉)
Engineering Department, Public University of Navarre (UPNA), Arrosadia Campus, 31006 Pamplona, Spain
e-mail: laura.carlosena@unavarra.es

A. Carlosena
Faculty of Medicine and Health, University of New South Wales (UNSW), Sydney, NSW, Australia

N. Aghamohammadi and M. Santamouris (eds.), *Urban Overheating: Heat Mitigation and the Impact on Health*, Advances in Sustainability Science and Technology,
https://doi.org/10.1007/978-981-19-4707-0_16

Keywords Urban overheating · Heatwaves · Spain · Heat-related deaths · Co-morbidity · Mitigation strategies

1 Introduction

More people live in urban areas than in rural areas. In 2018, 55% of the world's population resided in urban zones and, by 2050, the population is expected to increase by 2.5 billion, and 90% will concentrate in Asia and Africa [1]. Extensive urban expansion and population growth have aggravated environmental problems such as local climate, resource depletion, and quality of air [2]. Moreover, this phenomenon can significantly alter the local microclimate, producing enduring high temperatures [3].

The Urban Heat Island Effect (UHI) has been studied as the temperature differences between rural and urban locations. The first reference to point out this phenomenon dates back to 1833, in which the evidence of higher temperatures in cities was reported [4]. These higher urban temperatures have a severe effect on (i) energy consumption requirements for cooling buildings, (ii) outdoor pollution levels, (iii) heat-related mortality and morbidity, (iv) urban ecological footprint, and (v) survival levels [5]. Overheating sources include the released anthropogenic heat, high absorption of solar radiation by the urban materials and structures, decreased airflow and urban ventilation, reduced evapotranspiration, and limited radiative losses [6]. Envelope materials of buildings and urban structures significantly influence the urban thermal balance; traditional construction materials absorb solar and infrared radiation increasing their surface temperature and the ambient temperature. Moreover, neighborhoods with higher proportions of concrete and higher dense infrastructure exacerbate urban heat [7] despite being more walkable.

Heatwaves (HW) aggravate UHI by increasing mortality risks during prolonged periods of extreme heat compared to the community's usual climate [8]. The World Health Organization (WHO) acknowledges the effect of excess heat and the link to vulnerability and mortality, especially for the elderly, infants and children, pregnant women, outdoor and manual workers, athletes, and the poor [9]. In addition, a recent study noted that even though cold waves are more frequently reported in the Emergency Events Database, heatwaves were the primary cause of temperature-related deaths in Europe and the Russian Federation [10].

Research has focused, to some extent, on determining the effect of these extreme weather events on the population. During heatwaves in Greece, the inhabitants of low-income housing were during 85% of the hot period exposed to indoor temperatures over 30 °C [11]. A study of 15 Mediterranean cities showed that an increase of 1°C in the maximum outdoor temperature above the threshold results in a 3% increase in mortality [12]. During the heatwave of 2003, the total associated excess deaths in Spain were 8% (43,212 observed deaths compared with 40,046 expected deaths) and were only observed in those aged 75 years and over (15% more deaths than expected for the age group 75–84 and 29% for those aged 85 or over) [13]. A study on the impact of high temperatures on hospital admissions in Madrid showed that the temperature in which hospital admissions soar coincides with the temperature limit above which mortality sharply rises [14]. Moreover, they suggested that people die rapidly from circulatory diseases before being admitted to a hospital. A recent metanalysis [15] found positive associations between relative risk of cause-specific mortality and duration and excess of hot nights. Studies have calculated up to a 36% increase in discomfort hours social housing stock in southern Spain in 2050 [16].

This chapter overviews the current state of heat mortality trends described from research published in the last 25 years. Firstly, population trends, climate, heat-waves, and built environment in Spain will be described to provide a context-specific overview. The second part will cover the heat-related mortality trend found in research and analyze the data of heat-specific deaths from the Spanish National Institute of Statistics (INE, Instituto Nacional de Estadística, Spanish acronym). Finally, mitigation strategies and techniques will be presented, and the main conclusions will be drawn.

2 Population, Climate, and Future Projections for Spain

2.1 Population Trends

Spain is in Southwestern Europe and has 47,394,223 inhabitants (January 2021) and an average population density of 93 people per km^2. Nevertheless, its population is heterogeneously distributed. A recent study divided several countries into areas of 1 km^2 and determined that from a total of 505,000 areas in Spain, only 13% are lived in. As a result, the density in those areas is 737 people per km^2 [17]. For instance, Madrid and Barcelona have 5512 and 16,420 inhabitants per km^2, respectively [18, 19]. According to the INE, in the next 15 years, Spain will increase its population by a million inhabitants and by more than 3 million by 2070, following the current trends. Moreover, by 2035 it is expected that 26.5% of the population will be 65 years of age or older.

Advances in medicine and technology, such as infection control, pharmacology, or medical devices, have made possible the increase in life expectancy over the last century [20]. Nowadays, it is possible to operate, treat, and manage many diseases than once it was only thought off. Consequently, this has led to an aging population and the need to care and manage for chronic diseases, for example, hypertension or diabetes mellitus [21]. As a result, life expectancy in Spain has substantially extended; between 1999 and 2019, it increased from 75.4 to 80.9 for males and 82.3 to 86.2 for females.

The aging population and disease management have increased the burden in the healthcare system [22]. In Spain, the health expenditure percentage of gross domestic product (GDP) grew from 6.8% in 2000 to 9% in 2018 [23]. Nevertheless, the public per capita expenditure in health care [24] varies from autonomous communities

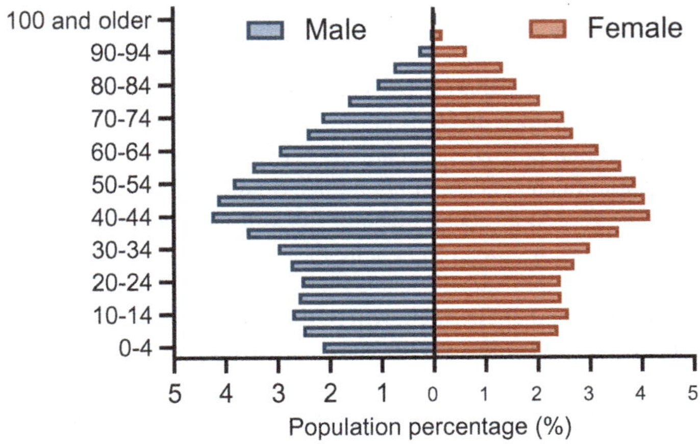

Fig. 1 Population pyramid Spain in 2020. Elaborated by authors with data from the National Institute of Statistics

due to the healthcare decentralization in Spain. The aging population is increasing rapidly (Fig. 1); the current national aging population index for 2021 according to INE is 129.17%, with regions such as Asturias with values up to 231%. Exposure to heatwaves increases mortality and morbidity, affecting the elderly population the most [25]. Regions with specially aged population will have to deal with a more demanding scenario.

2.2 *Climate in Spain*

The climate in Spain is diverse due to its geographical location and condition. Using the Köppen-Geiger [26] classification, the Meteorological State Agency of Spain (AEMET) classified the Iberian Peninsula into 13 climatic areas [27], shown in Fig. 2. Most of the Peninsula is covered by temperate with dry or hot summer climate (Csa), occupying approximately 40% of its surface, located in the southern central plateau region, and the Mediterranean coastal regions, except for the arid zones in the southeast. The different kinds of arid climates (BWh, BWk, BSh, and BSk) are present in the southeast of the Peninsula, and the Ebro Valley, and to some extent in the southern central plateau region and the Balearic Islands.

Other subtypes of temperate climates (Csb, Cfa, and Cfb) can be observed in the northeast of the Peninsula, surrounded by numerous mountains and part of the northern central plateau region and significant extensions of the Pyrenees. Finally, high altitude areas present continental climates (Dsb, Dsc, Dfb, and Dfc); they can be found in the Pyrenees and some small areas at high altitude in the Cantabrian and Iberian Mountain Ranges. Finally, the tundra (ET) is present in small areas on the highest plains of the Central Pyrenees [27].

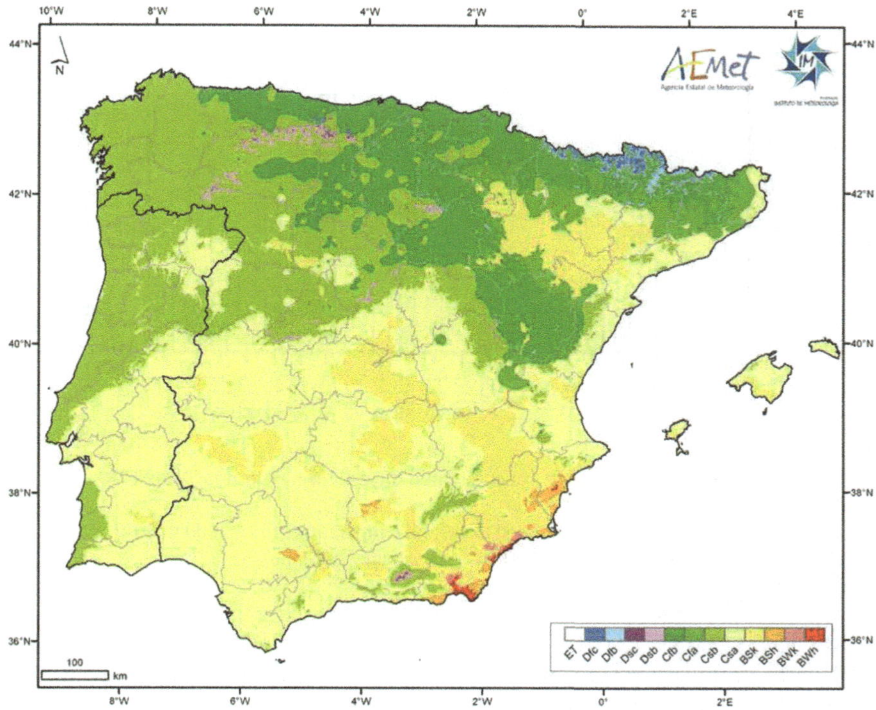

Fig. 2 Köppen-Geiger classification for the iberian peninsula end the Balearic Islands. From: Meteorological State Agency of Spain [27]

Future projections estimate that the mean value of maximum daily temperatures will rise, in relation to the reference period (2000–2009), by 1.6 °C across the period 2021–2050 and by 3.3 °C across the period 2051–2100 [28].

2.3 *Heatwaves*

Heatwaves are defined by the AEMET as an episode of at least three consecutive days, in which at least 10% of the meteorological stations considered registered maximum temperatures above the 95% percentile in their series of daily maximum temperatures in July and August for the years 1971–2000 [29]. The chosen parameters are (i) their duration, (ii) the number of affected regions, (iii) the heatwave maximum temperature, and (iv) the anomaly of the wave.

According to the data shown in Fig. 3, the most critical heatwave took place in 2015 and is the longest recorded in the country since 1975. It lasted 26 days, from the 27th to the 22nd of July, with a heat temperature maximum of 37.6 °C, 30 regions were affected. The second most extended heatwave was in 2003, with thousands of

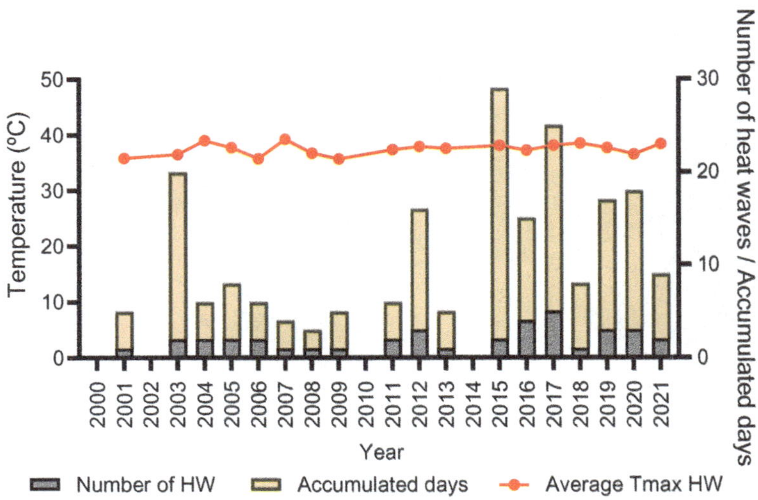

Fig. 3 Number of heatwaves, accumulated days, and temperatures from 2000 to 2021. Elaborated by authors with data from the Meteorological State Agency of Spain

victims across Europe. In 2017, five heatwaves took place, adding up to 25 days, and in 2012, the heat wave covered the greatest number of regions up to 40 provinces on 10th of August, followed by the year 2003 with 38 provinces.

Simulations from the EURO-CORDEX [30] project can help analyze different climate change scenarios. For example, it was projected that between 2021 and 2050 there will be an increase in intensity, frequency, duration, and spatial coverage of health waves in the Iberian Peninsula by 104%. The most considerable changes will occur in the eastern-central region, rising to 150% for the Mediterranean coast and the Pyrenees. This prediction is consistent with the different projections made by AEMET (Fig. 4).

2.4 *Spanish Built Environment*

Alongside appropriate heating during the wintertime, the problem of overheating in summer causes severe challenges in some countries. According to data from Eurostat, almost 20% of dwellings in European Union (EU) countries cannot be ventilated, air-conditioned, or cooled to achieve a relative comfort level during the summer [31]. This phenomenon is more significant in southern countries with substantially higher temperatures and longer summer periods.

In the case of Spain (Fig. 5), the building sector is aged; 55% of the built stock was constructed before 1980. The first construction regulation that included some minimal energy efficiency criteria was introduced in 1979 with the NBE-CT-79 [32].

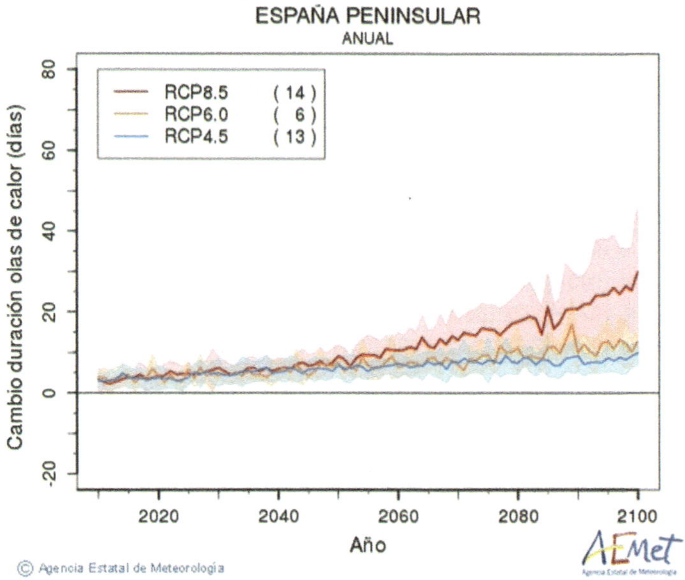

Fig. 4 Projections of change in the duration of heat waves in days according to three different scenarios. Vertical axis: change in day duration of heatwaves. From: Meteorological State Agency of Spain

From 2006 onward, "Código Técnico de la Edificación" transposed European directives, and the last modification in 2020 pushes achieving near-zero emissions buildings. These new regulations emphasize insulation that improves the buildings' envelope. Nevertheless, in summer, when the gains (internal or solar) are not adequately controlled, there is a tendency toward more summer discomfort as the envelope insulation increases [33–35].

Using the different criteria established by the Chartered Institution of Building Services Engineers (CIBSE), Escandón et al. [36] evaluated the possible comfort risks due to overheating under the current scenario and a climate change scenario. They concluded that 38% of the evaluated cases are already at risk of overheating as they fail to meet two of the three adaptive criteria set in TM52; by 2050, this figure will increase to almost 100%.

Urban vegetation or green infrastructure enhances the urban climate and thermal comfort, improves pollutant removal, and improves residents' health [37–39]. As a result, it is among the most considered urban heat mitigation techniques. Furthermore, studies have shown that the percentage of green space inside a one-kilometer and a three-kilometer radius is significantly related to perceived general health [40]. Nevertheless, the literature debates how best to measure green space's quality, quantity, and accessibility [41]. For example, the European Union has deployed a series of quality-of-life indicators that include the building stock state, pollution, noise, access to green areas, leisure, and natural and living environment.

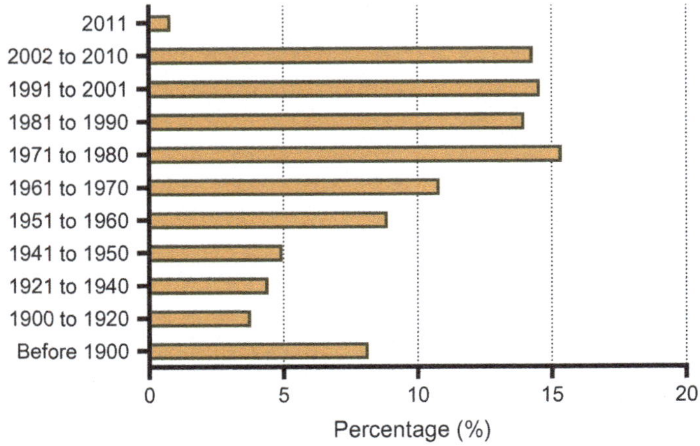

Fig. 5 Age distribution of Spanish residential building stock. Elaborated by authors with data from the National Institute of Statistics (2011)

Figure 6 shows the Land Use of five cities in Spain and the national average, according to the parameters defined by the European Union. As can be seen, the national average percentage dedicated to "green urban areas, sports, and leisure facilities is 0.21%. Depending on the urban settlement, this parameter varies greatly. In Spain, the average satisfaction with green areas and recreational areas is 6.6 out of 10 [42], whereas in the European Union (EU-28) is 7.1 out of 10.

3 Heat-Related Mortality in Spain

3.1 Study Cases in Spain

After a review was carried out in the Scopus database with the keywords *Heat, mortality, Spain,* and *heatwave,* forty papers were selected. After reading the abstract and conclusions, some were excluded since they studied only cold-related mortality or drought-related mortality; finally, 27 were analyzed. Table 1 summarizes the primary studies on heat-related mortality in Spain. In addition, the methodology employed, the location, dates of the study, and the observed variables are presented.

Among all the studies, only five studies considered all regions in Spain [46, 52, 56–58]. Some studies have dealt only with specific regions and provinces of Spain such as Castilla-La Mancha [43, 44, 54], Galicia [50], Madrid [48, 61, 63, 66], and Sevilla [48]. In Castilla-La Mancha, the effects on heat circulatory-cause mortality declined, but respiratory-cause mortality was maintained. The study that compared Madrid and Seville over a 36-year observed period found that both locations adapt to the increase in temperatures, finding women more susceptible to heat than men

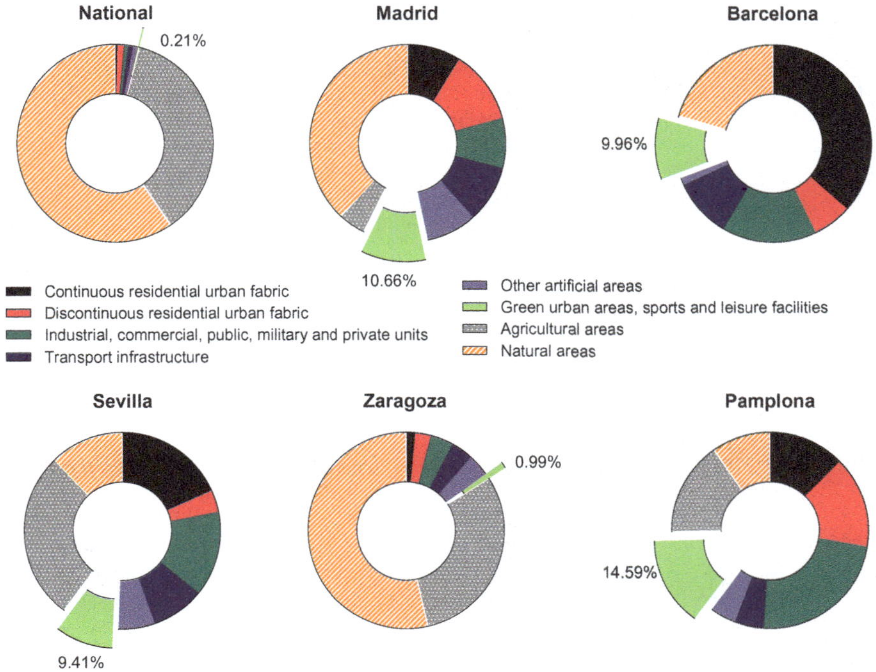

Fig. 6 Land use in 6 Spanish cities elaborated by authors with data from the National Institute of Statistics based on Copernicus Urban atlas data

[48]. Others have focused on single cites or municipalities such as Barcelona [51, 55, 64], Bilbao [51], Madrid [14, 49, 51, 62], Sevilla [51], Valencia [67], and Zaragoza [67]. Comparing four cities, the intensity of a heat wave is an important mortality risk indicator during heatwave days [51].

Most studies refer to a reduction of heat-related mortality; nevertheless, the causes are unclear, while others are signaling toward the effect of adaptation and the effectiveness of the Heat Health Prevention Plan (HHPP). Results pointed to a reduction in the relative risks of cause-specific and cause-sex mortality across the whole range of summer temperatures. These reductions, in turn, explained the observed downward trends in heat-attributable deaths, with the only exceptions of respiratory diseases for women and both sexes together [46]. The average difference between weather station temperature and ERA-5 reanalysis data is 1.4 °C. These differences are more remarkable for the minimum mortality temperature, where 80% of the cities show a higher threshold using weather station data [52].

After the 2003 heatwave in Europe, a HHPP was implemented in Spain. The HHPP implementation rendered that there was an increase in mortality attributable to temperature from 0.38% (1993–2002) to 1.21% (2004–2013) for moderate heat. Nevertheless, the highest mortality attributable fraction reductions were detected

Table 1 Main studies on heat-related mortality in Spain

References	Dates	Methodology	Locations	Observed variable
[43]	1975–2003	Autoregressive integrated moving average (ARIMA) models (Box-Jenkins) and cross-correlation functions (CCF)	Castile-La Mancha 5 regions	Daily organic-cause mortality (all except external causes) $T_{\text{max day}}$ $T_{\text{mean day}}$ $T_{\text{min day}}$
[44]	1975–2008	ARIMA models	Castile-La Mancha 5 regions	Organic cause mortality T_{max} (June–September)
[45]	1975–2008	ARIMA models	Castile-La Mancha 5 regions	Specific-cause mortality $T_{\text{mean day}}$ $P_{\text{mean day}}$
[46]	1980–2015	Standard quasi-Poisson regression models Time-varying distributed lag nonlinear models (DLNMs)	47 major cities	circulatory, respiratory, both T_{mean} from meteorological stations Study follows STROBE guidelines
[47]	1983–2013	Generalized linear model (GLM) methodology with the Poisson regression link	10 Spanish Provinces	Daily mortality data natural causes except accidents T_{max}
[48]	1983–2018	Cubic curvilinear estimation	Provinces of Seville and Madrid	Daily mortality data natural causes $T_{\text{max day}}$
[49]	1986–1997	Univariate ARIMA modeling	Madrid municipal area (city)	Mortality due to all causes except accidents, and circulatory and respiratory $T_{\text{max day}}$ $T_{\text{mean day}}$ $T_{\text{min day}}$ Relative humidity
[50]	1987–2006	Box–Jenkins stochastic procedures (ARIMA)	Galicia (4 Spanish provinces)	Daily mortality data $T_{\text{max day}}$

(continued)

Table 1 (continued)

References	Dates	Methodology	Locations	Observed variable
[51]	1990–2014	DLNM models	Barcelona, Bilbao, Madrid, and Seville (4 cities)	Daily mortality data natural causes except accidents, respiratory, and cardiovascular causes $T_{\text{mean day}}$ (3-day average) Relative humidity Wind speed (m/s),
[52]	1990–2014	A standard time-series quasi-Poisson	52 capitals	Daily mortality data natural causes $T_{\text{mean day}}$ (European climate assessment and dataset ERA-5 reanalysis data)
[53]	1991–1993	Autoregressive Poisson regression	Valencia (city)	Daily number of deaths $T_{\text{mean day}}$ Relative humidity
[54]	1991–2008	ARIMA model	Castile-La Mancha (5 capital provinces cities)	Daily mortality natural causes except accidents $T_{\text{max day}}$
[55]	1992–2015	Time-stratified case-crossover design	Barcelona	Natural mortality data during (summer months, excluding deaths < 13 years old) $T_{\text{mean day}}$ (source Urb Clim)
[56]	1993–2013	DLNM models	50 provinces (excluding Ceuta and Melilla)	Daily number of deaths $T_{\text{max day}}$
[14]	1995–2000	ARIMA modeling	Madrid, hospital admissions (city)	Hospital admissions organic and circulatory $T_{\text{max day}}$ $T_{\text{min day}}$ Relative humidity at 7am Daily air-pollutant concentrations
[57]	1995–2004	GEE for Poisson regression	52 capitals	Total daily mortality $T_{\text{max day}}$ $T_{\text{min day}}$
[58]	2000–2009	ARIMA and GLM with the Poisson regression	52 capitals	Daily mortality natural, circulatory, and respiratory $T_{\text{max day}}$

(continued)

Table 1 (continued)

References	Dates	Methodology	Locations	Observed variable
[59]	2000–2009	From [60]	52 capitals	Daily mortality natural causes $T_{\text{max day}}$ $T_{\text{min day}}$
[61]	2000–2009	GLM models with the Poisson link	Five geographical areas of the Madrid region	Daily mortality natural, circulatory, and respiratory $T_{\text{max day}}$ $T_{\text{min day}}$
[62]	2001–2009	GLM with the Poisson regression link	Madrid (city)	Daily deaths Parkinson disease (PD.) Daily PD-related emergency admissions $T_{\text{max day}}$
[63]	2001–2009	GLM with the Poisson regression	Madrid (Autonomous region)	Daily emergency hospital admissions natural, respiratory, and circulatory causes $T_{\text{max day}}$
[13]	2003–2003	Poisson regression model	50 provincial capitals	Observed versus expected deaths
[64]	2003–2013	DLNM Effects of night heat: a quasi-Poisson regression with GAM	Barcelona region	Daily mortality natural causes cardiovascular and respiratory diseases T_{hourly}
[65]	2010–2013	Poisson regression models	Madrid (17 districts)	Annual aggregated data on daily mortality due to natural causes T_{heat} $T_{\text{max day}}$
[66]	2013–2017	Time series regression models	Madrid autonomous region	Incidence of ST-segment elevation myocardial infarction T_{max}
[67]	2014–2021	ARIMA models with exogenous variables	Zaragoza (city)	Daily mortality natural causes $T_{\text{max day}}$
[28]	2021–2100	GLM with the Poisson scenario RCP8.5	52 provincial capitals	Population attributable risk (PAR) Mortality rate T_{max} $T_{\text{threshold}}$ annual average number of heatwave days

among the elderly, in mortality for cardiovascular causes, and in towns with socioeconomic vulnerability [56]. The threshold temperature is another critical aspect when determining heat mortality; a study concluded that western locations and low mean summer temperatures were associated with higher relative risks, suggesting thresholds may have been set too high in those areas [57].

Another study on extreme temperatures in Spain compared the daily heat mortality per provincial capital versus the cold-related mortality (3 deaths/day vs. 3.48 deaths/day), as a result, concluding that making prevention plans against low-temperature days was more "profitable" than prevention plans against high-temperature days in terms of avoidable mortality. Finally, after analyzing heat-attributable mortality across 52 provinces, attention should be drawn to the high mortality registered in Madrid and Barcelona compared to the remaining Spanish provinces [59]. However, a study comparing the effect of cold and heat in Castilla-La Mancha concluded the opposite: heat is statistically more significant than the effect of cold [44]. In that study, the variable, heatwave duration, was important in all-cause and respiratory-cause mortality, with wave persistence being related to a mean 3.5% increase in daily organic-cause mortality.

Considering a temperature increase of 1.6 °C across 2021–2050 and 3.3 °C for 2051–2100 [28] and no heat-adaptation process, overall annual mortality attributable to high temperatures in Spain would amount to 1414 deaths/year for 2021–2050, rising to 12,896 deaths/year in the period 2051–2100 [28]. However, under a heat-adaptation process scenario, the authors calculate that the annual mortality would decrease to 651 deaths/year for 2021–2050 and 931 deaths/year for 2051–2100.

3.2 *Heat-Specific Deaths in Spain*

As established and maintained by the WHO, the International Classification of Diseases (ICD) is a disease coding system [68]. It is used for mortality data; death certificates include both underlying and contributory causes of deaths through an ICD code.

To illustrate the Spanish scenario, population and mortality data obtained from the INE alongside the heat waves identified by the AEMET are presented. This approach allows a more context-specific analysis. In this section, heat-related deaths are those in which the death certificate mentions heat-related ICD-10 codes, see Table 2 captures heat-related ICD codes.

Publicly available data from INE only provides information on the underlying cause of death; therefore, deaths coded under contributory factor T67 could not be obtained. For this chapter, X30 codes are included as the main focus is heat produced naturally. Deaths due to exposure to heat of man-made origin are not appropriate concerning heatwaves. No deaths due to urticaria due to heat have been registered between 1999 and 2019.

Figure 7 provides a visual representation of heat-related deaths in Spain in total number and ratios combined with HW days. Evaluating the 20-year time frame, from

Table 2 ICD-10 codes related to heat exposure

ICD-10 code	Description
L50.2	Urticaria due to cold and heat
T67[a]	Effects of heat and light
X10–X19	Contact with heat and hot substances
X30	Exposure to excessive natural heat (including excessive heat as the cause of sunstroke exposure to heat)
W92	Exposure to excessive heat of man-made origin

According to the ICD-10, T67 can only be considered a contributing cause

the heat-related deaths (X30), 62% were people over 65 years old, and 58% were men. As seen in a 2015 study, 1.1% of deaths in Spain can be attributed to heat [69]. Many studies have suggested that the mortality exceeded mainly was attributed to people over 65 years old [50].

As mentioned earlier, heat is also an aggravating factor to other co-morbidities. Data from deaths due to other causes such as cardiovascular or respiratory triggered by heatwaves has not been captured, as has been already considered in the research presented in Table 1. Access to both attributable and contributing causes could provide a more comprehensive analysis of the effect of heatwaves in Spain

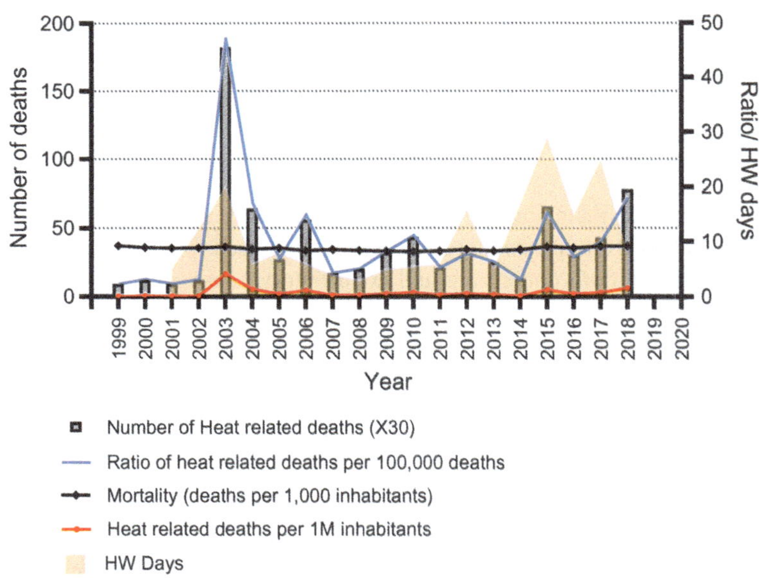

Fig. 7 Evolution of the number of heat-related deaths, the ratio of heat-related deaths per 100,000 deaths, heat-related deaths per 1 million inhabitants, mortality, and heatwave (HW) number of days. Elaborated by authors with data from the National Institute of Statistics and the Meteorological State Agency of Spain

from 1999 to 2019, codes T67. Similarly, accessing the monthly death rates and data on heat-related ICD codes would have allowed analysis focused on the warmer season (June to September). It is important to know that other confounders need to be considered, such as age, geographical location, co-morbidities, the month of death, living conditions, or socioeconomic status.

4 Mitigation Techniques

An increase in temperature is a health hazard, especially to vulnerable populations. Human physiology is tested when exposed to heat. People suffering from obesity, respiratory, cardiovascular diseases, diabetes, or renal failures have a decreased ability to adapt to environmental challenges [70]. As seen in Spain [71] and internationally, healthcare systems are overwhelmed by the increase in Emergency presentations [72] and hospital admissions [73] during heatwaves.

Collaboration between meteorological, government, and health institutions [74] is vital to ensure that weather forecast provides valuable information in anticipation of heatwaves. Using a proactive approach, weather forecasts can be quickly disseminated in media and social platforms to ensure the community is aware and prepared by early warning systems. Healthcare systems can equip themselves with resources and workforce planning to ensure staff levels are appropriate to meet the surge in demand [75]. Due to the time-sensitive nature of heat-related conditions [76], healthcare professionals looking after patients in emergency medical services need to be familiar and competent in rapid recognition and treatment. Regular training, adequate protocols, and guidelines for heat-related illnesses management should be implemented across healthcare systems [76]. Similarly, nursing homes and care facilities can set up plans to ensure residents' exposure to heat is minimized by installing a cooled or air-conditioned room.

Public health should set up a broad approach with different strategies, initiatives, and public campaigns to reduce the heat-related risks [77]. Awareness and educational campaigns targeted to the population can help increase awareness and knowledge on preparing, preventing, and managing heat-related conditions as they have shown promising results [78]. These campaigns should be done regularly at the beginning of summer and if heat waves are expected, focusing on those more vulnerable. Distribution and broadcast of this messaging could be done in primary care, pharmacy, retail, education centers, and places of work.

The mitigation techniques mentioned above need to be suited for the purpose. In addition, context-specific adaptation [79] is needed due to Spanish climate, housing, healthcare services, socioeconomic, and cultural diversity. Another critical element is ensuring the surveillance system works properly and provides valuable information. As demonstrated in France in the 2003 European heatwaves, authorities could not quickly assess the magnitude of the problem due to their surveillance [74]. As a result, people died due to existing diseases aggravated by the heat.

To be successful, healthcare mitigation is a multifaceted approach requiring prediction and prevention programs combined with emergency responses and accurate surveillance. After the 2003 heatwave, many European countries, including Spain, introduced HHPP, and the WHO developed a guide to encourage adoption. Multiple before and after studies have provided strong results. In the Spanish context, there is national HHPP, but within different autonomous communities, there are different characteristics [56]. The plan includes weather forecasts and the definition of four levels of risk depending on the number of days of HW. This plan is also combined with preventive information dissemination to the general population, a targeted focus on vulnerable groups, and a hotline and emergency services activation. Moreover, in 2008 the WHO developed a EuroHeat project made up of 8 elements: a heat-health action plan [80].

In Spain, the implementation of HHPP, as seen in a study [56] comparing the before and after HHPP implementation over ten years in eight regions, showed a positive outcome effect and reduction in extreme heat–related mortality. The more elements of the HHPP national plan incorporated, the better the resulting effects. Interestingly, the greatest effect in mortality reduction was observed in the elderly population. However, due to the variability of factors, such as temperature, heat adaptation, and access to air conditioner among many, those factors could be cofounders. Another vital factor to consider is the threshold chosen to issue heat warnings. Initially, it was set by a climatological criterion (95th percentile) [58] without considering the context-specific characteristics as social-economic and demographic. However, due to those characteristics, it is known that heat habituation by population leads to different mortality threshold temperatures; therefore, the threshold should consider a whole array of attributes that characterize a society.

Several cities have implemented different measures to alleviate the excess heat in the summer, especially during a heatwave. For example, Barcelona's city Hall enabled 155 "climate refuges" directed toward vulnerable collectives [81]. These spaces could be interior or exterior with shadow, water, and resting areas. With this measure, 37.5% of the population had access to a refuge at a 5-min walk and 87.6% at a 10-min walk.

A range of strategies are being implemented currently in different cities. Madrid has proposed the plan "Madrid + Natural" and designated 17 different strategies to mitigate the climate, such as green façades, green roofs, shading devices, cool roofs, and vegetation in streets [82]. Sevilla City Hall has partnered with the University of Sevilla to create "Sombras/shades" project to find comfort indicators, analyze passive strategies, and monitor their results. Many locations (Sevilla, Ecija, Utrera, Granada, and Valencia) install canvases during the summer months to shade streets in the city centers to reduce direct solar radiation toward the pavement. Nevertheless, currently, there is no available data on the effectiveness of the different proposed mitigation techniques.

5 Conclusions

This chapter presents the trend of heat-related mortality research in Spain. A context-specific overview is offered with information on population trends, climate, heat-waves, and built environment in Spain. The current scenario of an increasing and aging population combined with rising temperatures and extreme weather events present severe challenges for the near future for the Spanish population.

The general conclusion is that there are strong links between overheating and the occurrence of heatwaves on the number of deaths; there has been a decline in the number of heat-related deaths. The origin of the decline could be due to the implementation of heat-health prevention plans or intrinsic characteristics of the regions. However, the situation is very heterogeneous due to different climates, expenditures on the healthcare system, and urban morphologies.

The published research on the effect of heatwaves and mortality dealt with different population samples, regions of study, different health conditions, and climatic parameters. Thus, research should focus on standardizing some aspects. For example, the causes of deaths are considered; some authors exclude respiratory and cardiac causes, while others consider all-natural deaths. In addition, there is no consensus on the temperature variable or variables they considered (daily maximum temperature, daily mean temperature, daily minimum temperature). Moreover, very few studies considered relative humidity as a parameter that could reduce mortality. Finally, it has to be noted that most of the meteorological data in all studies were located nearby airports, so there might be a deviation when calculating threshold temperatures and other derived parameters.

This review chapter provides an updated overview of the effect of heat on death in Spain and offers some insights on the mitigation strategies that might aid in diminishing the effect of overheating. Research on quantifying the effect of mitigation is a promising field with great potential to improve the overall conditions of the population.

References

1. United Nations, Department of Economic and Social Affairs, Population Division (2019) World urbanization prospects: the 2018 revision
2. Yin C, Yuan M, Lu Y et al (2018) Effects of urban form on the urban heat island effect based on spatial regression model. Sci Total Environ 634:696–704. https://doi.org/10.1016/j.scitotenv.2018.03.350
3. Fabiani C, Pisello AL, Bou-Zeid E et al (2019) Adaptive measures for mitigating urban heat islands: The potential of thermochromic materials to control roofing energy balance. Appl Energy 247:155–170. https://doi.org/10.1016/j.apenergy.2019.04.020
4. Oke TR (1982) The energetic basis of the urban heat island. Q J R Meteorol Soc 108:1–24. https://doi.org/10.1002/qj.49710845502
5. Santamouris M (2020) Recent progress on urban overheating and heat island research. Integrated assessment of the energy, environmental, vulnerability and health impact. Synergies with the global climate change. Energy Build 207:109482. https://doi.org/10.1016/j.enbuild.2019.109482

6. Santamouris M (2015) Analyzing the heat island magnitude and characteristics in one hundred Asian and Australian cities and regions. Sci Total Environ 512–513:582–598. https://doi.org/10.1016/j.scitotenv.2015.01.060
7. O'Brien GA, Ross NA, Strachan IB (2019) The heat penalty of walkable neighbourhoods. Int J Biometeorol. https://doi.org/10.1007/s00484-018-01663-0
8. Brooke Anderson G, Bell ML (2011) Heat waves in the United States: mortality risk during heat waves and effect modification by heat wave characteristics in 43 U.S. Communities. Environ Health Perspect 119:210–218. https://doi.org/10.1289/ehp.1002313
9. Heat and Health. https://www.who.int/news-room/fact-sheets/detail/climate-change-heat-and-health. Accessed 27 Nov 2021
10. Brennenstuhl H, Will M, Ries E et al (2021) Patterns of extreme temperature-related catastrophic events in Europe including the Russian Federation: a cross-sectional analysis of the Emergency Events database. BMJ Open 11. https://doi.org/10.1136/bmjopen-2020-046359
11. Sakka A, Santamouris M, Livada I et al (2012) On the thermal performance of low income housing during heat waves. Energy Build 49:69–77. https://doi.org/10.1016/j.enbuild.2012.01.023
12. Baccini M, Biggeri A, Accetta G et al (2008) Heat effects on mortality in 15 European cities. Epidemiology 19:711–719. https://doi.org/10.1097/EDE.0b013e318176bfcd
13. Simon F, Lopez-Abente G, Ballester E, Martinez F (2005) Mortality in Spain during the heat waves of summer 2003. Eurosurveillance 10:9–10. https://doi.org/10.2807/esm.10.07.00555-en
14. Linares C, Díaz J (2008) Impact of high temperatures on hospital admissions: comparative analysis with previous studies about mortality (Madrid). Eur J Public Health 18:317–322. https://doi.org/10.1093/eurpub/ckm108
15. Royé D, Sera F, Tobías A et al (2021) Effects of hot nights on mortality in Southern Europe. Epidemiol Camb Mass 32:487–498. https://doi.org/10.1097/EDE.0000000000001359
16. Escandón R, Suárez R, Sendra JJ et al (2019) Predicting the impact of climate change on thermal comfort in a building category: the case of linear-type social housing stock in Southern Spain. Energies 12:2238. https://doi.org/10.3390/en12122238
17. Rae A (2018) Think your country is crowded? These maps reveal the truth about population density across Europe. In: The Conversation. http://theconversation.com/think-your-country-is-crowded-these-maps-reveal-the-truth-about-population-density-across-europe-90345. Accessed 21 Nov 2021
18. Densidad de población de Municipio de Madrid. http://www.madrid.org/desvan/desvan/AccionDatosUnaSerie.icm?codTema=201092&codMun=ZZ01. Accessed 3 Dec 2021
19. (2020) Idescat. Densidad de población. https://www.idescat.cat/pub/?id=aec&n=250&lang=es. Accessed 27 Nov 2021
20. Boudoulas KD, Triposkiadis F, Stefanadis C, Boudoulas H (2017) The endlessness evolution of medicine, continuous increase in life expectancy and constant role of the physician. Hellenic J Cardiol 58:322–330. https://doi.org/10.1016/j.hjc.2017.05.001
21. van Weel C (1996) Chronic diseases in general practice: the longitudinal dimension. Eur J Gen Pract 2:17–21. https://doi.org/10.3109/13814789609161652
22. Dall TM, Gallo PD, Chakrabarti R et al (2013) An aging population and growing disease burden will require a large and specialized health care workforce by 2025. Health Aff (Millwood) 32:2013–2020
23. (2018) Current health expenditure (% of GDP)—Spain|Data. https://data.worldbank.org/indicator/SH.XPD.CHEX.GD.ZS?locations=ES. Accessed 29 Nov 2021
24. Cappellaro G, Fattore G, Torbica A (2009) Funding health technologies in decentralized systems: a comparison between Italy and Spain. Health Policy 92:313–321. https://doi.org/10.1016/j.healthpol.2009.05.004
25. Campbell S, Remenyi TA, White CJ, Johnston FH (2018) Heatwave and health impact research: a global review. Health Place 53:210–218. https://doi.org/10.1016/j.healthplace.2018.08.017
26. Kottek M, Grieser J, Beck C et al (2006) World Map of the Köppen-Geiger climate classification updated. Meteorol Z 259–263. https://doi.org/10.1127/0941-2948/2006/0130

27. Departamento de Producción de la Agencia Estatal de Meteorología de España, Departamento de Meteorologia e Clima, Instituto de Meteorologoa de Portugal (2011) Atlas Climático Ibérico/Iberian Climate Atlas
28. Díaz J, Sáez M, Carmona R et al (2019) Mortality attributable to high temperatures over the 2021–2050 and 2051–2100 time horizons in Spain: adaptation and economic estimate. Environ Res 172:475–485. https://doi.org/10.1016/j.envres.2019.02.041
29. Olas de calor en España desde (1975) Documento actualizado en enero 202. https://aemetblog.es/2020/06/25/olas-de-calor-en-espana-desde-1975-2/. Accessed 20 Nov 2021
30. Lorenzo N, Díaz-Poso A, Royé D (2021) Heatwave intensity on the Iberian Peninsula: future climate projections. Atmos Res 258:105655. https://doi.org/10.1016/j.atmosres.2021.105655
31. (2016) Problem of overheating in summer. In: Energy—European Communication. https://ec.europa.eu/energy/content/problem-overheating-summer_en. Accessed 7 Jul 2021
32. Residencial, comercial e institucional. https://www.miteco.gob.es/es/cambio-climatico/temas/mitigacion-politicas-y-medidas/edificacion.aspx. Accessed 28 Nov 2021
33. Chvatal KMS, Corvacho H (2009) The impact of increasing the building envelope insulation upon the risk of overheating in summer and an increased energy consumption. J Build Perform Simul 2:267–282. https://doi.org/10.1080/19401490903095865
34. Fosas D, Coley DA, Natarajan S et al (2018) Mitigation versus adaptation: does insulating dwellings increase overheating risk? Build Environ 143:740–759. https://doi.org/10.1016/j.buildenv.2018.07.033
35. Pyrgou A, Castaldo VL, Pisello AL et al (2017) On the effect of summer heatwaves and urban overheating on building thermal-energy performance in central Italy. Sustain Cities Soc 28:187–200. https://doi.org/10.1016/j.scs.2016.09.012
36. Escandón R, Suárez R, Alonso A, Mauro GM (2022) Is indoor overheating an upcoming risk in southern Spain social housing stocks? Predictive assessment under a climate change scenario. Build Environ 207:108482. https://doi.org/10.1016/j.buildenv.2021.108482
37. Bartesaghi Koc C, Osmond P, Peters A (2018) Evaluating the cooling effects of green infrastructure: a systematic review of methods, indicators and data sources. Sol Energy 166:486–508. https://doi.org/10.1016/j.solener.2018.03.008
38. Jayasooriya VM, Ng AWM, Muthukumaran S, Perera BJC (2017) Green infrastructure practices for improvement of urban air quality. Urban For Urban Green 21:34–47. https://doi.org/10.1016/j.ufug.2016.11.007
39. Selmi W, Weber C, Rivière E et al (2016) Air pollution removal by trees in public green spaces in Strasbourg city, France. Urban For Urban Green 17:192–201. https://doi.org/10.1016/j.ufug.2016.04.010
40. Maas J, Verheij RA, Groenewegen PP et al (2006) Green space, urbanity, and health: how strong is the relation? J Epidemiol Community Health 60:587–592. https://doi.org/10.1136/jech.2005.043125
41. Barlow P, Lyons S, Nolan A (2021) How Perceived adequacy of open public space is related to objective green space and individuals' opinions of area-level quality. Sustainability 13:8516. https://doi.org/10.3390/su13158516
42. (2013) INEbase/Encuesta de condiciones de vida. In: INE. https://www.ine.es/dyngs/INEbase/es/operacion.htm?c=Estadistica_C&cid=1254736176807&menu=resultados&idp=1254735976608#!tabs-1254736194824. Accessed 29 Nov 2021
43. Miron IJ, Criado-Alvarez JJ, Diaz J et al (2008) Time trends in minimum mortality temperatures in Castile-La Mancha (Central Spain): 1975–2003. Int J Biometeorol 52:291–299. https://doi.org/10.1007/s00484-007-0123-6
44. Linares C, Diaz J, Tobías A et al (2015) Impact of heat and cold waves on circulatory-cause and respiratory-cause mortality in Spain: 1975–2008. Stoch Environ Res Risk Assess 29:2037–2046. https://doi.org/10.1007/s00477-014-0976-2
45. Miron IJ, Linares C, Montero JC et al (2015) Changes in cause-specific mortality during heat waves in central Spain, 1975–2008. Int J Biometeorol 59:1213–1222. https://doi.org/10.1007/s00484-014-0933-2

46. Achebak H, Devolder D, Ballester J (2018) Heat-related mortality trends under recent climate warming in Spain: a 36-year observational study. PLOS Med 15:e1002617. https://doi.org/10.1371/journal.pmed.1002617
47. Díaz J, Carmona R, Mirón IJ et al (2018) Time trend in the impact of heat waves on daily mortality in Spain for a period of over thirty years (1983–2013). Environ Int 116:10–17. https://doi.org/10.1016/j.envint.2018.04.001
48. Follos F, Linares C, Vellón JM et al (2020) The evolution of minimum mortality temperatures as an indicator of heat adaptation: the cases of Madrid and Seville (Spain). Sci Total Environ 747. https://doi.org/10.1016/j.scitotenv.2020.141259
49. Díaz J, Jordán A, García R et al (2002) Heat waves in Madrid 1986–1997: effects on the health of the elderly. Int Arch Occup Environ Health 75:163–170. https://doi.org/10.1007/s00420-001-0290-4
50. deCastro M, Gomez-Gesteira M, Ramos AM et al (2011) Effects of heat waves on human mortality, Galicia, Spain. Clim Res 48:333–341. https://doi.org/10.3354/cr00988
51. Royé D, Codesido R, Tobías A, Taracido M (2020) Heat wave intensity and daily mortality in four of the largest cities of Spain. Environ Res 182. https://doi.org/10.1016/j.envres.2019.109027
52. Royé D, Íñiguez C, Tobías A (2020) Comparison of temperature–mortality associations using observed weather station and reanalysis data in 52 Spanish cities. Environ Res 183:109237. https://doi.org/10.1016/j.envres.2020.109237
53. Ballester F, Corella D, Pérez-Hoyos S et al (1997) Mortality as a function of temperature. A study in Valencia, Spain, 1991–1993. Int J Epidemiol 26:551–561. https://doi.org/10.1093/ije/26.3.551
54. Linares C, Sánchez R, Mirón IJ, Díaz J (2015) Has there been a decrease in mortality due to heat waves in Spain? Findings from a multicity case study. J Integr Environ Sci 12:153–163. https://doi.org/10.1080/1943815X.2015.1062032
55. Ingole V, Marí-Dell'Olmo M, Deluca A et al (2020) Spatial variability of heat-related mortality in Barcelona from 1992–2015: a case crossover study design. Int J Environ Res Public Health 17. https://doi.org/10.3390/ijerph17072553
56. Martínez-Solanas È, Basagaña X (2019) Temporal changes in temperature-related mortality in Spain and effect of the implementation of a heat health prevention plan. Environ Res 169:102–113. https://doi.org/10.1016/j.envres.2018.11.006
57. Tobias A, Armstrong B, Zuza I et al (2012) Mortality on extreme heat days using official thresholds in Spain: a multi-city time series analysis. BMC Public Health 12:133. https://doi.org/10.1186/1471-2458-12-133
58. Díaz J, Carmona R, Mirón IJ et al (2015) Geographical variation in relative risks associated with heat: update of Spain's heat wave prevention plan. Environ Int 85:273–283. https://doi.org/10.1016/j.envint.2015.09.022
59. Carmona R, Díaz J, Mirón IJ et al (2016) Mortality attributable to extreme temperatures in Spain: a comparative analysis by city. Environ Int 91:22–28. https://doi.org/10.1016/j.envint.2016.02.018
60. Díaz J, Carmona R, Mirón IJ et al (2015) Comparison of the effects of extreme temperatures on daily mortality in Madrid (Spain), by age group: the need for a cold wave prevention plan. Environ Res 143:186–191. https://doi.org/10.1016/j.envres.2015.10.018
61. López-Bueno JA, Díaz J, Linares C (2019) Differences in the impact of heat waves according to urban and peri-urban factors in Madrid. Int J Biometeorol 63:371–380. https://doi.org/10.1007/s00484-019-01670-9
62. Linares C, Martinez-Martin P, Rodríguez-Blázquez C et al (2016) Effect of heat waves on morbidity and mortality due to Parkinson's disease in Madrid: a time-series analysis. Environ Int 89–90:1–6. https://doi.org/10.1016/j.envint.2016.01.017
63. Díaz J, López IA, Carmona R et al (2018) Short-term effect of heat waves on hospital admissions in Madrid: analysis by gender and comparision with previous findings. Environ Pollut 243:1648–1656. https://doi.org/10.1016/j.envpol.2018.09.098

64. Royé D (2017) The effects of hot nights on mortality in Barcelona, Spain. Int J Biometeorol 61:2127–2140. https://doi.org/10.1007/s00484-017-1416-z
65. López-Bueno JA, Díaz J, Sánchez-Guevara C et al (2020) The impact of heat waves on daily mortality in districts in Madrid: the effect of socio demographic factors. Environ Res 190. https://doi.org/10.1016/j.envres.2020.109993
66. García-Lledó A, Rodríguez-Martín S, Tobías A et al (2020) Heat waves, ambient temperature, and risk of myocardial infarction: an ecological study in the community of Madrid. Rev Esp Cardiol 73:300–306. https://doi.org/10.1016/j.recesp.2019.05.009
67. Roldán E, Gómez M, Pino MR et al (2016) The effect of climate-change-related heat waves on mortality in Spain: uncertainties in health on a local scale. Stoch Environ Res Risk Assess 30:831–839. https://doi.org/10.1007/s00477-015-1068-7
68. World Health Organization (2004) International statistical classification of diseases and related health problems, 10th revision (ICD–10). World Health Organization, Geneva, Switzerland
69. Gasparrini A, Guo Y, Hashizume M et al (2015) Mortality risk attributable to high and low ambient temperature: a multicountry observational study. Lancet 386:369–375. https://doi.org/10.1016/S0140-6736(14)62114-0
70. Kenny GP, Yardley J, Brown C et al (2010) Heat stress in older individuals and patients with common chronic diseases. CMAJ 182:1053–1060
71. Cajoto V, JA DP (2005) Health impact of 2003 heat wave at Hospital de Riveira (A Coruna). In: Anales de medicina interna (Madrid, Spain: 1984), pp 15–20
72. Heaviside C, Macintyre H, Vardoulakis S (2017) The urban heat island: implications for health in a changing environment. Curr Environ Health Rep 4:296–305. https://doi.org/10.1007/s40572-017-0150-3
73. Hansen AL, Bi P, Ryan P et al (2008) The effect of heat waves on hospital admissions for renal disease in a temperate city of Australia. Int J Epidemiol 37:1359–1365. https://doi.org/10.1093/ije/dyn165
74. Kirch W, Menne B, Bertollini R (2005) Extreme weather events and public health responses
75. Runkle J, Svendsen ER, Hamann M et al (2018) Population health adaptation approaches to the increasing severity and frequency of weather-related disasters resulting from our changing climate: a literature review and application to Charleston, South Carolina. Curr Environ Health Rep 5:439–452
76. Rublee C, Dresser C, Giudice C et al (2021) Evidence-based heatstroke management in the emergency department. West J Emerg Med 22:186–195. https://doi.org/10.5811/westjem.2020.11.49007
77. Kovats RS, Hajat S (2008) Heat stress and public health: a critical review. Annu Rev Public Health 29:41–55
78. Hasan F, Marsia S, Patel K et al (2021) Effective community-based interventions for the prevention and management of heat-related illnesses: a scoping review. Int J Environ Res Public Health 18:8362. https://doi.org/10.3390/ijerph18168362
79. Dixon-Woods M (2014) The problem of context in quality improvement. Perspect Context Lond Health Found 87–101
80. Matthies F, Bickler G, Hales S, Marín NC (2008) Heat-health action plans: guidance. World Health Organization
81. Barcelona A de (2021) 155 refugios climáticos repartidos por toda la ciudad para hacer frente al calor|Info Barcelona|Ayuntamiento de Barcelona. https://www.barcelona.cat/infobarcelona/es/tema/emergencia-climatica/mas-de-160-refugios-climaticos-repartidos-por-toda-la-ciudad-para-hacer-frente-al-calor-4_1083949.html. Accessed 27 Nov 2021
82. Madrid A de (2019) Cambio Climático - Madrid + Natural - Ayuntamiento de Madrid. https://www.madrid.es/portales/munimadrid/es/Inicio/El-Ayuntamiento/Medio-ambiente/Energia-y-cambio-climatico?vgnextoid=0ca36936042fc310VgnVCM1000000b205a0aRCRD&vgnextchannel=4b3a171c30036010VgnVCM100000dc0ca8c0RCRD&idCapitulo=8617429. Accessed 27 Nov 2021

Chapter 17
Impact of Urban Overheating and Heat-Related Mortality in Mexico

M. E. González-Trevizo, K. E. Martínez-Torres, A. Luna-León, J. F. Armendáriz-López, and J. Sandoval-Félix

Abstract Latin America and the Caribbean is the region with the largest population concentrated in megacities, 81% of the population live in urban areas, and by 2050 the proportion will hover at 89%. México, with its inequitable healthcare access and almost 100 million people living in 4,189 urban settlements faces a sustainable city agenda for climate action and good health and well-being. This chapter addresses morbidity and mortality pre-pandemic data related to heat-related diseases from the National Morbidity Yearbook and the National Mortality Census to identify and characterize trends by state, age, gender, and environmental conditions under the ICD-10 system. Evidence presented in a municipal case study indicates that ischemic heart disease counts for ≈33% of total deaths in the urban context, being an elderly female population above 55 years old the most vulnerable.

Keywords Urban overheating · Health · Heat-related mortality · Morbidity · Energy

M. E. González-Trevizo (✉) · K. E. Martínez-Torres
Facultad de Ingeniería, Arquitectura y Diseño, Universidad Autonóma de Baja California, Ensenada, México
e-mail: eduardo.gonzalez35@uabc.edu.mx

K. E. Martínez-Torres
e-mail: karen.martinez24@uabc.edu.mx

A. Luna-León
Facultad de Arquitectura y Diseño, Universidad Autonóma de Baja California, Mexicali, México
e-mail: anibal@uabc.edu.mx

J. F. Armendáriz-López
Facultad de Ciencias de la Ingeniería y La Tecnologia, Universidad Autonóma de Baja California, Tijuana, México
e-mail: farmendariz@uabc.edu.mx

J. Sandoval-Félix
Urban Lab: Laboratorio de Urbano y de Políticas Públicas, Ensenada, México
e-mail: javier.sandoval@urbanlab.city

N. Aghamohammadi and M. Santamouris (eds.), *Urban Overheating: Heat Mitigation and the Impact on Health*, Advances in Sustainability Science and Technology,
https://doi.org/10.1007/978-981-19-4707-0_17

1 Introduction

The IPCC's Special Report on the Impacts of 1.5 °C Global Warming [1] underlined the significant difficulty for cities in developing adaptation measures that allow for the maintenance of favorable circumstances for the population. The rise in sea level poses a risk of flooding in around 130 port towns worldwide with populations over one million, resulting in a slew of disputes related to population displacement and migration. However, there is evidence that climate change poses significant challenges to water, energy, food insecurity, and transportation infrastructure. Similarly, the increased frequency of heatwaves, hurricanes, forest fires, and changes in ecosystems endanger various tourist activities critical to the inhabitants' economic activity. Although Reyes et al., [2] assert that new research approaches have increased awareness of the complexity that climate change implies for sustainable development, it will remain necessary to address fundamental issues such as increasing the certainty of environmental data monitoring [3], which aid in comprehending the problem. Additionally, it will be required to develop protocols that facilitate comparisons between investigations [4] to facilitate a more expansive decision-making process.

The Urban Heat Islands (UHI) refer to the increase in temperature experienced by cities concerning the surrounding non-urban areas. This is a typical occurrence regardless of the city's size or climate type. The UHI's impact fields have been classified along seven axes: energy, health-morbidity, health-mortality, urban environmental quality, risk and vulnerability, human comfort, and demography [5]. It is critical to emphasize that two decades ago, UHI were demonstrated to have a considerable effect on building energy use [6]. Indeed, climate variability now plays a critical role in the development of methods for gaining greater control over the energy use of buildings [7]. Similarly, It is critical to optimize and establish criteria for a more effective inquiry design to collect more relevant data [8].

Urban overheating has several harmful implications on human health. Temperature increases drive greater air pollution levels and the duration of heatwaves, which can aggravate the health of people with chronic conditions [9]. Climate change increases respiratory and allergy illnesses because of the population's overexposure to ozone, smoke, pollen, mold, and diseases derived from water and food pollution [10]. In this regard, overheating can amplify the negative consequences of climate change on urban health. On the other hand, it has been calculated that temperature rises have a direct effect on mortality during the hot season, reaching an average estimate of 37%. However, the degree of effect varies significantly between countries and subregions [11]. As a result, additional research must define more precisely the boundaries of the impact of UHI on human health [12]. A greater understanding of inhabitants' thermal comfort will enable more effective urban design [13], which should ideally reach neighborhood-scale [14], involving wind fluxes and vegetation morphology [15].

Elderly people, children, women, and those with chronic diseases, as well as those taking particular medications will be the most vulnerable to fluctuations in the environment's temperature and humidity, as well as to infections spread by mosquito bites [1]. Water pollution is particularly hazardous to pregnant women, in fact, women are historically highly vulnerable to natural disasters. For their part, children under the age of five account for 88% of morbidity due to climate change, as they are more prone to become ill as a result of diverse environmental conditions. Indigenous peoples are also a vulnerable group, as they cannot access food, water, and shelter. It is critical to emphasize that those who work outdoors are directly exposed to poor air quality and various climatic occurrences [10]. Air pollution, in general, can have adverse effects on the respiratory system, heart, kidneys, eyes, skin, brain, bones, blood, and blood vessels. Air pollution has been shown to impair sleep, reproduction, and metabolism. Additionally, it can induce or worsen allergies, diabetes, and certain types of cancer [16, 17].

2 Motivation and Justification

In cities, high and extreme temperatures stimulate the concentration of pollutants. Different contaminants alter rainfall patterns and degrade water quality, wreaking havoc on the subsurface layers and aquatic ecosystems' health. As a result, these pollutants amplify the effect of heat islands by retaining radiation in and around cities [18]. Prolonged exposure to pollutants influences the corrosion of metals, minerals, and vegetation; thus, it is critical to use Geographic Information Systems (GIS) to interpolate data that aids in decision-making to resolve the problem [19]. The shape of flora can help mitigate the effects of UHIs [20], yet the intensities of UHI are lower in cities with arid or semi-arid climates than in green places [21]. As a result, comprehending these dynamics is critical when considering the design or morphology of a city.

Thus, Latin America is the region with the largest population concentrated in megacities, 14.2% reside in six cities with 10 million inhabitants [22] challenged by global warming that exacerbates the risk of heat-related mortality. For decades, there has been a warning about the effects in Mexico and its relationship with the etiology of vector-borne, acute respiratory, and diarrheal diseases, or heat stroke due to extreme weather events on the health of the Mexican population [23]. Therefore, it is essential to understand patterns of mortality rates in urban areas as a matter of public policy interest [24]. This study reports morbidity and heat-related diseases in terms of the ICD-10 system for a pre-pandemic scenario in Mexico (2019), including sociodemographic, age, and environmental aspects in the national context, and through the selection of a local case study, the building cooling demands of two different locations can be observed.

3 Method

This study analyzed morbidity and mortality pre-pandemic data (2019) related to heat-related diseases to identify and characterize components by state, age, gender, and environmental conditions under the ICD-10 system. This study analyzes collected data from the Mexican health system under the criteria described below, in addition, a Case Study of Ensenada city from ongoing research is briefly explored to illustrate energy aspects regarding land surface temperature (LST) on contrasting cooling energy demand for two nodes with a different urban condition:

- **Taxonomy**. Defined by the International Statistical Classification of Diseases and Related Health Problems 10th Revision (ICD-10), a widely used diagnostic tool for epidemiology, health management, and therapeutic purposes worldwide [25].
- **Data sources**. On health and demographic aspects, Observations from the vital statistical data published in censuses and surveys of The National Institute of Statistics, Geography (INEGI, Census 2020) [26, 27], monthly cases per month of occurrence from the Morbidity Yearbook (period 1984–2019) of Mexico's Secretariat of Health [28]. On the environment, normalized climate data from the National Meteorology Service (SMN, period 1981–2010), interpolated data through the Meteonorm® software, which uses the climatic information values from the database of the Global Energy Balance File (GEBA) of the World Meteorological Organization (WMO) [29], and thermal maps from the Informatics Unit for Atmospheric and Environmental Sciences (UNIATMOS, period 1902–2015) [30].
- **Study Area**. (a) National level. Mexico is a federation of 32 states and is bounded on the north by the USA, and on the southeast by Guatemala and Belize. It is the world's eleventh most populated country, with over 124 million citizens, 77% of whom live in metropolitan areas. The population is young, with roughly 27% of residents under the age of 15 and only 7% over 65. Inequality of income and poverty are ongoing health problems. Mexico has the highest rate of income inequality in the OECD [31]. (b) Municipal and city level (case study). The municipality of Ensenada is bounded at 28° and 32° N and −112.8° and −116.9 W, with its 52.51 thousand km^2 of territory, represents 74.1% of the state of Baja California. For its part, Ensenada is a coastal city situated on Todos Santos Bay, a mid-rise building city that presents a mild semi-arid climate (Köppen BSk) [32].
- **LST Retrieval**. An automated mapping algorithm created in ERDAS IMAGINE 2014 compatible with LANDSAT 8 was applied in Ensenada city. For data processing, it uses Bands 10, 4, and 5. The emissivity corrected land surface temperature T_s was calculated as follows using brightness temperature (BT), λ is the wavelength of emitted radiance, and ε_λ is the emissivity calculated [33]:

$$T_S = \frac{\text{BT}}{\left\{1 + \left[\left(\frac{\lambda \text{BT}}{\rho}\right) \ln \varepsilon_\lambda\right]\right\}}$$

- **Energy simulation**. Estimation of energy loads on typical housing in Ensenada city was conducted using TRNSYS® 17, a simulation software used in the fields of building energy performance to simulate the behavior of transient systems [34].

4 Results

Morbidity is recognized as a prelude to mortality, health systems define a picture of heat-related diseases under the ICD-10 system, the same illnesses and symptoms summarized by several researchers [35–37]. Table 1 and Fig. 1 report hospitalizations during the year of study (2019) concerning gender, month, and state. It is important to note that J18.2, J13, and J14, non-acute, care, or self-care cases were not included. As seen below, women are susceptible to respiratory diseases and heatstroke with a representativity of 56% and 54%, while men report vulnerability to ischemic and cerebrovascular diseases, 59% and 53%, respectively.

The map shows the 32 Federal Entities (FE) referred to through blue bubbles that indicate City of Reference (CoR), which concentrates the largest population of each FE, the total population of Mexico is 126,014,024 inhabitants, the most populated state is the Estado de México with 16,992,418 people and the least populated is Colima with 731,391 people. The national average of demographic concentration in CoR is 22.9% with a minimum of 1% in the state of Tlaxcala, up to Aguascalientes with ±60.6%, Mexico City, although it concentrates ±79% of its population, is distributed in its 16 delegations, of which it concentrates ±19% in Iztapalapa with 1,835,486 residents. The population with access to health services (white bubbles) in each CoR has a national average of 76%, the minimum CoR is Tuxtla Gutierrez in Chiapas at 59.1%, and the maximum is La Paz, Baja California Sur at 86.5%. The contrast in the diameter of the white and blue bubbles indicates the concentration of each state, regarding their access to health.

Even though the highest number of hospitalizations in México occurs in January due to the high prevalence of respiratory diseases, pneumonia, and influenza, the CoR's predominant climate is Cwb (subtropical highland climate) [38], with the warmest month being May. The map depicts the may thermal distribution of the

Table 1 Heat-related hospitalizations with respect to gender in México (2019)

ICD-10 Code	Disease	Female	Male	Total
J00-J06, J20, J21	Acute respiratory infections	13,285,490	10,492,948	23,778,438
T67, X30	Exposure to excessive natural heat	2422	2052	4474
I20-I25	Ischemic heart diseases	28,647	41,530	70,177
I60-I67, I69	Cerebrovascular diseases	23,446	26,686	50,132
J09-J11	Influenza	48,608	41,750	90,358
J12-J18	Pneumonia	67,863	68,918	136,781
Total				24,130,360

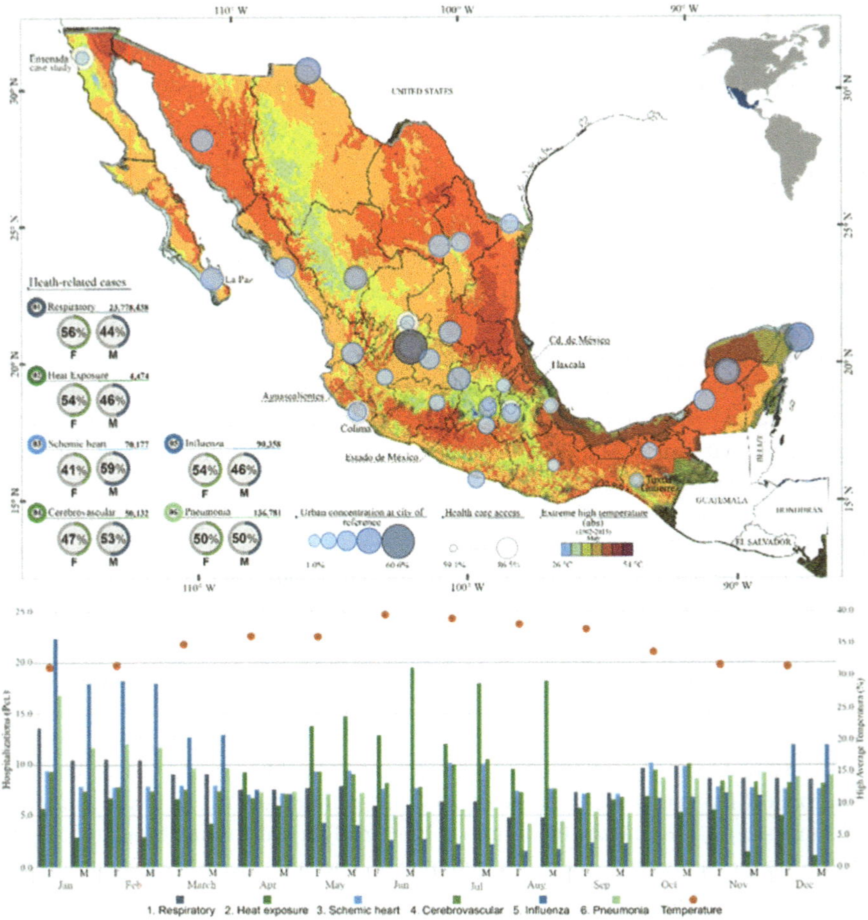

Fig. 1 Heat-related hospitalizations in México based on UNIATMOS map (period: 1902-2015) (2019)

absolute extreme maximum temperatures reported by the UNIATMOS with historical datasets from 1902–2015, and temperatures ranging from 15.3 to 52.5 °C. The preceding is evident in the inverse trend that cases of respiratory diseases present concerning those of heat exposure, which increase in percentage terms during May, especially in the female population (14.8%), June (19.5%), July (18%), and August (18.2%). Similar behavior can be seen in cerebrovascular diseases, whose distribution is sustained and regular throughout the year, but with significant months where the male population stands out, such as July (10.6%), August (7.7%), and October (10.1%).

Referring back to the research aims, this national prelude on morbidity set an opposite reference to address mortality aspects in a more detailed resolution. The mortality rate connected with heat-related disorders was observed in the municipality

Table 2 ICD-10 mortality cases in Ensenada (2019)

ICD-10 Code	Diseases	Urban	Rural
E86.0	Dehydration	2	0
I20-I25	Ischemic heart diseases	381	23
I30-I5A	Other forms of heart disease	154	3
I60-I69	Cerebrovascular diseases	184	8
I10–I16, I26–I28, I70–I99	Other circulatory diseases	161	7
J09-J18	Influenza and pneumonia	139	2
J40-J47	Chronic lower respiratory diseases	81	3
J20–J22; J60-J86; J95-J99	Other respiratory causes	43	2
Total		1,145	48

Note: Some Diseases were not presented, X30 Exposure to excessive natural heat, R50.9 Fever unspecified

of Ensenada, Baja California. 87,249 people live in urban regions and 43,474 individuals live in rural areas. The number of people who died in urban and rural areas in 2019 is shown in Table 2. According to the data, approximately 95.9% (1,145) of the deceased resided in urban regions and 4.1% (48 people) in rural areas. According to the data in this table, the major cause of circulatory death in both locations was ischemic heart disease, followed by cerebrovascular illness. However, influenza and pneumonia had a greater mortality rate attributable to respiratory causes in the urban context. Interestingly, the data in this table indicate that most of the population died from circulatory reasons.

The four major causes of mortality are depicted in Fig. 2. The findings indicate that the highest death rate was observed in a population over the age of 55. According to this figure, women above the age of 70 have an elevated death rate. Additionally, the data indicates that 60% of deaths occurred among men and 40% occurred among women. Additionally, the findings indicated an increase in mortality among men (in blue) due to ischemic heart (I20-I25), with a notable increase between 35 and 39 years -productive age- and 60–64 years and a similar but, diluted pattern in cerebrovascular disease (I60-I69) records. Contrary to what was indicated by the number of hospitalizations nationwide, where it was appreciated that the female population was more susceptible through a clinical picture related to respiratory diseases. As for conditions of hyperthermia, women have upward trends and peak values in older ages, in all four cases of diseases, this can be seen from 7074 years with peak values in the last stage of life, 85 and more.

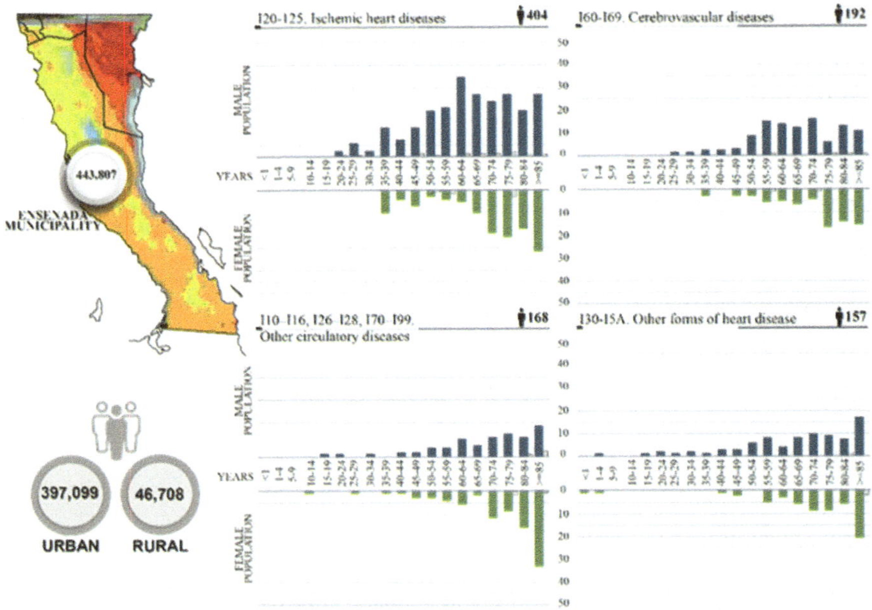

Fig. 2 Mortality rate in Ensenada by gender, urban/rural context, and age based on UNIATMOS map & Pyrgou A & Santamouris M (2018)

4.1 A Brief Exploration of the Urban Overheating and Energy

Due to the escalation of Urban Overheating as a result of climate change, new issues have arisen in addressing the dangers and energy needs associated with meeting building thermal conditions. This section briefly explores the importance of paying attention to the vulnerability of inhabitants of temperate climates because of the thermal adaptability during the warm season (August). Remote sensing techniques were used to correlate LST Overheating criteria with energy consumption of a prototypical residential building and two dissimilar urban nodes in the CoR of Ensenada, Baja California, a port city with BSk climate condition, 330,652 residents, and 105,736 housing units [27] is situated on 31.86 N and −116.60 W [32].

An algorithm for automated mapping of the Land Surface Temperature using Landsat eight thermal infrared sensor Band 10 was performed, as seen in Fig. 3, is useful to observe the urban thermal environment of the ground surface across the city layout. In addition, an accurate and representative hourly time series for a weather typical year using a stochastic model through Meteonorm software was calculated. For this purpose, the location situation was set as "city" and statistical uncertainty is 8% on Global Radiation (W/m^2), 7% on Beam Radiation (W/m^2) and 1.1 °C on Dry Bulb Temperature (°C). Each location is related to a Basic Geostatistical Area

(AGEB), which constitutes the basic unit of the INEGI´s National Geostatistical Framework (Table 3).

A numerical analysis was developed using a mathematical subroutine through Type 56 of TRNSYS® 17 software, a typical urban housing with a construction area of 38.5 m^2 and a volume of 104 m^3 was analyzed, which is built with concrete block walls of 0.12 m thick with a "U" value of 3.918 W/m^2K, a joist and filler block roof system of 0.17 m thick and a "U" value of 1.98 W/m^2K, 3 mm single glass window with U Value of 5.73/m^2K reinforced concrete floor 0.10 m thick with "U" value of 4.68 W/m^2K, wooden door 2.47 W/m^2K.

Internal Building loads were defined for a typical family of four occupants with the following scheduled occupancy, one working parent, one dedicated to household duties, and two children with student activities. The occupant´s internal gains were defined as standing light work (185 W per occupant). Internal gain due to equipment

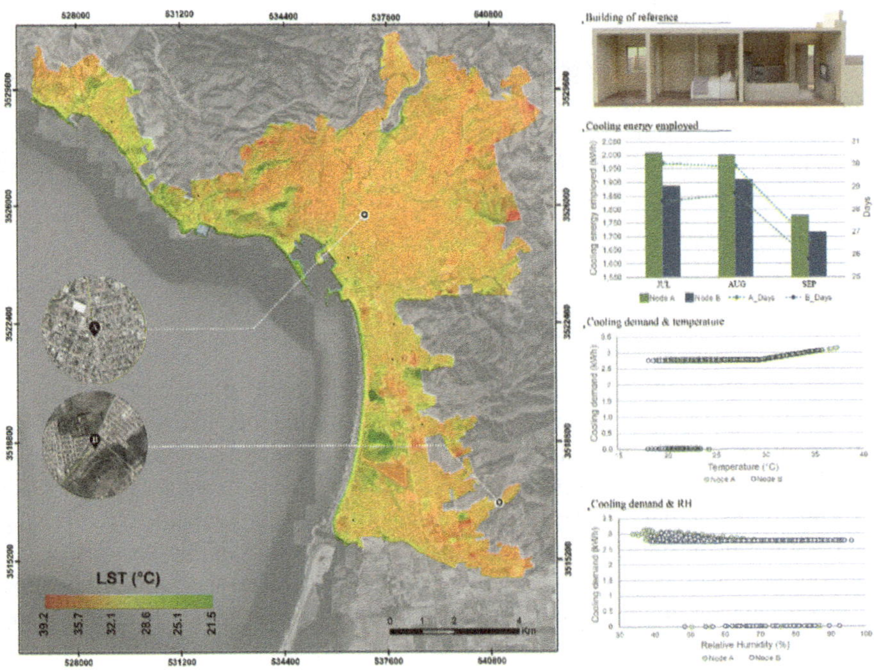

Fig. 3 A brief exploration of urban overheating and energy. Imagen Landsat 8, retrieval date: Aug 20th, 2019, 6:16 pm

Table 3 Urban nodes of analysis

Node	AGEB	Latitude	Longitude	Elevation
A	AGEB 0740	31.86	−116.60	16 masl
B	AGEB 9814	31.78	−116.56	81 masl

considered with a radiative gain peak load of 5.52 kWh, convective peak loads of 1.61 kWh, artificial lightning gains of 5 W/m^2, and finally a constant air change per hour rate (ACH) of 1.0, as infiltration/ventilation. The cooling system´s thermal and energy performance were pre-calculated to study the cooling demand of a real single package of 13 SEER, 1200 CFM, and a total net capacity of 35,000 BTU [39].

Figure 3 shows that July and August are the warmest months with the highest cooling energy demand, from September the temperatures present a decrease. However, when analyzing the relative humidity of the two nodes, it can be observed that node B presents a drier climate compared to the A values. Thus, to determine the energy consumption of the city of Ensenada's warmest and coldest nodes, an energy simulation of these two nodes was conducted. It was discovered that the climate of node A consumes more energy during the summer, which is associated with the low humidity content of the environment, as illustrated in the figure. Contrary to node A, the B node has a greater number of operational days than July. This is because the average temperature in August is higher, which results in a more significant number of operational days.

The operating efficiency of the cooling system has a close relationship with the site´s climatic conditions since the cooling unit has refrigerant condensation problems. In the same figure, it can be seen that the cooling unit is affected when temperatures are greater than 30 C. While, in the case of relative humidity, it affects when you have lower relative humidity.

5 Discussion

Mexico is second only to the United States in terms of obesity prevalence; in terms of life expectancy, women live an average of 77.7 years, and men live to the age of 72. After diabetes mellitus, the only other heat-related chosen cause of death is ischemic heart disease (72.45 per 1000 total deaths). As in Pyrgou and Santamouris [37], the difficulty of predicting the adaptation potential of the population in the face of rising temperatures in the future is highlighted. Likewise, the results of this work also agree with the findings of Liu, et al. [36] about the importance of generating mitigation plans for vulnerability to hyperthermia conditions and managing emergency plans that favor timely actions. Ideally, local governments should implement strategies based on Geographic Information Systems (GIS), to solve the lack of information. However, in Mexico, there are few cases in which there is an uprising of this type throughout the cities.

Mexico is also geographically extensive, has a continental area of 1,959,248 km^2, with 21,000 primary care units operated by public institutions, and has a great variety of climates that hinder specific preventive medicine strategies, to mention, in the 32 CoR (Af, As, Aw, Bsh, BSk, Bwh, Bwk, Cfa, Cwb) with warm months that differ from April to August, there are localities with historical average maximum temperatures of 39.6 °C in Hermosillo, Sonora and extreme temperatures up to 52.0 °C in

Mexicali, Baja California. However, the recurrence and magnitude of heatwaves are increasingly studied in temperate climates [40].

Additionally, the SARS-CoV-2 health crisis imposes a new constraint on the standardization of records, due to the federal government's hospital conversion edict. The most significant constraint on conducting mortality research is the unavailability of high-resolution geographic resources. The national basic geostatistical units (AGEBS) do not include mortality data, and the reasons for death are not linked to specific climatic zone scenarios. The development of protocols for reporting and monitoring claims related to the health of building occupants through Building Information Modeling (BIM) may represent a significant step forward; although in Mexico only the standard NMX-C-527–1-ONNCCE-2017 has been published [41], representing an alternative that is difficult to implement in the short or medium term.

6 Conclusion

Morbidity results in this study showed that the impact on women's health by presenting higher rates than men due to heat exposure (54%), respiratory failure (56%), and influenza (54%). However, the evidence indicates that the impact of pneumonia is similar between men and women. In contrast, men are significantly more affected by schematic heart diseases (59%) and cardiovascular accidents (53%). The latter is worrying for the public health system, given the high rate of obesity that leads to diseases such as diabetes mellitus and cerebrovascular disease.

In terms of Urban Overheating, there is a marked trend in hospitalization for reasons related to heat exposure. The three types of climates presented in 19 Urban areas of reference around México with the highest incidence are Subtropical highland climate (Cwb), Cold semi-arid climate (BSk), and Tropical Savana climate (Aw), they represent 60%.

The present analysis has substantial constraints as a result of the poor level of integration of information management and recording in the national health sector, which is composed of a mix of social insurance schemes, a voluntary public program for the uninsured, and private insurance systems. Among the existing limitations and conclusions, the following can be mentioned:

- Urban overheating is a differential phenomenon that promotes risks in extreme heat events in urban areas of adverse sociodemographic characteristics of the city.
- Acute respiratory infections in autumn/winter and circulatory diseases in spring/summer are dominant risk factors and significant concerns due to income inequality, gentrification, health inequities, and access to health system services.
- Health access systems must develop preventive healthcare programs geared around demographic factors such as gender and age distribution during specific periods of urban overheating exacerbation on specific local climate zone.

- Integrating health statistics into high-resolution geographic systems is critical for establishing preventive measures for chronic conditions in developing countries. [42].
- The evidence refers to the need for alert protocols that connect heat-related pathologies to environmental and atmospheric conditions by seasons, such as temperature, humidity, and air quality.
- A robust methodological strategy for smart cities should consider a tracking data philosophy [43] of relevant tools currently used such as Satellite Remote Sensing, stationary and mobile weather stations, online and print media, prospective scenarios, sectors by density, and landscape typology (residential, offices, industry, commerce, etc.) and implement protocols based on dynamic platforms, i.e., BIM, IoT, RFID, and/or GIS also by sectors depending on their sociodemographic aspects and national regulations [41].

Acknowledgements The time dedicated to this research was possible due to the UABC internal financing research projects, and its Library system data base, Programa para el Desarrollo Profesional Docente (PRODEP) and research activities supported by Consejo Nacional de Ciencia y Tecnología (CONACyT) through its Sistema Nacional de Investigadores (SNI). In addition, the authors would like to gratefully acknowledge B. Eng. Randy Martinez for his valuable orientation and our undergraduate students Alina Vasquez, Linette Carrillo, Lidia Ornelas, Melissa Salgado, and Erick Sanchez for their enthusiasm and assistance with data collection details.

References

1. Roberts D, Pidcock R, Chen Y, Connors S, Tignor M (2019) 2018: global warming of 1.5 °C. An IPCC special report on the impacts of global warming of 1.5 °C above pre-industrial levels and related global greenhouse gas emission pathways, in the context of strengthening the global response to the threat of climate c. Nairobi
2. Reyers B, Folke C, Moore ML, Biggs R, Galaz V (2018) Social-ecological systems insights for navigating the dynamics of the anthropocene. Annu Rev Environ Resour 43:267–289
3. Xi DDZ, Taylor SW, Woolford DG, Dean CB (2019) Statistical models of key components of wildfire risk. Annu Rev Stat Appl 6:197–222
4. Otto FEL (2017) Attribution of weather and climate events. Annu Rev Environ Resour 42:627–646
5. Gonzalez-Trevizo ME, Martinez-Torres KE, Armendariz-Lopez JF, Santamouris M, Bojorquez-Morales G, Luna Leon A (2021) Research trends on environmental, energy and vulnerability impacts of Urban Heat Islands: an overview. Energy Build 246(111051):1–27
6. Santamouris M et al (2001) On the impact of urban climate on the energy consuption of building. Sol Energy 70(3):201–216
7. Li H, Wang Z, Hong T (2021) A synthetic building operation dataset. Sci Data 8(1):1–14
8. Kim SW, Brown RD (2021) Urban heat island (UHI) variations within a city boundary: A systematic literature review. Renew Sustain Energy Rev 148(August 2020):111256
9. United States Environmental Protection Agency (2021) "Heat Island." Heat Island Impacts
10. Levy BS, Patz JA (2017) The impact of climate change on public health, human rights, and social justice, vol 1–5. Elsevier Inc
11. Vicedo-Cabrera AM et al (2021) The burden of heat-related mortality attributable to recent human-induced climate change. Nat Clim Chang 11(6):492–500

12. Leal Filho W et al (2021) Addressing the urban heat islands effect: a cross-country assessment of the role of green infrastructure. Sustain 13(2):1–20
13. Pigliautile I, Pisello AL, Bou-Zeid E (2020) Humans in the city: Representing outdoor thermal comfort in urban canopy models. Renew Sustain Energy Rev 133(August):110103
14. Martinez S, Machard A, Pellegrino A, Touili K, Servant L, Bozonnet E (2021) A practical approach to the evaluation of local urban overheating—a coastal city case-study. Energy Build. 253:111522
15. Di Bernardino A et al (2021) Urban microclimate modelling for comfort and energy studies. In Palme M, Salvati A (eds) Urban microclimate modelling for comfort and energy studies, 1st ed., Cham
16. Schraufnagel DE et al (2019) Air pollution and noncommunicable diseases: a review by the forum of international respiratory societies' environmental committee, part 1: the damaging effects of air pollution. Chest 155(2):409–416
17. Schraufnagel DE et al (2019) Air pollution and noncommunicable diseases: a review by the forum of international respiratory societies' environmental committee, part 2: air pollution and organ systems. Chest 155(2):417–426
18. Mirzaei PA (2015) Recent challenges in modeling of urban heat island. Sustain Cities Soc 19:200–206
19. Vidal F, Vicente R, Mendes Silva J (2019) Review of environmental and air pollution impacts on built heritage: 10 questions on corrosion and soiling effects for urban intervention. J Cult Herit 37:273–295
20. Dugord PA, Lauf S, Schuster C, Kleinschmit B (2014) Land use patterns, temperature distribution, and potential heat stress risk—the case study berlin, Germany. Comput Environ Urban Syst 48:86–98
21. Shahraiyni HT, Sodoudi S, El-Zafarany A, El Seoud TA, Ashraf H, Krone K (2016) A comprehensive statistical study on daytime surface urban heat island during summer in urban areas, case study: Cairo and its new towns. Remote Sens 8(8):1–20
22. ECLAC—United Nations (2021) Regional urban statistics: Urban and cities platform. Urban and cities platform of Latin America and the Caribbean [Online]. Available: https://plataformaurbana.cepal.org/en/regional-urban-statistics. [Accessed 01 Dec 2021]
23. Instituto Nacional de Ecología and Instituto Nacional de Salud Publica (2006) Estudio Diagnostico sobre los efectos del cambio climático en la salud humana de la poblacion de México. Mexico
24. Baca-López K, Fresno C, Espinal-Enríquez J, Flores-Merino MV, Camacho-López MA, Hernández-Lemus E (2021) Metropolitan age-specific mortality trends at borough and neighborhood level: the case of Mexico City. PLoS One 16(1 January):1–26
25. World health organization (2016) International statistical classification of diseases and related health problems, 1st edn. WHO Press, Switzerland
26. Instituto Nacional de Estadística y Geografía (INEGI) (2021) Mortalidad. Registros administrativos—estadísticas vitales [Online]. Available: https://www.inegi.org.mx/programas/mortalidad/#Tabulados. [Accessed: 15 Nov 2021]
27. Instituto Nacional de Estadística y Geografía (INEGI) (2021) SCINCE 2020. Sistema para la Consulta de Información Censal. [Online]. Available: https://gaia.inegi.org.mx/scince2020/. [Accessed: 01 Nov 2021]
28. Gobierno de México (2021) Casos de enfermedad por mes de ocurrencia (2019). Anuario de morbilidad 1984–2019. [Online]. Available: https://epidemiologia.salud.gob.mx/anuario/html/casos_mes.html. [Accessed: 15 Nov 2021]
29. Genossenschaft Meteotest (2020) Handbook part I: Software. Bern
30. Fernández Eguiarte A et al. Mapa de temperatura máxima extrema absoluta (1902-2015), Umbrales de exposición. Atlas climático de México, Unidad de Informática para las Ciencias Atmosféricas y Ambientales (UNIATMOS), Instituto de Ciencias de la Atmósfera y Cambio Climático, UNAM. [Online]. Available: http://uniatmos.atmosfera.unam.mx/ACDM/servmapas. [Accessed: 01 Dec 2021]

31. Gonzalez-Block MA, Reyes Morales H, Cahuana-Hurtado L, Balandran A, Mendez E (2020) Mexico health system review. Health Syst Transit 22(2)
32. XXII Ayuntamiento de Ensenada (2017) Anuario estadistico municipal de Ensenada, Ensenada
33. Avdan U, Jovanovska G (2016) Algorithm for automated mapping of land surface temperature using LANDSAT 8 satellite data. J Sensors 2016
34. Thermal Energy System Specialists (2021) What is TRNSYS?. [Online]. Available: http://www.trnsys.com/. [Accessed: 06 Nov 2021]
35. Aghamohammadi N, Fong CS, Idrus MHM, Ramakreshnan L, Sulaiman NM (2021) Environmental heat-related health symptoms among community in a tropical city. Sci Total Environ 782:146611
36. Liu X, Yue W, Yang X, Hu K, Zhang W, Huang M (2020) Mapping urban heat vulnerability of extreme heat in Hangzhou via comparing two approaches. Complexity 2020
37. Pyrgou A, Santamouris M (2018) Increasing probability of heat-related mortality in a mediterranean city Due to urban warming. Int J Environ Res Public Health 15(8):1–14
38. Beck HE, Zimmermann NE, McVicar TR, Vergopolan N, Berg A, Wood EF (2018) Present and future köppen-geiger climate classification maps at 1-km resolution. Sci Data 5:1–12
39. TRANE (2008) TRANE product data. Single packaged convertible gas/electric 13 SEER. [Online]. Available: http://www.climatesolutionsinc.com/wp-content/uploads/Trane-2YCC3-Brochure.pdf. [Accessed: 18 Nov 1BC]
40. Molina MO, Sánchez E, Gutiérrez C (2020) Future heat waves over the mediterranean from an Euro-CORDEX regional climate model ensemble. Sci Rep 10(1):1–10
41. Gobierno de México, *Declaratoria de vigencia de la Norma Mexicana NMX-C-527–1-ONNCCE-2017.* México: Diario Oficial de la Federación, 2017, p. 1.
42. United States Global Change Research Program (2021) National integrated heat health information system (NIHHIS) | U.S. climate resilience toolkit [Online]. Available: https://toolkit.climate.gov/tool/national-integrated-heat-health-information-system-nihhis. [Accessed: 28 Nov 2021]
43. California Department of Public Health and the Centers for Disease Control's (CDC) (2021) Tracking California information action for healthier communities. Views, maps and data [Online]. Available: https://trackingcalifornia.org/main/maps-and-data. [Accessed: 01 Nov 2021]

Correction to: Impact of Urban Overheating and Heat-Related Mortality in Hong Kong

Junyi Hua, Yuan Shi, Chao Ren, Kevin Ka-Lun Lau, and Edward Yan Yung Ng

Correction to:
Chapter 14 in: N. Aghamohammadi and M. Santamouris (eds.), *Urban Overheating: Heat Mitigation and the Impact on Health*, Advances in Sustainability Science and Technology,
https://doi.org/10.1007/978-981-19-4707-0_14

In the original version of the book, the affiliation of Hua Junyi has been changed to "School of International Affairs and Public Administration, Ocean University of China". Email address: huajunyi@ouc.edu.cn.

The correction chapter and the book has been updated with the changes.

The updated original version of this chapter can be found at
https://doi.org/10.1007/978-981-19-4707-0_14

N. Aghamohammadi and M. Santamouris (eds.), *Urban Overheating: Heat Mitigation and the Impact on Health*, Advances in Sustainability Science and Technology,
https://doi.org/10.1007/978-981-19-4707-0_18

The manufacturer's authorised representative in the EU is Springer Nature Customer Service Centre GmbH, Europaplatz 3, 69115 Heidelberg, Germany. If you have any concerns regarding our products, please contact ProductSafety@springernature.com

Printed and bound by CPI Group (UK) Ltd, Croydon, CR0 4YY

30/09/2026

02225242-0014